Code Check Complete

An Illustrated Guide to the Building, Plumbing, Mechanical, and Electrical Codes

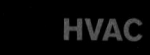

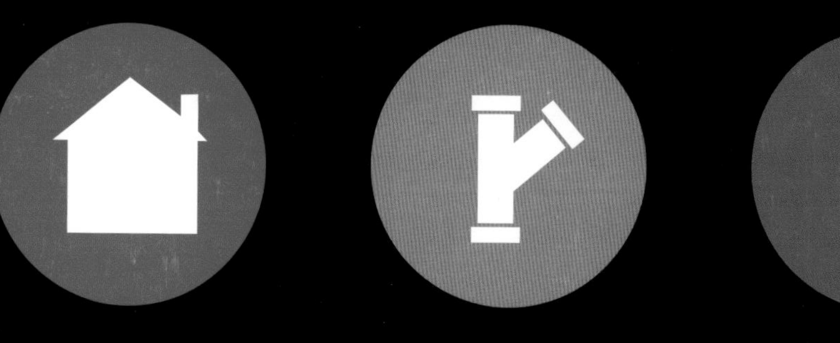

BUILDING PLUMBING HVAC ELECTRICAL

Redwood Kardon / Douglas Hansen / Michael Casey / Illustrated by Paddy Morrissey

Illustrations and Layout: Paddy Morrissey
 ISBN 978-1-56158-911-1
 Printed in China | 10 9 8 7 6 5 4 3 2 1

INTRODUCTION

Code☑Check Complete is a compilation of the four individual Code Check field guides to the building codes: Building, Plumbing, HVAC, and Electrical.

This book is intended for building inspectors, design professionals, plan reviewers, contractors, home inspectors, educators, and do-it-yourself home owners. The primary code used in the first three sections (Building, Plumbing, and HVAC) is the 2006 International Residential Code®, the most widely adopted residential code in the United States. The electrical section is based on the National Electrical Code®, the oldest and most widely used electrical code. Several other codes are referenced in the Plumbing and HVAC sections, including the Uniform Codes that are used in many parts of the country.

The major codes are updated on a three-year cycle. Since the first edition, the Code Check field guide (flip–chart format) has been updated with each code cycle, and the current "combination" book is in its 5th edition. Building jurisdictions do not always adopt a code during the calendar year of publication, and some actually lag several years behind. These books can still be used in areas where older codes apply, thanks to the "code changes" featured at the end of each section. Significant changes are highlighted throughout the text and summarized there, and users of older codes are alerted to items for which the most current code does not apply.

For further information on codes, or to contact the authors, please visit **www.codecheck.com.**

CONTENTS

Code ✓Check® Building Second Edition

BY DOUGLAS HANSEN, REDWOOD KARDON, AND MICHAEL CASEY

Based on Chapters 1–11 of the 2006 International Residential Code®

For more information on the Building, Electrical, Mechanical, and Plumbing codes; valuable resources; and why Benjamin Franklin is featured in the Code Check series, visit: ***www.codecheck.com***

Code Check Building is a condensed guide to the building portions of the *2006 International Residential Code for One- & Two-Family Dwellings.* Most building jurisdictions around the country have either adopted the International Residential Code (IRC) or a code that is based on it. The IRC is prescriptive and simpler to use than the International Building Code (IBC), or the legacy codes that preceded the IBC. For example, the IBC has different occupancy categories for garages and for dwellings, and users must move back and forth between different chapters to find all the rules for fire-resistive construction to separate the two occupancies. The IRC simply prescribes the covering that must be on the garage side.

KEY TO USING CODE CHECK BUILDING

Each text line ends with an IRC code reference in brackets, ex:

☐ Ceiling height in habitable spaces min 7ft ____ [305.1]

The rule for minimum ceiling heights in habitable spaces is found in section 305.1. Note: In the IRC, the number is actually R305.1. We have omitted the letter at the beginning to save space and include more information on each line.

A colored code citation followed by a superscript number indicates a change from the 2003 edition of the IRC, ex:

☐ Unrated walls min 5ft to property line ____ [302.1][1]

This rule changed in the 2006 edition. The change is explained as item #1 on the list of significant code changes on page 60.

T1 refers to Code Check Building Table 1, ex:

☐ Determine Seismic Design Category **T1** ____ [301.2.2.1]

F3 refers to Code Check Building Figure 3, ex:

☐ Setbacks & clearances to slopes >1 vert: 3 horiz **F3** ____ [403.1.7]

The word OR at the end of the line signals that an alternative rule or rules will be listed in the following line or lines without check boxes preceding them.

The word EXC at the end of the line means that an alternative rule or rules will be listed in the following line or lines without check boxes preceding them.

☐ Foundation wall to extend min 6in above finish grade EXC ____ [404.1.6]
4in OK if masonry veneer is used ____ [404.1.6]

The basic rule is for the foundation wall to extend at least 6 inches above grade. Only 4 inches is required if a masonry veneer is used.

REFERENCE DOCUMENTS

The IRC is part of the suite of codes published by the ***International Code Council.*** It is limited to one- and two-family dwellings and townhouses not more than three stories above grade. It is a prescriptive document containing rules and instructions. When aspects of a building exceed the scope of the IRC, the ***International Building Code*** is a more comprehensive document, containing engineering regulations for structural design. It is acceptable to use any of the specific performance-based provisions of the International Codes as an alternative to the prescriptive rules in the IRC.

The American Concrete Institute publishes two documents that can be used as alternates to the prescriptive rules of the IRC. These are: *ACI 318–Building Code Requirements for Structural Concrete*, and *ACI 530–Building Code Requirements for Masonry Structures.*

The Truss Plate Institute publishes *TPI 1–National Design Standard for Metal Plate Connected Wood Truss Construction*, and contributes to *BCSI 1-03 Guide to Good Practice for Handling, Installing & Bracing of Metal Plate Connected Wood Trusses.*

SEQUENCE OF CODE CHECK BUILDING

The IRC building portions are arranged in the following "from the ground up" sequence: (1) Administration, (2) Definitions, (3) Planning, (4) Foundations, (5) Floors, (6) Wall Construction, (7) Wall Coverings, (8) Roof/Ceiling Construction, (9) Roof Assemblies, and (10) Chimneys and Fireplaces. Code Check Building follows the same basic sequence, with more detailed sections near the end on egress, fire protection, and safety glazing.

SEISMIC DESIGN CATEGORIES

The IRC assigns all buildings a seismic design category (SDC) from A to E. SDC A is the least likely to experience seismic damage, and E is the most vulnerable. Category D is broken into three subparts, D_0, D_1, and D_2. The D_0 designation is new in the 2006 IRC. Buildings in SDC E must be designed to the International Building Code. In some cases, the local building official can designate them as belonging to SDC D_2, provided shear walls are in one plane from the foundation to the uppermost story, there are no cantilevers, and the building has a regular shape.

ABBREVIATIONS

# = number
ACI = American Concrete Institute
AMI = in accordance with manufacturer's instructions
ASTM = American Society for Testing & Materials
BO = building official
cfm = cubic feet per minute
CMU = concrete masonry units
ex = example
exc = except
EXC = exception to rule will follow in the next line
ft = foot/feet
gal = gallon(s)
horiz = horizontal
hr = hour
IBC = International Building Code
in = inch(es)
lb = pound(s)
L&L = listed and labeled
max = maximum
min = minimum
o.c. = on center
No. = number
PL = property line
PT = pressure treated
psf = pounds per square foot
psi = pounds per square inch
req = require
req'd = required
reqs = requires
SDC = Seismic Design Category
sq = square
TJI® = manufactured I-joists*
vert = vertical
w/ = with
w/o = without

** TJI® is a trademark of Trus Joist™, a Weyerhaeuser business*

WHY BENJAMIN FRANKLIN?

Benjamin Franklin was chosen as the main character for our Code Check illustrations for a number of reasons. The "First American's" insatiable curiosity, scientific genius, and civic mindfulness drove him to study fire safety, safe exiting, public sanitation, improved heating methods, and of course, electricity.

Franklin made major contributions to each of the four main disciplines of building inspection: Building, Plumbing, Mechanical, and Electrical. Franklin's first attempt to safeguard the public through building codes came in 1735 with his call for minimum standards in the design of fireplace hearths, hearth extensions, and combustible material clearance. The principles Franklin proposed are codified in all the modern building codes, which prescribe these clearances in detail. In 1735, Franklin organized the first volunteer Fire Department in Philadelphia, which still remains the model for our modern fire departments. He also understood the importance of building design in slowing the spread of a fire and was proud that his final home–built after his return from France in 1785–did not have concealed spaces where fire could spread. Thus, by judicious use of plaster, Franklin anticipated the fireblocking rules in today's codes. He also took an interest in designing stairways that were the proper pitch. One of Franklin's early inventions was the Franklin Stove, still called by his name today. By the mid-1740s, firewood was already scarce in the Philadelphia area, and air pollution was beginning to be a problem. Franklin's design retained more heat than the fireplaces or stoves then in use, thus reducing the demand for wood. Ever the altruist, he refused to patent his inventions and insisted on giving them to the world. Franklin never lost interest in the subject. In 1785, concerned over the polluting effects of coal, he designed another stove that burned soft coal and consumed its own smoke.

An unusual aspect of Franklin's personality, at least by the standards of the 18th century, was his concern for personal hygiene. From his youth he was an avid swimmer, and in his adult life he bathed as often as once a week–a practice considered scandalous at the time. Franklin not only brought the first bathtub to America but also improved its design and spent much of his time reading and writing while soaking. His concern for sanitation went beyond his own personal needs, and he was instrumental in creating America's first public sewer system in Philadelphia.

If you were to be transported back to any major European city in Franklin's time, you would probably find yourself overwhelmed with the stench of daily life. Franklin, a great man of reason, was not content with this fetid mess. Despite the fact that germ theory was still a century off, Franklin saw the virtues of separating urban dwellers from their waste products. Thanks, in no small part, to his civic mindfulness Philadelphia became internationally noted for its cleanliness.

Of all Franklin's scientific accomplishments, the best known is his lightning rod. After correctly hypothesizing the nature of lightning in 1750, his famous kite experiment in 1752 confirmed that lightning is a form of electricity and can be directed to the earth through rods and conductors. The principles of grounding all date back to him. To this day, passive lightning protection systems are referred to as "Franklin" systems. As luck would have it, his own house was hit by lightning and saved by his invention.

In 1736, after an extensive fire in Philadelphia, Benjamin Franklin created the first fire department–a fire brigade named The Union Fire Company.

Building and plumbing codes exist to safeguard persons and property. At Code Check, we feel that purpose to be a continuation of the work of Benjamin Franklin. His ideas are still alive in today's building codes and are carried on by code-making organizations and the people who practice those codes.

CODE CHECK BUILDING CONTENTS

FIG. 1

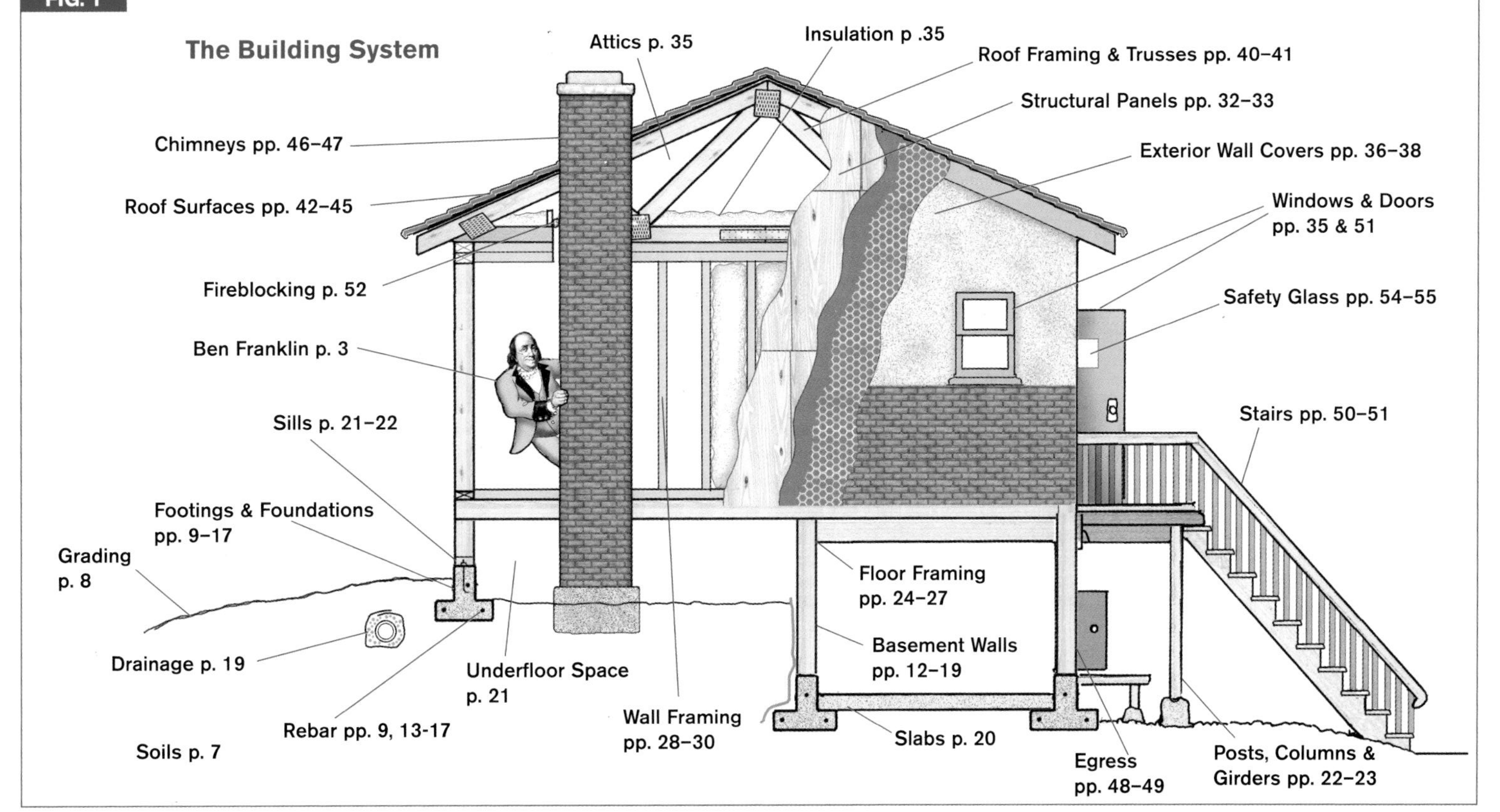
The Building System
Attics p. 35
Insulation p .35
Roof Framing & Trusses pp. 40–41
Structural Panels pp. 32–33
Chimneys pp. 46–47
Exterior Wall Covers pp. 36–38
Roof Surfaces pp. 42–45
Windows & Doors pp. 35 & 51
Fireblocking p. 52
Safety Glass pp. 54–55
Ben Franklin p. 3
Sills p. 21–22
Stairs pp. 50–51
Footings & Foundations pp. 9–17
Grading p. 8
Floor Framing pp. 24–27
Drainage p. 19
Underfloor Space p. 21
Basement Walls pp. 12–19
Rebar pp. 9, 13-17
Wall Framing pp. 28–30
Slabs p. 20
Soils p. 7
Egress pp. 48–49
Posts, Columns & Girders pp. 22–23

THE BUILDING SYSTEM

PLANNING, PERMITS, & BUILDING LOCATION

Before beginning a building or project, plans must be approved by the local building department. The building official determines the climatic and geographic design criteria that apply. The plans must include setbacks from the property lines and adjacent slopes.

Planning

- ☐ If engineered design req'd for nonconventional elements, use IBC ___ [301.1.3]
- ☐ Determine climatic & geographic design criteria **T1** ___ [301.2]
- ☐ Determine SDC **T1** ___ [301.2.2.1]
- ☐ Consider wind limitations in bldg design **T1** ___ [301.2.1.4]
- ☐ Consider snow loads in bldg design **T1** ___ [301.2]
- ☐ Establish flood elevation **T1** ___ [324.1.3]
- ☐ Are special inspection reports from outside agencies req'd? **T2** ___ [109.1.5]
- ☐ Determine soil type **T3** ___ [401.4]
- ☐ Are tests needed to determine expansive soils? ___ [401.4]

TABLE 1 CLIMATIC & GEOGRAPHIC CONDITIONS[A] [T301.2(1)]

Ground Snow Load	Wind Speed (mph)	Seismic Design Category	Is Property Subject to Damage from			
			Weathering[B]	Frost Line Depth	Termites	Flood Hazards[C]

A. Fill in w/ information obtained from local building department.
B. Choose negligible, moderate, or severe–affects strength of concrete & grade of CMU.
C. Reference the flood hazard map adopted by community & date of adoption.

Room Dimensions

- ☐ Min 1 room ≥120sq.ft ___ [304.1]
- ☐ Habitable rooms: min 70sq.ft, exc kitchens ___ [304.2]
- ☐ Habitable rooms: min dimension 7ft ___ [304.3]
- ☐ Ceiling height in habitable spaces: min 7ft ___ [305.1]
- ☐ Kitchens, baths, & hallways min ceiling height: 7ft EXC ___ [305.1]
  6ft 8in OK in baths over fixture & front clearance area ___ [305.1]
- ☐ Ceiling height unfinished basements: min 6ft 8in ___ [305.1]

Location & Setbacks

- ☐ Unrated walls: min 5ft to PL ___ [302.1][1]
- ☐ Openings in walls <5ft to PL: max 25% of wall area ___ [T302.1][2]
- ☐ No openings in walls <3ft to PL EXC ___ [302.1]
  Roof projections <5ft to PL: 1-hour construction (min setback 2ft) ___ [302.1]
- ☐ Foundation vents OK & openings in walls perpendicular to PL OK ___ [302.1]

Permits & Inspections

- ☐ Approved plans & permit card on site ___ [106.3.1]
- ☐ Inspection & approval before covering any work ___ [109.4]
- ☐ Permits not req'd for the following: ___ [105.2]
  - Sheds <120sq.ft[3]
  - Fences ≤6ft, retaining walls ≤4ft & w/ no surcharge
  - Water tanks on grade ≤5,000 gal & height/width ratio ≤2:1
  - Sidewalks & driveways[4]
  - Painting & other finish work
  - Prefab pools <24in deep, swings, playground eqpmt

TABLE 2 SPECIAL INSPECTIONS [109.1.5] & IBC CHAPTER 17

Inspection Type	Agency	Continuous	Periodic	Obtained
Fire-resistance rated construction				
Steel – welding				
Steel – high-strength bolts				
Concrete – reinforcement				
Concrete – bolts				
Concrete – slump test				
Application of pre-stressing force				
Grouting of bonded tendons				
Masonry – welding reinforcement				
Masonry – cold weather protection				
Soils				
Special cases				

SOILS

All building loads must be transferred to the foundation footings, which in turn must be placed in competent soil to prevent building movement. The Unified Soil Classification System attempts to predict the properties and behavior of soils based on simple field tests. The required thickness of footings and width of foundation walls depends on the soil type.

Soil Tests & Load Values

- ☐ BO may req soil test in areas likely to have expansive, shifting, or compressible soils **F2** ______ [401.4]
- ☐ BO may req soil investigation if soils <1,500psf likely to be present _[T401.4.1]
- ☐ BO may allow presumptive values of **T3** in lieu of geotechnical evaluation [401.4.1]
- ☐ Compressible or shifting soils: (peat) must be removed ______ [401.4.2]

FIG. 2 Graded Soils

Well-graded soil

Poorly graded soils

Soil tests determine the percentages of sand, silt, & clay, as well as the plasticity index & the gradation based on a sieve test. Check w/ the local BO to determine your site classification & whether testing will be necessary.

TABLE 3 SOIL TYPES & PRESUMPTIVE LOAD-BEARING VALUES [401.4.1]

Abbreviation	Class of Material[A,B]	Load-Bearing Pressure
	Sedimentary & foliated rock	4,000 lb./sq.ft.
GW[C]	Well-graded gravel & gravel-sand mixtures	3,000 lb./sq.ft.
GP[C]	Poorly graded gravel & sandy gravel mixtures	
SW[C]	Well-graded sands	2,000 lb./sq.ft.
SP[C]	Poorly-graded sands	
SM[C]	Silty sands & poorly graded sand-silt mixtures	
GM[C]	Siltey gravel & poorly graded gravel/sand-silt mixtures	
SC	Clayey sands & poorly graded sand-clay mixtures	
GC	Clayey gravels & poorly graded sand-clay mixtures	
CL	Clay w/ low compressibility	1,500 lb./ sq.ft
ML	Silt w/ low compressibility	
CH	Clay w/ high compressibility	
MH	Silt w/ high compressibility	

A. Some soils w/ more than one characteristic must be represented w/ a dual symbol, such as SM-SC.
B. Soils are well graded when they contain a uniform mix of particle sizes. They are poorly graded when all particles are the same size or if there are gaps in the distribution of particle size.
C. These soils are considered Group I w/ good drainage characteristics per [T405.1].

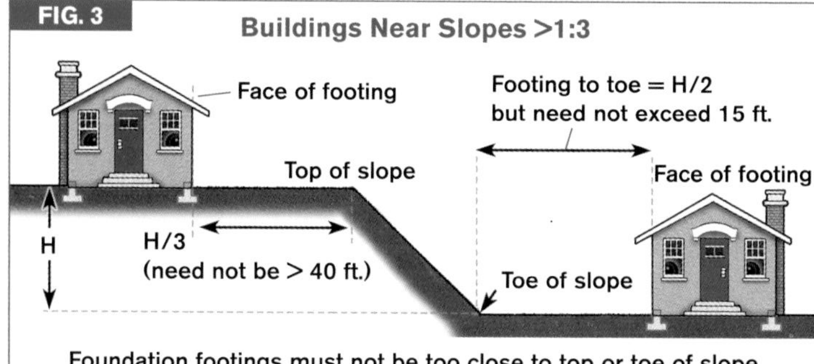

FIG. 3 Buildings Near Slopes >1:3

Foundation footings must not be too close to top or toe of slope.

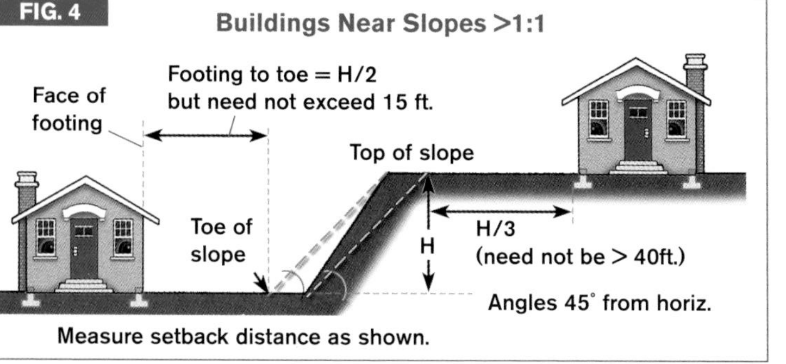

FIG. 4 Buildings Near Slopes >1:1

Measure setback distance as shown.

SITE GRADING

Site grading must establish a foundation height that will allow positive drainage to slope away from the building, while at the same time not creating a situation in which the foundation is placed on fill, other than engineered and compacted fill. The basic rule is to have positive slope for 10 ft. in all directions from the building, with surface drainage directed to a storm sewer or equivalent. A new rule in the 2006 IRC allows an exception for situations in which the building is close to the property line or if there are physical barriers preventing a positive slope for 10 ft. In those cases, the grading can have a 5% slope to a swale.

Slope/Grade

- ☐ Surface graded away from foundation: min 6in/10ft **F5** EXC ________ [401.3]
  5% slope to swale OK when barrier prevents 6in/10ft ________ [401.3X][5]
- ☐ Swale: min slope 2% when closer than 10ft to bldg ________ [401.3X]
- ☐ Setbacks & clearances to slopes >1: vert:3 horiz **F3** ________ [403.1.7]
- ☐ Setbacks & clearances to slopes >1: vert:1 horiz **F4** ________ [403.1.7]
- ☐ Alternate setbacks permitted w/ approval of BO ________ [403.1.7.4]
- ☐ Graded site–top of foundation: min 12in + 2% above street drain **F5** ________ [403.1.7.3]

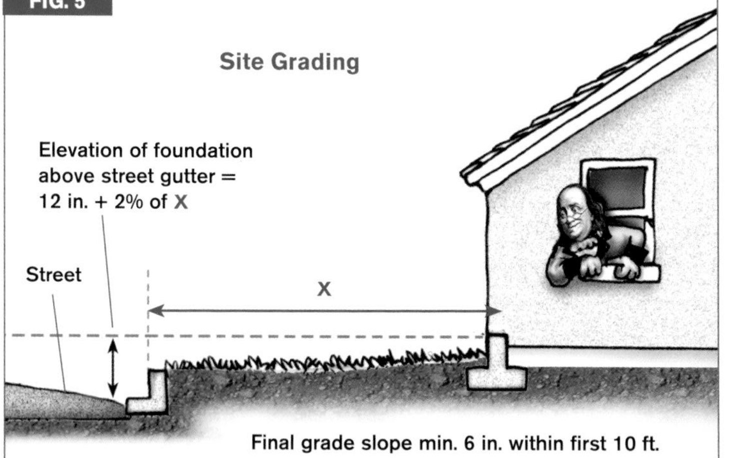

FIG. 5 Site Grading

Final grade slope min. 6 in. within first 10 ft.

FOOTINGS

The foundation footing must be capable of supporting all loads transmitted to it, including the dead loads, live loads, and environmental loads. Footings must be supported on previously undisturbed material or on engineered fill; care must be taken in grading a site to ensure that footings are not cast on unconsolidated soil. The footing is typically cast monolithically with the foundation on houses with crawlspaces and "inverted T" foundations, or slabs-on-grade.

Footings

- ☐ Depth: min 12in into undisturbed native soil or engineered fill **F6** [403.1.4, 403.1]
- ☐ Must extend below frost line or be frost protected [403.1.4.1]
- ☐ Width: per **T5** [403.1.1]
- ☐ Min thickness: 6in [403.1.1]
- ☐ Min projection past foundation: 2in each side, max projection ≤ footing thickness **F6** [403.1.1]
- ☐ SDC D_0, D_1 & D_2 monolithic slab interior footings below bearing or braced walls min 12in deep from top of slab [403.1.4.2][6]
- ☐ Foundations & slabs on expansive soil req design to prevent uplift [403.1.8]
- ☐ Top surface of all footings level [403.1.5]
- ☐ Bottom surface of footings: max slope 1:10 (create steps where necessary)

TABLE 4 — REINFORCING STEEL COVER[A]

Foundation Surface	Min. Cover ≤# 5 bars	Min. Cover ≥# 6 bars
Concrete cast against & permanently exposed to earth	3 in.	3 in.
Concrete exposed to earth or weather	1 in.	2 in.
Not exposed to weather, ex.–top of indoor slab	3/4 in.	3/4 in. up to # 11 bars

A. The cover in this table is recommended by *ACI 318–Building Code Requirements for Structural Concrete.* This document is the industry standard for proper practices. These #s are not specifically a part of the IRC.

TABLE 5 — MIN. WIDTH OF CONCRETE OR MASONRY FOOTINGS (IN INCHES) [T403.1]

Construction Type	# of Stories	Load Bearing Value of Soil[A] (psf) 1,500	2,000	3,000	≥4,000
Conventional light-frame construction	1	12	12	12	12
	2	15	12	12	12
	3	23	17	12	12
4-in. brick veneer over frame or 8-in. hollow-concrete masonry	1	12	12	12	12
	2	21	16	12	12
	3	32	24	16	12
8-in. solid or fully grouted masonry	1	16	12	12	12
	2	29	21	14	12
	3	42	32	21	16

A. See **T3** for vert. load-bearing values of different soil types.

FORMS

Foundation forms contain concrete while it is still in its fluid state and provide protection for reinforcing steel and other elements that are to be permanently embedded. Forms must not be removed until the concrete is sufficiently cured, and all traces of wood forms must be removed to prevent any potential wood-destroying organisms from entering the building.

Forms

- ☐ Size: per tables **T7,9,10** [404.1.1,2]
- ☐ Pipe penetrations must be sleeved [2603.5]
- ☐ Excavation free of debris & roots [408.5, 506.2]
- ☐ Wood beam connections: 1/2in air space on 3 sides **F7** [319.1]
- ☐ Foundation wall to extend min 6in above finish grade EXC [404.1.6] 4in OK if masonry veneer is used [404.1.6]

FIG. 6

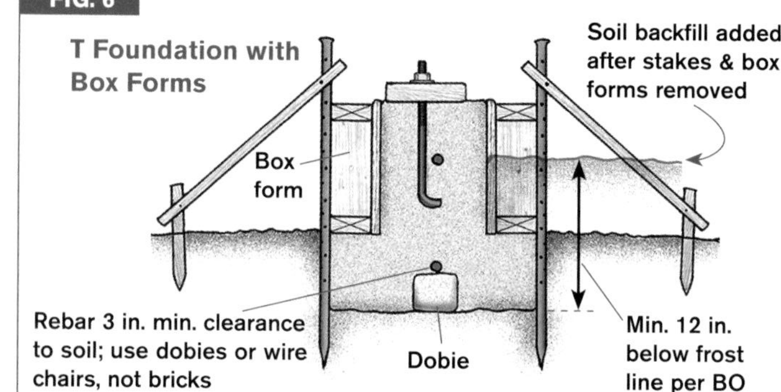

FIG. 7

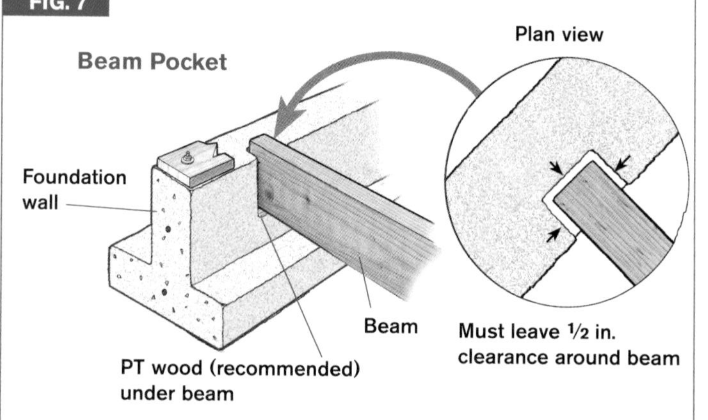

CONCRETE

Concrete must have sufficient strength for the application and environment. Hi-strength concrete (>2,500 psf) might require special inspections (T2). In general, the more northern the latitude, the more severe the weathering potential. ACI 318 defines the testing standards for concrete.

Mixing & Strength

- ☐ Min compressive strength: per **T6** [402.2]
- ☐ Materials & testing to conform to ACI 318 [402.2]

TABLE 6 — MIN. COMPRESSIVE STRENGTH OF CONCRETE (AT 28 DAYS) IN PSI [T402.2]

Type or Location of Concrete Construction	Weathering Potential		
	Negligible	Moderate	Severe
Basement walls, foundations, & other concrete not exposed to the weather	2,500	2,500	2,500[A]
Basement slabs & interior slabs on grade, exc. garage floor slabs	2,500	2,500	2,500[A]
Basement & foundation walls, exterior walls, & other vert concrete exposed to weather	2,500	3,000[B]	3,000[B]
Porches, carport slabs, & steps exposed to the weather & garage floor slabs	2,500	3,000[B,C]	3,500[B,C]

A. Must be air-entrained if exposed to freeze-thaw during construction.
B. Air-entrainment req'd. Air content between 5% & 7% by volume of concrete.
C. Garage floor slab air-entrainment may be reduced to 3% if strength increased to 4,000 psi.

ANCHOR BOLTS

Anchor bolts resist lateral forces that could cause a building to lift or slide off the foundation. Anchor bolts must have sufficient embedment to resist pullout and must be spaced properly to secure the sill in place. Washers must be capable of distributing a load across the sill without it cracking or splitting.

General

- ☐ Min 2 bolts per piece of sill material EXC [403.1.6]
  Walls: 24in connecting offset braced wall panels req only 1 bolt [403.1.6X2][7]
- ☐ Walls: 12in in length connecting offset braced wall panels no bolt req'd [403.1.6X3][8]
- ☐ Sill plates under braced wall panels req bolts [403.1.6]
- ☐ 1/2in bolts: min 7in embedment max 6ft spacing EXC [403.1.6]
  Straps placed at intervals providing equivalent anchorage as 1/2in bolts [403.1.6X1]
- ☐ Max distance: 12in from end of sill **F9** [403.1.6]
- ☐ Min distance: 7 bolt diameter from end of sill [403.1.6]
- ☐ SDC C, D_0, D_1, & D_2 reqs 2in plate washers [403.1.6.1]
- ☐ Sufficient distance from edge for washer to fully seat on sill **F8,9** [403.1.6.1]
- ☐ SDC C, D_0, D_1, & D_2 4ft o.c. for >2 stories [403.1.6.1]

FIG. 8

Anchor Bolt Holder

Anchor bolt holders help keep the bolt correctly aligned & reduce the likelihood of the foundation cracking around the bolt.

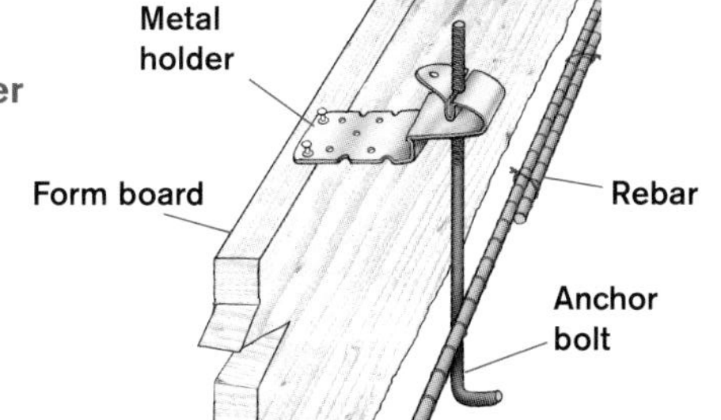

FIG. 9

Anchor Bolts & Hold-Downs

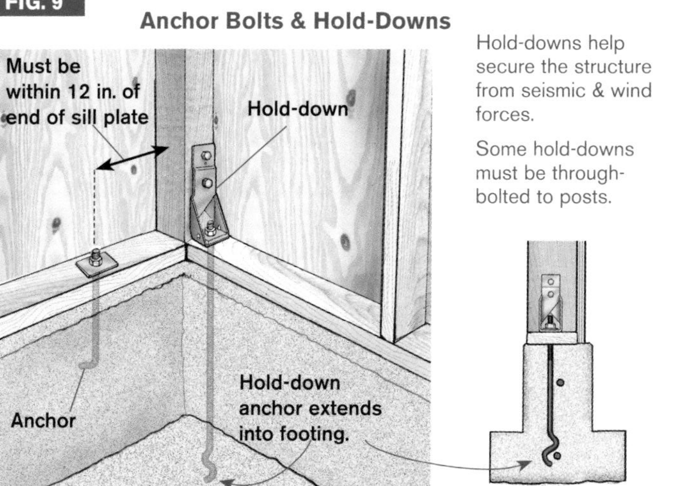

Hold-downs help secure the structure from seismic & wind forces.

Some hold-downs must be through-bolted to posts.

FOUNDATIONS

Rubble Stone Masonry Foundations

- ☐ Min thickness: 16in, max unbalanced backfill 8ft [404.1.8]
- ☐ Soil lateral pressure must be ≤30 psf (types GW, GP, SW, & SP) [404.1.8]
- ☐ Not allowed in SDC D_0, D_1, & D_2 [404.1.8]

Insulating Concrete Form (ICF) Foundations

- ☐ Flat ICF foundations: min 5 1/2in thick [404.4.2]
- ☐ Waffle & screen grid ICF: min 6in thick [404.4.3,4]
- ☐ Max slump: 6in, max aggregate: 3/4in or AMI [404.4.5]
- ☐ Limited to buildings not more than 60ft in plan dimension [404.4.1]

Insulating Concrete Form (ICF) Foundations (cont.)

- ☐ Floor clear spans: max 32ft, roof clear span: max 40ft [404.4.1]
- ☐ Steel coverage req'd to comply w/ **T5** EXC [404.4.6.1]
  Reduction to 3/4in coverage OK when forms left in place [404.4.6.1X]
- ☐ Provide protection against termite hazards [404.4.7.2]

Wood Foundations

- ☐ All fasteners: stainless, hot-dipped galvanized, silicon bronze, or copper [402.1.1]
- ☐ All lumber: PT & marked by accredited agency [402.1.2]
- ☐ Field-treat all cut ends [402.1.2]
- ☐ Footings: well-graded washed gravel min 6in thick if crawlspace, 8in basement [403.2]
- ☐ Drain system to the porous layer & basement sump pump req'd [405.2.3]

Frost-Protected Shallow Foundations

- ☐ Only applies for foundations of heated areas (not garages or utility areas) [403.3]
- ☐ Insulation: R-value per air-freezing index (check w/ local BO) [T403.3]
- ☐ Horiz insulation below ground reqs paving protection for 24in from foundation [403.3.2]
- ☐ Drain rock below insulation: drain to daylight or approved sewer [403.3.3]

BASEMENT FOUNDATION WALLS

In addition to being the proper size for the vertical load, foundation walls must be designed to resist lateral forces imposed by the soils and the other forces acting on the building. A properly anchored floor system helps restrain these walls within the limitations imposed by this section.

Foundation Walls

- ☐ Design req'd for walls subject to hydrostatic pressure [404.1.3]
- ☐ Design req'd for walls w/o lateral support at top or bottom if >48in unbalanced backfill **F10** [404.1.3]

Plain Concrete or Masonry

- ☐ Plain concrete or masonry thickness per **T7, T9** [404.1.1&2]
- ☐ Only SDC A, B, & C unless <8ft high & ≤4ft unbalanced backfill [404.1.4]

SDC D Requirements

- ☐ SDC D_0, D_1, & D_2 requirements: [404.1.4]
  - Min thickness 7½in EXC
  - 6in thickness OK if height not > 4ft 6in
- ☐ Plain masonry in SDC D_0, D_1, & D_2 min thickness 8in [404.1.4]
- ☐ Max height: 8ft, max unbalanced backfill: 4ft [404.1.4]
- ☐ Plain concrete or masonry in SDC D_0, D_1, & D_2: min 1 #4 bar in upper 12in [404.1.4]
- ☐ Plain masonry reqs min 1 vert #3 bar 4ft o.c. & tied to horiz footing reinforcement [404.1.4]

Concrete or Masonry

- ☐ Reinforced concrete & masonry per **T9,10** [404.1.1,2][9]
- ☐ Table reinforcement assumes grade 60 rebar **T9,10** [T404.1.19(2,3,4)]
- ☐ Vert reinforcement: min distance from soil side **T8** [404.1.1,2]
- ☐ Min height above finished grade: 6in, 4in if veneer [404.1.6]
- ☐ Thickness of foundation: not less than thickness of wall supported EXC [404.1.5]
  8in foundation can support 10in veneered or cavity wall to 20ft high [404.1.5]

Pier & Curtain Wall Foundations

- ☐ Limited to light-frame construction ≤2 stories in height [404.1.5.1]
- ☐ Continuous concrete footings req'd under all load bearing-walls [404.1.5.1]
- ☐ Max height of 4in masonry supporting wood frame: 4ft [404.1.5.1]
- ☐ Max height of unbalanced fill: 24in for solid masonry, 12in for hollow [404.1.5.1]
- ☐ Sill must be anchored (see **p. 11**) [404.1.5.1]
- ☐ Reinforcement req'd in SDC D_0, D_1, & D_2 [404.1.5.1]

TABLE 7 PLAIN (UNREINFORCED) MASONRY FOUNDATION WALLS [T404.1.1(1)]

Max. Wall Height (ft.)	Max. Unbalanced Backfill Height (ft.)	Minimum Wall Thickness[A] (in.)		
		GW, GP, SW, & SP Soils	GM, GC, SM-SC, & ML Soils	ML-CL, SC, MH, & Inorganic CL Soils
5	4	6 solid or 8	6 solid or 8	6 solid or 8
	5	6 solid or 8	8	10
6	4	6 solid or 8	6 solid or 8	6 solid or 8
	5	6 solid or 8	8	10
	6	8	10	12
7	4	6 solid or 8	8	8
	5	6 solid or 8	10	10
	6	10	12	10 solid
	7	12	10 solid	12 solid
8	4	6 solid or 8	6 solid or 8	8
	5	6 solid or 8	10	12
	6	10	12	12 solid
	7	12	12 solid	See footnote B
	8	10 solid	12 solid	See footnote B
9	4	6 solid or 8	6 solid or 8	8
	5	8	10	12
	6	10	12	12 solid
	7	12	12 solid	See footnote B
	8	12 solid	See footnote B	See footnote B
	9	See footnote B	See footnote B	See footnote B

A. The term *solid* refers to either grouted hollow units or solid masonry units.
B. For walls in this range, refer to **T10** for reinforcement requirements.

*The load-bearing capacity of the soil (**T3**) is used to determine the minimum footing width (**T5**).*

*The Seismic Design Category in **T1** and the type and thickness of wall system supported by the foundation will determine the minimum thickness of the masonry wall foundation.*

*In **T7**, the soil type and unbalanced backfill determine the foundation thickness.*

*In **T9** and **T10**, the soil type and unbalanced backfill determine the reinforcement requirements.*

TABLE 8 MIN. DIST. OF VERT. STEEL FROM SOIL SIDE OF WALL

Thickness of wall	8 in.	10 in.	12 in.
Distance face of soil to center of steel	5 in.	$6^3/_4$ in.	$8^3/_4$ in.

*The reinforcing steel req'd. in **T10** must be on the tension side of the wall to resist lateral soil pressure. This table provides the min. distance from the soil side of the wall. Steel in concrete foundation walls (**T9**), must be $1^1/_4$ in. + $^1/_2$ the bar diameter from the inner face.*

CONCRETE FOUNDATION WALLS

TABLE 9 — CONCRETE FOUNDATION WALLS [R404.1.1(5)]

Max. Wall Height	Max. Unbalanced Backfill Height	Min. Vert. Reinforcement Size & Spacing (DR = design req'd. per ACI-318, PC = plain concrete) — Soil Classes & Design Lateral Soil											
		GW, GP, & SP • 30				GM, GC, SM,-SC, & ML • 45				SC, ML-CL, & inorganic CL • 60			
		Minimum Wall Thickness											
		$5^1/_2$ in.	$7^1/_2$ in.	$9^1/_2$ in.	$11^1/_2$ in.	$5^1/_2$ in.	$7^1/_2$ in.	$9^1/_2$ in.	$11^1/_2$ in.	$5^1/_2$ in.	$7^1/_2$ in.	$9^1/_2$ in.	$11^1/_2$ in.
5 ft.	4 ft.	PC	PC	PC	PC	PC	PC	PC	PC	PC	PC	PC	PC
	5 ft.	PC	PC	PC	PC	PC	PC	PC	PC	PC	PC	PC	PC
6 ft.	4 ft.	PC	PC	PC	PC	PC	PC	PC	PC	PC	PC	PC	PC
	5 ft.	PC	PC	PC	PC	PC	PC	PC	PC	#4@35 in.	PC	PC	PC
	6 ft.	PC	PC	PC	PC	#5@48 in.	PC	PC	PC	#5@36 in.	PC	PC	PC
7 ft.	4 ft.	PC	PC	PC	PC	PC	PC	PC	PC	PC	PC	PC	PC
	5 ft.	PC	PC	PC	PC	PC	PC	PC	PC	#5@47 in.	PC	PC	PC
	6 ft.	PC	PC	PC	PC	#5@42 in.	PC	PC	PC	#6@43 in.	#5@48in.	PC	PC
	7 ft.	#5@46 in.	PC	PC	PC	#6@42 in.	#5@46 in.	PC	PC	#6@34 in.	#6@48in.	PC	PC
8 ft.	4 ft.	PC	PC	PC	PC	PC	PC	PC	PC	PC	PC	PC	PC
	5 ft.	PC	PC	PC	PC	#4@38 in.	PC	PC	PC	#5@43 in.	PC	PC	PC
	6 ft.	#4@37 in.	PC	PC	PC	#5@37 in.	PC	PC	PC	#6@37 in.	#5@43 in.	PC	PC
	7 ft.	#5@40 in.	PC	PC	PC	#6@37 in.	#5@41 in.	PC	PC	#6@34 in.	#6@43 in.	PC	PC
	8 ft.	#6@43 in.	#5@47 in.	PC	PC	#6@34 in.	#6@43 in.	PC	PC	#6@27 in.	#6@32 in.	#6@44 in.	PC

TABLE 9 CONTINUED ON NEXT PAGE

TABLE 9 CONCRETE FOUNDATION WALLS [R404.1.1(5)] – (cont.)

Max. Wall Height	Max. Unbalanced Backfill Height	Min. Vert. Reinforcement Size & Spacing (DR = design req'd. per ACI-318, PC = plain concrete)											
		Soil Classes & Design Lateral Soil											
		GW, GP, & SP • 30				GM, GC, SM, SC, & ML • 45				SC, ML-CL, & inorganic CL • 60			
		Minimum Wall Thickness (in.)											
		5 1/2 in.	7 1/2 in.	9 1/2 in.	11 1/2 in.	5 1/2 in.	7 1/2 in.	9 1/2 in.	11 1/2 in.	5 1/2 in.	7 1/2 in.	9 1/2 in.	11 1/2 in.
9 ft.	4 ft.	PC	PC	PC	PC	PC	PC	PC	PC	PC	PC	PC	PC
	5 ft.	PC	PC	PC	PC	#4@35 in.	PC	PC	PC	#5@40 in.	PC	PC	PC
	6 ft.	#4@34 in.	PC	PC	PC	#6@48 in.	PC	PC	PC	#6@36 in.	#5@39 in.	PC	PC
	7 ft.	#5@36 in.	PC	PC	PC	#6@34 in.	#5@37 in.	PC	PC	#6@33 in.	#6@38 in.	#5@37 in.	PC
	8 ft.	#6@38 in.	#5@41 in.	PC	PC	#6@33 in.	#6@38 in.	#5@37 in.	PC	#6@24 in.	#7@39 in.	#6@39 in.	#4@48 in.
	9 ft.	#6@34 in.	#6@46 in.	PC	PC	#6@26 in.	#7@41 in.	#6@41 in.	PC	#6@19 in.	#7@31 in.	#7@41 in.	#6@39 in.
10 ft.	4 ft.	PC	PC	PC	PC	PC	PC	PC	PC	PC	PC	PC	PC
	5 ft.	PC	PC	PC	PC	#4@33 in.	PC	PC	PC	#5@38 in.	PC	PC	PC
	6 ft.	#5@48 in.	PC	PC	PC	#6@45 in.	PC	PC	PC	#6@34 in.	#5@37 in.	PC	PC
	7 ft.	#6@47 in.	PC	PC	PC	#6@34 in.	#6@48 in.	PC	PC	#6@30 in.	#6@35 in.	#6@48 in.	PC
	8 ft.	#6@34 in.	#5@38 in.	PC	PC	#6@30 in.	#7@47 in.	#6@47 in.	PC	#6@22 in.	#7@35 in.	#7@48 in.	#6@45 in.
	9 ft.	#6@34 in.	#6@41 in.	#4@48 in.	PC	#6@23 in.	#7@37 in.	#7@48 in.	#4@48 in.	DR	#6@22 in.	#7@37 in.	#7@47 in.
	10 ft.	#6@28 in.	#7@45 in.	#6@45 in.	PC	DR	#7@31 in.	#7@40 in.	#6@38 in.	DR	#6@22 in.	#7@30 in.	#7@38 in.

MASONRY FOUNDATION WALLS

TABLE 10 MIN. VERT. REINFORCEMENT OF 8-IN. , 10-IN., & 12-IN. MASONRY FOUNDATION WALLS [T404.1.1(2)(3)(4)]

Wall Height	Unbalanced Backfill Height	Bar Size & O.C. Spacing								
		GW, GP. SW, & SP Soils			GM, GC, SM, SM-SC, & ML Soils			SC, ML-CL, & Inorganic CL Soils		
		8-in. wall	10-in. wall	12-in. wall	8-in. wall	10-in. wall	12-in. wall	8-in. wall	10-in. wall	12-in. wall
6 ft. 8 in.	≤4 ft.	#4@ 48 in.	#4@ 56 in.	#4@ 72 in.	#4@ 48 in.	#4@ 56 in.	#4@ 72 in.	#4@ 48 in.	#4@ 56 in.	#4@ 72 in.
	5 ft.	#4@ 48 in.	#4@ 56 in.	#4@ 72 in.	#4@ 48 in.	#4@ 56 in.	#4@ 72 in.	#4@ 48 in.	#4@ 56 in.	#4@ 72 in.
	6 ft. 8 in.	#4@ 48 in.	#4@ 56 in.	#4@ 72 in.	#5@ 48 in.	#5@ 56 in.	#4@ 72 in.	#6@ 48 in.	#5@ 56 in.	#5@ 72 in.
7 ft. 4 in.	≤4 ft.	#4@ 48 in.	#4@ 56 in.	#4@ 72 in.	#4@ 48 in.	#4@ 56 in.	#4@ 72 in.	#4@ 48 in.	#4@ 56 in.	#4@ 72 in.
	5 ft.	#4@ 48 in.	#4@ 56 in.	#4@ 72 in.	#4@ 48 in.	#4@ 56 in.	#4@ 72 in.	#4@ 48 in.	#4@ 56 in.	#4@ 72 in.
	6 ft.	#4@ 48 in.	#4@ 56 in.	#4@ 72 in.	#5@ 48 in.	#4@ 56 in.	#4@ 72 in.	#5@ 48 in.	#5@ 56 in.	#5@ 72 in.
	7 ft. 4 in.	#5@ 48 in.	#4@ 56 in.	#4@ 72 in.	#6@ 48 in.	#5@ 56 in.	#5@ 72 in.	#6@ 48 in.	#6@ 56 in.	#6@ 72 in.
8 ft.	≤4 ft.	#4@ 48 in.	#4@ 56 in.	#4@ 72 in.	#4@ 48 in.	#4@ 56 in.	#4@ 72 in.	#4@ 48 in.	#4@ 56 in.	#4@ 72 in.
	5 ft.	#4@ 48 in.	#4@ 56 in.	#4@ 72 in.	#4@ 48 in.	#4@ 56 in.	#4@ 72 in.	#4@ 48 in.	#4@ 56 in.	#4@ 72 in.
	6 ft.	#4@ 48 in.	#4@ 56 in.	#4@ 72 in.	#5@ 48 in.	#4@ 56 in.	#4@ 72 in.	#5@ 48 in.	#5@ 56 in.	#5@ 72 in.
	7 ft.	#5@ 48 in.	#4@ 56 in.	#4@ 72 in.	#6@ 48 in.	#5@ 56 in.	#5@ 72 in.	#6@ 40 in.	#6@ 56 in.	#6@ 72 in.
	8 ft.	#5@ 48 in.	#5@ 56 in.	#5@ 72 in.	#6@ 48 in.	#6@ 56 in.	#6@ 72 in.	#6@ 32 in.	#6@48 in.	#6@64
8 ft. 8 in.	≤4 ft.	#4@48 in.	#4@ 56 in.	#4@ 72 in.	#4@ 48 in.	#4@ 56 in.	#4@ 72 in.	#4@48 in.	#4@ 56 in.	#4@ 72 in.
	5 ft.	#4@48 in.	#4@ 56 in.	#4@ 72 in.	#4@ 48 in.	#4@ 56 in.	#4@ 72 in.	#5@48 in.	#4@ 56 in.	#4@ 72 in.
	6 ft.	#4@48 in.	#4@ 56 in.	#4@ 72 in.	#5@ 48 in.	#4@ 56 in.	#4@ 72 in.	#6@48 in.	#5@ 56 in.	#5@ 72 in.
	7 ft.	#5@48 in.	#4@ 56 in.	#4@ 72 in.	#6@ 48 in.	#5@ 56 in.	#5@ 72 in.	#6@40 in.	#6@ 56 in.	#6@ 72 in.
	8 ft. 8 in.	#6@48 in.	#5@ 56 in.	#5@ 72 in.	#6@ 32 in.	#6@48 in.	#7@ 72 in.	#6@ 24 in.	#6@48 in.	#6@48 in.

TABLE 10 CONTINUED ON NEXT PAGE

TABLE 10 MIN. VERT. REINFORCEMENT OF 8-IN. , 10-IN., & 12-IN. MASONRY FOUNDATION WALLS [T404.1.1(2)(3)(4)] – (cont.)

Wall Height	Unbalanced Backfill Height	Bar Size & O.C. Spacing								
		GW, GP, SW, & SP Soils			GM, GC, SM, SM-SC, & ML Soils			SC, ML-CL, & Inorganic CL Soils		
		8-in. wall	10-in. wall	12-in. wall	8-in. wall	10-in. wall	12-in. wall	8-in. wall	10-in. wall	12-in. wall
9 ft. 4 in.	≤4 ft.	#4@ 48 in.	#4@ 56 in.	#4@ 72 in.	#4@ 48 in.	#4@ 56 in.	#4@ 72 in.	#4@ 48 in.	#4@ 56 in.	#4@ 72 in.
	5 ft.	#4@ 48 in.	#4@ 56 in.	#4@ 72 in.	#4@ 48 in.	#4@ 56 in.	#4@ 72 in.	#5@ 48 in.	#4@ 56 in.	#4@ 72 in.
	6 ft.	#4@ 48 in.	#4@ 56 in.	#4@ 72 in.	#5@ 48 in.	#5@ 56 in.	#5@ 72 in.	#6@ 48 in.	#5@ 40 in.	#5@ 72 in.
	7 ft.	#5@ 48 in.	#4@ 56 in.	#4@ 72 in.	#6@ 48 in.	#5@ 56 in.	#5@ 72 in.	#6@ 40 in.	#6@ 56 in.	#6@ 72 in.
	8 ft.	#6@ 48 in.	#5@ 56 in.	#5@ 72 in.	#6@ 40 in.	#6@ 56 in.	#6@ 72 in.	#6@ 24 in.	#6@ 24 in.	#6@ 56 in.
	9 ft.4 in.	#6@ 40 in.	#6@ 56 in.	#6@ 72 in.	#6@ 32 in.	#6@ 40 in.	#6@ 48 in.	#6@ 16 in.	#6@ 16 in.	#6@ 40 in.
10 ft.	≤4 ft.	#4@ 48 in.	#4@ 56 in.	#4@ 72 in.	#4@ 48 in.	#4@ 56 in.	#4@ 72 in.	#4@ 48 in.	#4@ 56 in.	#4@ 72 in.
	5 ft.	#4@ 48 in.	#4@ 56 in.	#4@ 72 in.	#4@ 48 in.	#4@ 56 in.	#4@ 72 in.	#5@ 48 in.	#4@ 56 in.	#4@ 72 in.
	6 ft.	#4@ 48 in.	#4@ 56 in.	#4@ 72 in.	#5@ 48 in.	#5@ 56 in.	#5@ 72 in.	#6@ 48 in.	#5@ 56 in.	#5@ 72 in.
	7 ft.	#5@ 48 in.	#5@ 56 in.	#4@ 72 in.	#6@ 48 in.	#6@ 56 in.	#6@ 72 in.	#6@ 32 in.	#6@ 48 in.	#6@ 72 in.
	8 ft.	#6@ 48 in.	#5@ 56 in.	#5@ 72 in.	#6@ 32 in.	#6@ 48 in.	#6@ 72 in.	#6@ 24 in.	#6@ 40 in.	#6@ 48 in.
	9 ft.	#6@ 40 in.	#6@ 56 in.	#6@ 72 in.	#6@ 24 in.	#6@ 40 in.	#6@ 56 in.	#6@ 16 in.	#6@ 24 in.	#6@ 40 in.
	10 ft.	#6@ 32 in.	#6@ 48 in.	#6@ 64 in.	#6@ 16 in.	#6@ 32 in.	#6@ 40 in.	#6@ 16 in.	#6@ 24 in.	#6@ 32 in.

RETAINING WALLS

Retaining walls that are not part of a building are unrestrained–there is no basement slab or floor system to hold them in place–and they must incorporate design that will resist the lateral pressure of soils. Though the IRC does not require a building permit for retaining walls up to 4ft. in height, they do require proper design per IBC specifications for walls that are taller than 2ft. Retaining walls usually have drainage systems to relieve the hydrostatic pressure behind them.

Retaining Walls

- ☐ Design req'd for retaining walls w/o lateral support & retaining >24in unbalanced fill ______ [404.5][10]
- ☐ Safety factor of 1.5 against lateral sliding & overturning ______ [404.5][11]

FIG. 10

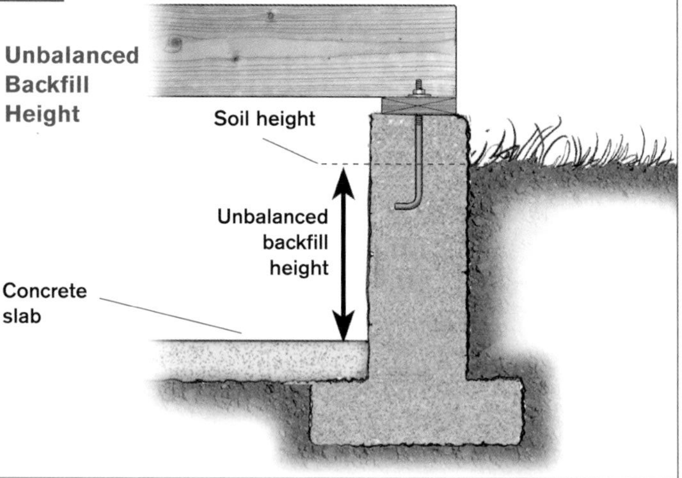

FIG. 11

Aspect Ratios for Unbalanced Foundations

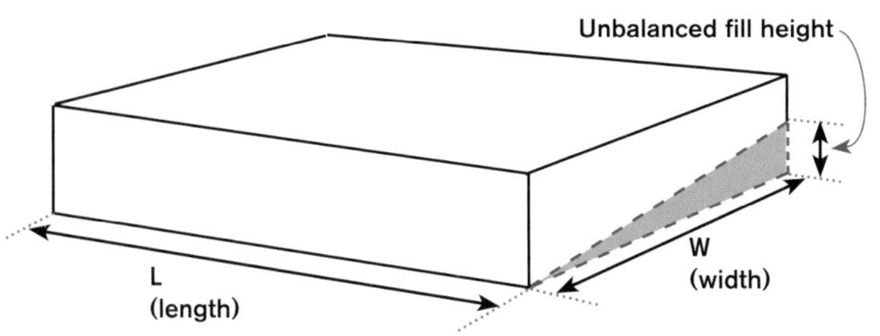

BASEMENT WALL LATERAL SUPPORT

Basement walls are restrained by the floor slab at the bottom and the floor system at the top. Unbalanced basements require special means of securing the floor system to transfer loads to the walls. New provisions in the 2006 IRC specify the minimum methods of attaching the floor to the walls, including anchor bolts and framing hardware. The aspect ratio limitations in **T11** are based on a plywood floor diaphragm with nailing 6 in. o.c.

Lateral Support

- ☐ Foundation walls req lateral support if >48in unbalanced backfill __ [404.1.4]
- ☐ Foundation walls are considered laterally supported if ALL of the following are met: ______ [404.1][12]
  - Basement floor 3½in slab tight against wall **F10**
  - Joist & blocking hardware connections to sill per **T11**
  - Sill bolt spacing per soil type & backfill height per **T11**
  - Joist blocking full-depth first 2 joist spaces parallel to foundation wall, flat 2×4 blocking for rest of joists
  - Unbalanced basements **F11** limited to values of **F11**

TABLE 11 REQUIREMENTS FOR LATERAL RESTRAINT OF BASEMENT WALLS [T404.1(1)(2)(3)]

Max. Wall Height (ft.)	Max. Unbalanced Backfill Height (ft.)	Connection Type (indicated by letter) Anchor Bolt Spacing (middle number, in inches) Max. L/W Aspect Ratio (decimal number)		
		Soil Class (see T3)		
		GW, GP, SW, & SP Soils	GM, GC, SM-SC, & ML Soils	SC, MH, ML-CL, & Inorganic CL Soils
7	4	A – 72 – 4.0	A – 58 – 4.0	A – 43 – 4.0
	5	A – 44 – 4.0	B – 30 – 3.4	B – 22 – 2.6
	6	B – 26 – 3.0	C – 17 – 2.0	C – 13 – 1.5
	7	C – 16 – 1.9	C – 11 – 1.2	D – 8 – 0.9
8	4	A – 72 – 4.0	A – 66 – 4.0	A – 50 – 4.0
	5	A – 51 – 4.0	B – 34 – 3.9	B – 25 – 2.9
	6	B – 29 – 3.4	B – 20 – 2.3	C – 15 – 1.7
	7	B – 18 – 2.1	C – 12 – 1.4	C – 9 – 1.1
	8	C – 12 – 1.4	C – 8 – 1.0	D – 6 – 0.7
9	4	A – 72 – 4.0	A – 72 – 4.0	A – 56 – 4.0
	5	A – 57 – 4.0	B – 38 – 4.0	B – 29 – 3.3
	6	B – 33 – 3.8	B – 22 – 2.6	C – 17 – 1.9
	7	B – 21 – 2.4	C – 14 – 1.6	C – 10 – 1.2
	8	C – 14 – 1.6	C – 9 – 1.1	D – 7 – 0.8
	9	C – 10 – 1.1	D – 7 – 0.8	D – 5 – 0.6

A = Three 8d per joist per **T14**.
B = One 20-gauge angle clip each joist w/ five 8d per leg.
C = 1 1/4-in.-thick steel angle w/ horiz. leg attached to sill bolt, vert. leg attached to joist/blocking w/ 1/2 in. min. diameter bolt.
D = 2 1/4-in.-thick steel angles, 1 on each side of joist/blocking, horiz. legs attached to sill bolts, vert. legs attached w/ 1/2in. min. diameter bolt through both angles.

DRAINAGE

Groundwater must be directed away from the foundation. If soils have good drainage characteristics, proper surface drainage and grading may be adequate. For poorly draining soils, drainage systems must be installed, ex: drain tiles or piping, filter fabric, and drainage rock. Retaining walls on the site should be designed in accord with the IBC.

Drainage

- ☐ Concrete & CMU fndns enclosing habitable or usable space below grade req drains EXC **F12,13** ______ [405.1]
  Drain system not req'd on well-drained ground or sand/gravel mixture [405.1X]
- ☐ Filter fabric req'd over top of drain field **F12** ______ [405.1]
- ☐ Min 2in of crushed rock under drain pipe **F12** ______ [405.1]
- ☐ Roof drain must discharge at least 5ft from footing or to approved drain system if soils are expansive or collapsible **F12** ______ [801.3]

FIG. 12 Foundation & Downspout Drainage

FIG. 13

Pipes Parallel to Footings

45°

Pipes or excavations not allowed within 45° angle of footings

Waterproofing, Dampproofing, & Backfill

- ☐ Walls retaining earth & enclosing interior space:
  - Dampproofing to finished grade OR ______ [406.1]
  - If high water table, waterproofing to finished grade ______ [406.2]
- ☐ CMU must be parged before dampproofing ______ [406.1]
- ☐ Joints in waterproofing lapped & sealed ______ [406.2]
- ☐ Do not install backfill until walls anchored to floor EXC ______ [404.1.7]
  Bracing not req'd for walls supporting <4ft of unbalanced backfill_ [404.1.7X]

CONCRETE SLABS

Concrete slabs that serve as floors require vapor retarders to prevent moisture from rising into the building and damaging the floors or creating indoor air-quality problems, such as mold. Slabs require reinforcement to resist cracking and upheaval.

Concrete Slabs

- ☐ Slab-on-ground floors: min 3½in thick ______ [506.1]
- ☐ Width & depth of footing under interior bearing walls: per **T4** ______ [F403.1(1)]
- ☐ BO may OK slab on expansive soil if adequate performance found in similar soil ______ [403.1.8]
- ☐ Remove all vegetation, top soil, & debris ______ [506.2]
- ☐ Fill compacted & free of vegetation & foreign material ______ [506.2.1]
- ☐ Treat soils in areas subject to termite infestation ______ [320.1]

Concrete Slabs (cont.)

- ☐ Base ¼in sand, gravel, or crushed stone <2in diameter EXC ______ [506.2.2]
  Base not req'd on Group I soils (good drainage–see **T3**) ______ [506.2.2X]
- ☐ Vapor retarder below slab, joints min 6in lap EXC ______ [506.2.3]
  Retarder not req'd under garages, walks, patios, unheated structures, or per BO ______ [506.2.3X]
- ☐ Reinforcing mesh reqs support between center & upper ⅓ of slab for duration of slab placement **F14** ______ [506.2.4][13]

SDC D_0, D_1, & D_2

- ☐ Slab w/ turned-down footing reqs #4 bars at top & bottom of footing [403.1.3.2]
- ☐ Monolithic slab w/ footing 1 #5 bar or 2 #4 in middle ⅓ of footing depth **F14** ______ [403.1.3.2X]

FIG. 14

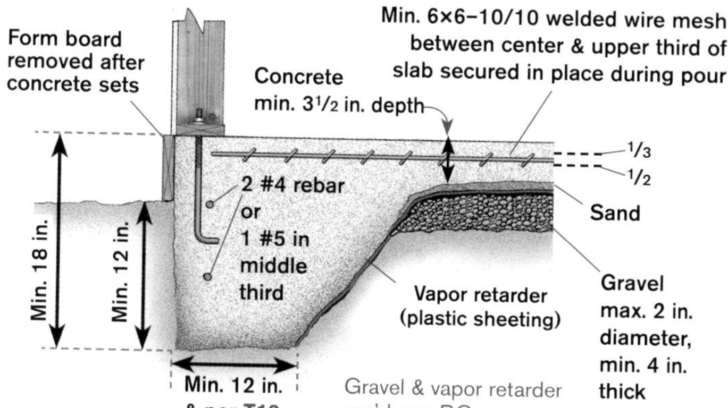

UNDERFLOOR

Houses with underfloor crawlspaces may require a plumbing and mechanical inspection before the floor can be installed. Also verify that proper clearances, ventilation, and access openings have been provided. Vent openings should not be blocked with insulation or mechanical equipment.

General

- ☐ Remove all vegetation, organic material, forms, & construction debris_ [408.5]
- ☐ Remove all wood forms **F6,9** ______ [408.5]

Access Openings

- ☐ Floor openings: min 18in × 24in ______ [408.4]
- ☐ Perimeter wall openings: min 16in high × 24in wide ______ [408.4]
- ☐ Provide full-depth exterior access well w/ footprint size at least 16in × 24in **F15** ______ [408.4]
- ☐ Perimeter wall opening *not* allowed under door ______ [408.4]
- ☐ Opening must allow removal of equipment in crawlspace ______ [1305.1.4]

Ventilation

- ☐ Min 1sq.ft per 150sq.ft of underfloor area ______ [408.1][14]
- ☐ Vent openings: within 3ft of each corner ______ [408.1]
- ☐ Each vent opening min 1sq.ft ______ [408.2]
- ☐ Vents must be covered w/ grates, grills, louvers, or corrosion-resistant screen ______ [408.2]
- ☐ Vents ≤3ft of PL OK ______ [302.1X3]
- ☐ Vents may be eliminated if continuous vapor retarder, lapped 6in, taped & 6in up stem wall & either: ______ [408.3][15]
  - Insulated perimeter walls & continuous mechanical ventilation, or
  - Insulated perimeter walls & conditioned air to underfloor area, or
  - Underfloor area used as a plenum

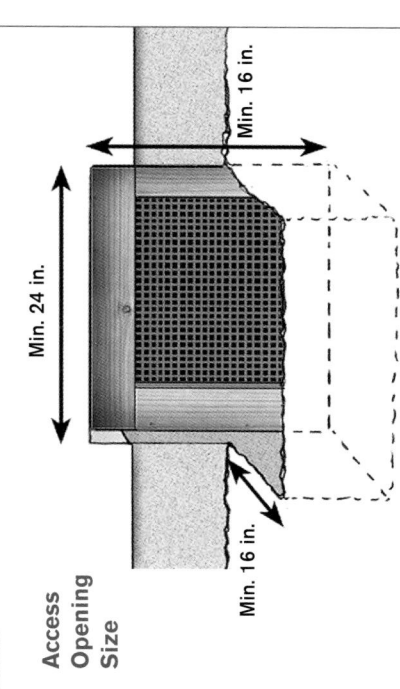

FIG. 15

PLATES & SILLS

Wood plates on top of foundations must be protected against termites and moisture. They must be secured to the foundation.

Sills

- ☐ Concrete or masonry foundation wall to extend 6in above finished grade, 4in OK if masonry veneer **F16** ______ [404.1.6]
- ☐ Sills & posts <8in above earth must be PT or naturally decay resistant **F17** ______ [319.1]
- ☐ Soil clearance to untreated wood joists: min 18in **F16** ______ [319.1]
- ☐ Soil clearance to untreated wood beams: min 12in **F16** ______ [319.1]
- ☐ Fasteners for treated wood to be corrosion-resistant exc bolts ≥$^1/_2$in ______ [319.3]

FIG. 16

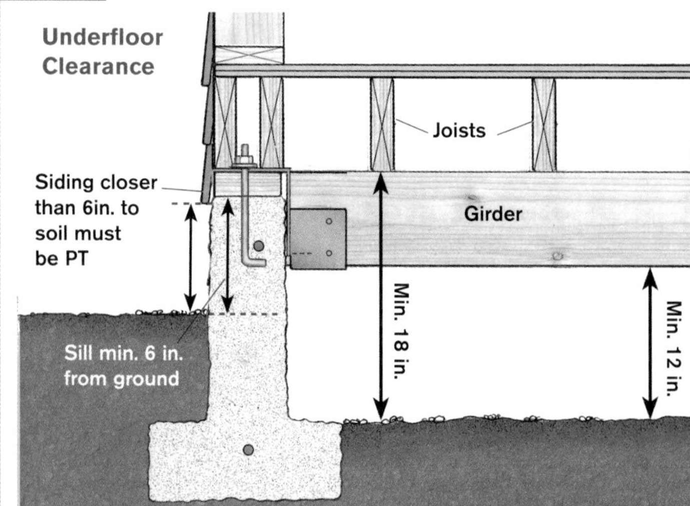

CRIPPLE WALLS

Cripple walls (short walls between the foundation and the lowest floor) must be braced to resist lateral loads.

Cripple Walls

- ☐ Cripple studs no less than size of studs above them ______ [602.9]
- ☐ Cripple walls >4ft height: sized as additional story ______ [602.9]
- ☐ Cripple wall <14in height: fully sheathed or solidly blocked ______ [602.9]
- ☐ Cripple wall bracing: per table in SDC D_2 (see **p. 32**) **T18** ______ [602.10.2.2]

Cripple Walls (cont.)

- ☐ Other than SDC D_2 brace as req'd for floor above & decrease max spacing to 18ft, increase bracing percentage by 15% ______ [602.10.2.1]
- ☐ OK to redesignate any cripple wall as 1st-story wall for bracing [602.10.2.3]

POSTS, COLUMNS & GIRDERS

Most houses with crawl spaces are built over concrete piers and posts that support wood girders. Clearance must be maintained between the soil and the wood that is not pressure-preservative treated. The placement of interior bearing walls must be considered in laying out the locations of piers, posts, and girders.

Posts & Columns

- ☐ Posts closer than 8in to ground must be PT ______ [407.1]
- ☐ Steel: min 3in diameter, wood columns min 4×4 **F17** ______ [407.3]
- ☐ Steel columns must be corrosion protected ______ [407.2]
- ☐ Bottom end of columns req restraint to prevent lateral displacement EXC_ [407.3]
  SDC A, B, & C columns enclosed by foundation exempt **F17** ______ [407.3X]
- ☐ Masonry piers max height 10× any dimension of pier ______ [606.6]
- ☐ Unfilled hollow piers limited to 4× any dimension of pier ______ [606.6]
- ☐ Hollow CMU piers req min 4in concrete cap or fill in top course **F28** _[606.6.1]

Girders

- ☐ Girder spans per **T12,16** ______ [502.5]
- ☐ Spacing per bearing wall location & joist span **T13** ______ [502.3]
- ☐ Girder end joints must be over supports ______ [502.6]
- ☐ Girder bearing on concrete min 3in ______ [502.6]
- ☐ Girder joints over supports must be tied **F17** ______ [502.9]
- ☐ Max 1 joist depth horiz under perpendicular bearing walls **F21** ______ [502.4]
- ☐ Holes max 1/3 girder depth **F18** ______ [502.8.1]
- ☐ Holes min 2in from top or bottom **F18** ______ [502.8.1]
- ☐ No notches in middle third of span or on tension side of ≥4in thick members except at ends ______ [502.8.1]

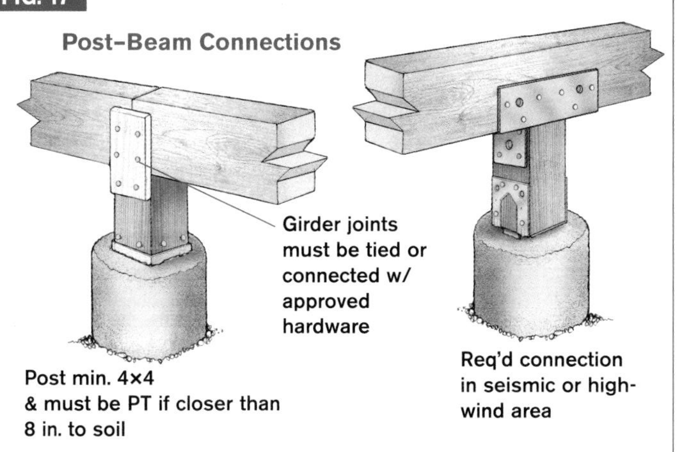
FIG. 17
Post–Beam Connections
Girder joints must be tied or connected w/ approved hardware
Post min. 4×4 & must be PT if closer than 8 in. to soil
Req'd connection in seismic or high-wind area
FIG. 18
Notching & Boring: Joists & Girders
No notching in middle third, holes OK
Notch depth max. 1/6 joist depth, notch length max. 1/3 joist depth
2 in. min.
Holes must be min. 2 in. from edges or other holes, max. size 1/3 depth.
Notch at end 1/4 of depth max.
Outer third
No notching at bottom if ≥4 in. thick, except at ends

TABLE 12 GIRDER & HEADER SPANS[A] IN FEET–INCHES [T502.5(1)]

Support	Min. Header Size	Building Width[B]		
		20 ft.	28 ft.	36 ft.
		Span	Span	Span
Roof & ceiling	2–2×4	3–6	3–2	2–10
	2–2×6	5–5	4–8	4–2
	2–2×8	6–10	5–11	5–4
	2–2×10	8–5	7–3	6–6
	2–2×12	9–9	8–5	7–6
Roof, ceiling, & 1 center–bearing floor	2–2×4	3–1	2–9	2–5
	2–2×6	4–6	4–0	3–7
	2–2×8	5–9	5–0	4–6
	2–2×10	7–0	6–2	5–6
	2–2×12	8–1	7–1	6–5
Roof, ceiling, & 1 clear–span floor	2–2×4	2–8	2–4	2–1
	2–2×6	3–11	3–5	3–0
	2–2×8	5–0	4–4	3–10
	2–2×10	6–1	5–3	4–8
	2–2×12	7–1	6–1	5–5
Roof, ceiling, & 2 center–bearing floors	2–2×4	2–7	2–3	2–0
	2–2×6	3–9	3–3	2–11
	2–2×8	4–9	4–2	3–9
	2–2×10	5–9	5–1	4–7
	2–2×12	6–8	5–10	5–3

A. Assuming #2 grade lumber & 30 psf ground snow load.
B. For widths between those shown, spans may be interpolated.
For header spans of interior bearing walls, use **T16**.

FLOOR FRAMING

Floor framing supports live and dead loads in a building and also transfers lateral loads to the wall structures. Horizontal framing members must have sufficient size to support these loads and must not be damaged by excessive or improper notching or boring.

General–Standard Dimensional Sawn Lumber

- ☐ Joist spans: not to exceed max numbers in table **T13** [502.3]
- ☐ Min end bearing: 3in on concrete, 1 1/2in on wood [502.6]
- ☐ Max 1 joist depth horiz under perpendicular bearing walls **F21** [502.4]
- ☐ Double joists under parallel bearing walls **F19** [502.4]
- ☐ Holes max size 1/3 of joist depth **F18** [502.8.1]
- ☐ Holes min 2in from top, bottom, notches, or other holes **F18** [502.8.1]
- ☐ No notches middle 1/3 of span **F18** [502.8.1]

Joist or Beam Bearing

- ☐ On concrete or masonry: min 3in [502.6]
- ☐ On wood plates: min 1 1/2in [502.6]
- ☐ Joists into side of girder req hanger, or min 2×2 ledger [502.6.2]
- ☐ Joist lap min 3in & 3 10d nails **F20** [502.6.1]

Joist Blocking

- ☐ Joists blocked at all ends [502.7]
- ☐ Blocking req'd at intermediate supports in SDC D_0, D_1, or D_2 [502.7X]
- ☐ Joists >2×12 blocked, bridged, or 1×3 backer strip @ 8ft o.c. [502.7.1]

Framing at Openings

- ☐ Single header & trimmer joists: OK to 4ft header span [502.10]
- ☐ Double header & trimmer when header spans >4ft [502.10]
- ☐ Header joists >6ft reqs joist hanger [502.10]
- ☐ Tail joists >12ft req joist hangers or min 2×2 ledger [502.10]

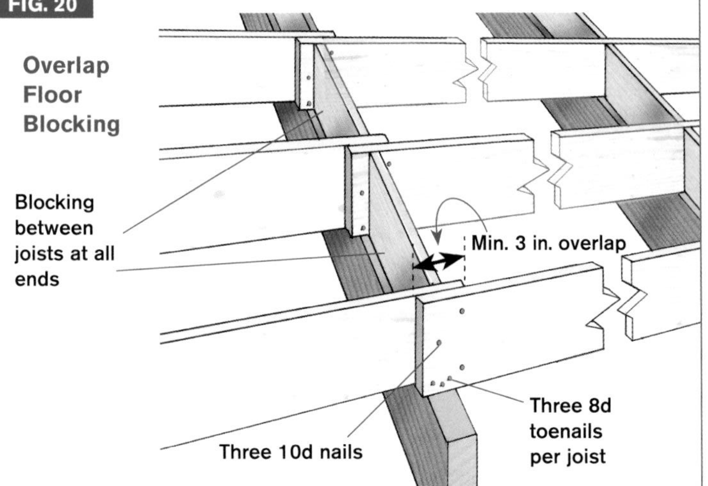

TABLE 13 — JOIST SPANS, NOTCHING, & BORING [T502.3.1(2)] & [502.8.1]

Floor Joist Spans—40 lb. Live Load				Notching		Boring
DF#2	12 in. o.c.	16 in. o.c.	24 in. o.c.	End	Outer 1/3	2 in. to edge
2×6	10 ft. 9 in.	9 ft. 9 in.	8 ft. 1 in.	1 3/8 in.	7/8 in.	1 1/2 in.
2×8	14 ft. 2 in.	12 ft. 7 in.	10 ft.3 in.	1 7/8 in.	1 1/2 in.	2 3/8 in.
2×10	17 ft. 9 in.	15 ft. 5 in.	12 ft. 7 in.	2 3/8 in.	1 1/2 in.	3 1/8 in.
2×12	20 ft. 7 in.	17 ft. 10 in.	14 ft. 7 in.	2 7/8 in.	1 7/8 in.	3 1/2 in.

FIG. 21 Bearing Wall Suppport

45° max.

Bearing walls should not offset more than the joist depth from the supporting girder below the floor.

STEEL FRAMING

Steel framing is becoming more popular as the costs of lumber increase and the search for straight lumber becomes more difficult. Many new homes are framed with dimensional and engineered lumber for exterior walls and steel for interior nonbearing walls. Steel can also be the sole framing material of the house. While not specified in the code, insulating pads on joists, studs, and trusses can greatly reduce noise in steel-framed houses. The IRC has extensive prescriptive requirements for steel framing.

Steel Framing

- ☐ Steel floor framing straight & free of defects ____ [505.1,603.1]
- ☐ Load-bearing steel must be marked w/ manufacturer's identification, thickness coating designation, & yield strength ____ [505.2.2, 603.2.2]
- ☐ Limited to buildings 60ft in length perpendicular to joist span ____ [505.1.1,603.1.1]
- ☐ Limited to buildings 40ft in length parallel to joist span ____ [505.1.1,603.1.1][16]
- ☐ Limited to buildings 2 stories in height ____ [505.1.1,603.1.1][17]
- ☐ Not OK in areas w/ designed wind speed >110mph ____ [505.1.1,603.1.1]
- ☐ Studs in line w/ joists, rafters, & trusses max 3/4in offset ____ [603.1.2]
- ☐ Bearing stiffeners req'd at all bearing locations of floor joists ____ [505.3.4]
- ☐ No splicing of studs or structural members ____ [505.3.8, 603.3.6]
- ☐ Steel-to-steel connections: min 1/2in to edge and 1/2in center to center ____ [505.2.4,603.2.4]
- ☐ Floor-sheathing screws: countersunk heads & min 3/8in from edge ____ [505.2.4]
- ☐ All screws: min 3 exposed threads through steel ____ [505.2.4,603.2.4]
- ☐ All exterior walls req wood structural panel sheathing ____ [603.7]
- ☐ Sheathing: fastened #8 min screws every 6in at edges12in field ____ [T603.3.2(1)]

FLOOR TRUSSES & WOOD I–JOISTS

Wood I-joists, laminated veneer lumber, metal-plate connected trusses, and other manufactured products are very commonly used in floor and ceiling systems. These have the advantage of dimensional uniformity and do not shrink after construction. It is essential that these engineered products be installed in accordance with manufacturer's instructions. They must not be altered except when permitted by the manufacturer's recommendations or as specified by a design professional.

General

- ☐ Trusses must conform to accepted engineering practice ________ [502.11.1]
- ☐ Bracing per drawings; if none shown, then follow BCSI 1-03 ____ [502.11.2]
- ☐ No alteration of trusses w/o approval of registered design professional ________ [502.11.3]
- ☐ No additional loads w/o verification of truss loading per drawings _ [502.11.3]
- ☐ Design drawings must be submitted to BO & w/ truss shipment __ [502.11.4]
- ☐ Cuts, notches, & holes in TJIs & other engineered lumber only where permitted by manufacturer's instructions or registered design professional **F22-25** ________ [502.8.2][18]

FIG. 22

Prefabricated I-Joists

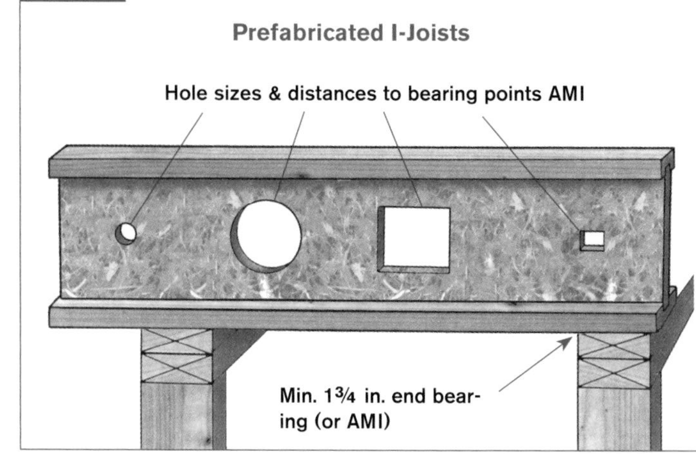

Prefabricated wood I-joists are shipped with their own span tables & installation instructions. They are designed to support floors. Point loads, such as bearing walls, require special support, such as the squash blocks shown here.

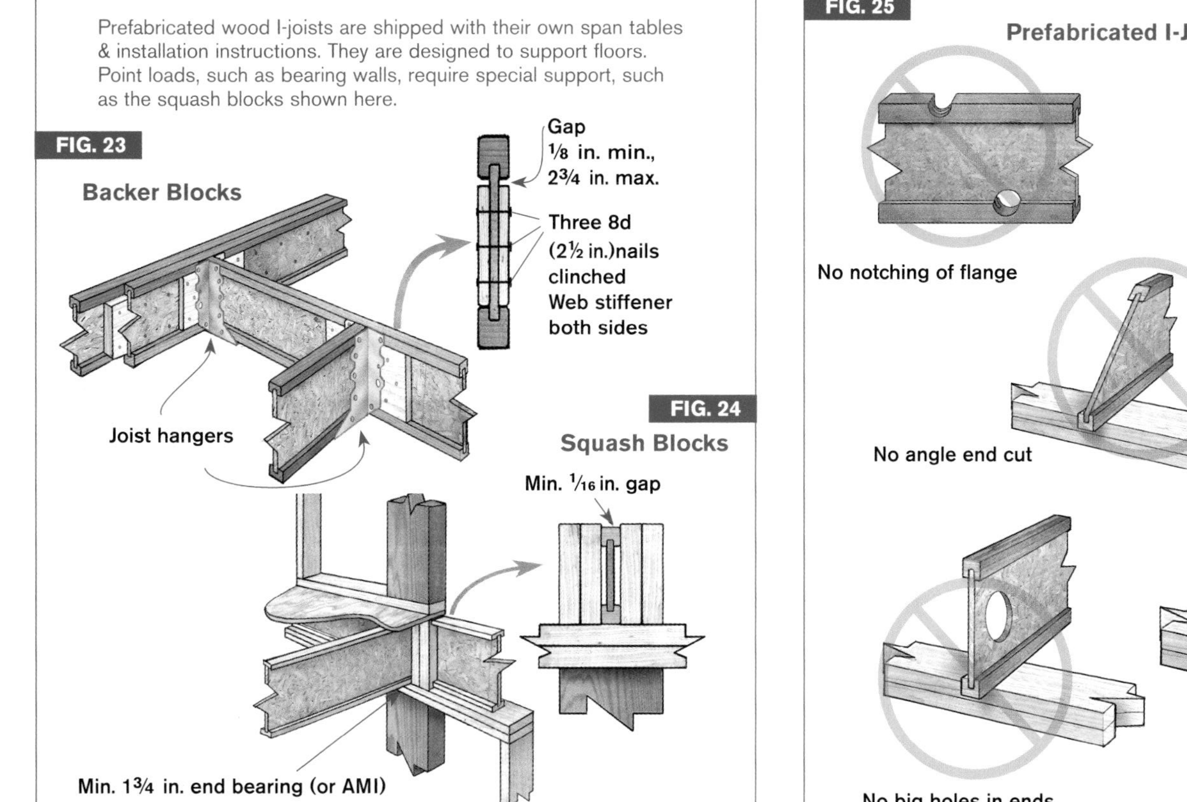

WOOD I-JOISTS

WALL FRAMING

The building frame must be capable of supporting the dead loads–the weight of the building and its fixed equipment–and the live loads–the weight of its occupants and furnishings. In addition, the frame must be capable of transmitting loads from wind or earthquakes to the building foundation.

Stud Framing

- ☐ Stud size & spacing: per **T15** _______ [602.3.1]
- ☐ End-jointed lumber OK if identified by grade mark _______ [602.1.1]
- ☐ Studs must fully bear on min 2in nominal thickness sole plate _______ [602.3.4]

FIG. 26

Notching and Boring Studs

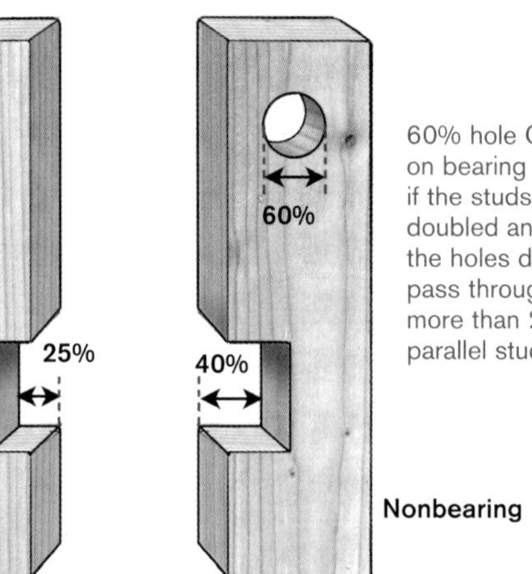
40%
25%
Bearing
60%
40%
Nonbearing
60% hole OK on bearing walls if the studs are doubled and the holes do not pass through more than 2 parallel studs.

Stud Notching & Boring

- ☐ Bearing or exterior wall: max notch 25% boring 40% **F26** _______ [602.6]
- ☐ 60% boring if doubled & ≤2 successive studs **F26** _______ [602.6]
- ☐ Nonbearing: max notch 40%, boring 60% **F26** _______ [602.6]
- ☐ Holes no closer than 5/8in to face of stud _______ [602.6]

Corner Framing

- ☐ Exterior corners: 3 studs min EXC **F27** _______ [F602.3(2)]
  2 studs OK if wood backup cleats, clips, or other approved material is installed as backing for facing materials _______ [F602.3(2)]
- ☐ Lap plates at corners _______ [602.3.2]

Top Plates

- ☐ Bearing wall intersections & corners must overlap _______ [602.3.2]
- ☐ End joints must offset 24in min _______ [602.3.2]
- ☐ Joints need not occur over studs _______ [602.3.2][19]
- ☐ Nailing: min 10d@24in o.c. **T14** _______ [602.3]
- ☐ Min 1 1/2in-wide 16 gauge straps over notches/holes ≥50% of plate width EXC _______ [602.6.1]
  If notch or cut is covered by structural panel _______ [602.6.1X]
- ☐ Strap must have at least 8 16d nails into plate on each side _______ [602.6.1]
- ☐ Single top plate OK if tied with 3in×6in steel plate w/ 6 8d nails & Joists/trusses/rafters land within 1in over stud _______ [602.3.2X]
- ☐ Openings <4ft in bearing walls req min 2 2×4 header _______ [602.7]
- ☐ Min header bearing: 1 1/2in _______ [T502.1(1) &(2)]
- ☐ Header spans: **T12,16** _______ [602.7, T502.5(1)&(2)]
- ☐ No header req'd for nonbearing walls–flat 2×4 OK up to 8ft _______ [602.7.2]

TABLE 14 NAILING SCHEDULE [T602.3(1)]

Connection	Nailing[A]
Joist to sill or girder, toe nail	3-8d
1×6 subfloor or less to each joist, face nail	2-8d or 2-1¾ staples
2 in. subfloor to joist or girder, blind & face nail	2-16d
Sole plate to joist or blocking, face nail	16d @16 in.o.c.
Top or sole plate to stud, end nail	2-16d
Stud to sole plate, toe nail	3-8d or 2-16d
Double studs, face nail	10d @ 24 in. o.c.
Double top plates, face nail	10d @ 24 in. o.c.
Sole plate to joist or blocking at braced wall panels	3-16d @ 16 in. o.c.
Doubled top plates, face nail of lap splice	8-16d
Blocking between joists or rafters to top plate, toe nail	3-8d
Rim joist to top plate, toe nail	8d 6 in. o.c.
Top plates, laps, & intersections, face nail	2-10d
Built-up or continuous header	16d @ 16 in. o.c. each edge
Ceiling joists to plate, toe nail	3-8d
Continuous header to stud, toe nail	4-8d
Ceiling joists, laps over partitions, face nail	3-10d
Ceiling joists to parallel rafters, face nail	3-10d
Rafter to plate, toe nail	2-16d
1 in. brace to each stud and plate, face nail	2-8d or 2-1¾ staples
1×6 sheathing to each bearing, face nail	2-8d or 2-1¾ staples
1×8 sheathing to each bearing, face nail	2-8d or 3- 1¾ staples
Wider than 1×8 sheathing to each bearing, face nail	3-8d or 4- 1¾ staples
Built-up corner studs	10d @ 24 in.o.c.
Built-up girders and beams, 2 in. lumber layers	10d each layer @ 32 in.o.c. top & bottom staggered & 2 nails at ends & each splice
2 in. planks	2-16d @ each bearing
Roof rafters to ridge, valley, or hip rafters, toe nail	4-16d
Roof rafters to ridge, valley, or hip rafters, face nail	3-16d
Rafter ties to rafters, face nail	3-10d
Collar ties to rafters, face nail	3-10d
Wood structural panels 5⁄16 in. to ½ in. on walls	6d 6 in. o.c. edges 6d 12 in. o.c. field[B]
Wood structural panels 5⁄16 in. to ½ in. on roofs	8d 6 in. o.c. edges 8d 12 in. o.c. field
Wood structural panels 19⁄32 in. to 1 in.	8d 6 in. o.c. edges 8d 12 in. o.c. field
Wood structural panels 1⅛ in. to 1¼ in.	10d common or 8d deformed shank 6 in. o.c. edges 12 in. o.c. field

A. Sheathing nails common except as noted, others smooth common, box, or deformed shank.
B. Gable ends req. 6 in. o.c. field nailing.

WALL FRAMING

FIG. 27

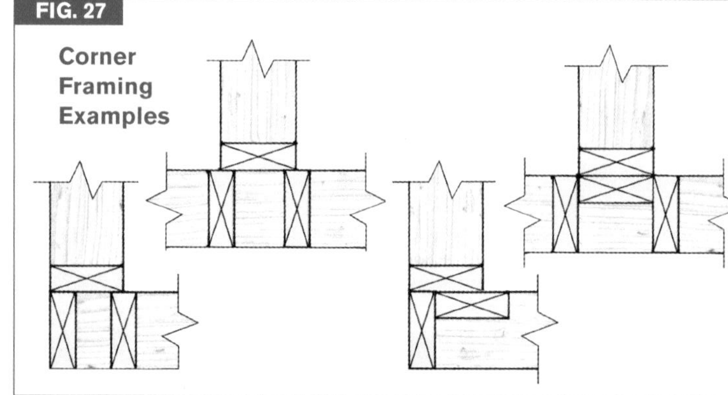

TABLE 15	STUD SIZING, SPACING, NOTCHING, & BORING [T602.3(5)] & [602.6]		
Stud Size	**2×4**	**3×4**	**2×6**
Bearing Walls (to 10ft. high)			
Supporting roof & ceiling	24 in. o.c.	24 in. o.c.	24 in. o.c.
Roof & ceiling + 1 floor	16 in. o.c.	24 in. o.c.	24 in. o.c.
Roof & ceiling + 2 floors	n/a	16 in. o.c.	16 in. o.c.
Notching **F26**	7/8 in.	7/8 in.	1 3/8 in.
Boring **F26**	1 3/8 in.	1 3/8 in.	2 3/16 in.
Boring 2 doubled consec.	2 in.	2 in.	3 1/4 in.
Nonbearing Walls			
Notching **F26**	1 3/8 in.	1 3/8 in.	2 3/16 in.
Boring **F26**	2 in.	2 in.	3 1/4 in.

TABLE 16	GIRDER SPANS FOR INTERIOR BEARING WALLS[A] (FT.-IN.) [T502.5(2)]			
# of Floors Supported	Min. Girder Size	Building Width[B]		
		20	28	36
1	2–2×6	4–6	3–11	3–6
	2–2×8	5–9	5–0	4–5
	2–2×10	7–0	6–1	5–5
	2–2×12	8–1	7–0	6–3
	3–2×8	7–2	6–3	5–7
	3–2×10	8–9	7–7	6–9
	3–2×12	10–2	8–10	7–10
2	2–2×6	3–2	2–9	2–5
	2–2×8	4–1	3–6	3–2
	2–2×10	4–11	4–3	3–10
	2–2×12	5–9	5–0	4–5
	3–2×8	5–1	4–5	3–11
	3–2×10	6–2	5–4	4–10
	3–2×12	7–2	6–3	5–7

A. Assuming #2 grade lumber.
B. For widths between those shown, spans may be interpolated.

T16 *applies to either headers or floor girders supporting interior nonbearing walls.*

T12 *is used for headers or girders supporting bearing walls.*

CONCRETE MASONRY UNITS (CMUS)

CMUs (or concrete blocks) are used for basement walls, foundation stem walls, piers, and walls above grade. In hurricane country, they are often used because the mass of the blocks provides a more naturally wind-resistant structure. CMUs require lateral support and reinforcement. Engineered design is not needed when following the prescriptive requirements of the IRC, or *ACI 530–Building Code Requirements for Masonry Structures*, published by the American Concrete Institute.

General CMU Wall Requirements

- ☐ 6in block OK for 1 story to 9ft + 6ft of gable ____ [606.2.1]
- ☐ 8in block if more than 1 story or >9ft ____ [606.2.1]
- ☐ Beam connections: 1/2in air space on 3 sides **F7** ____ [319.1]
- ☐ Lateral support req'd in vert or horiz direction ____ [606.9]
- ☐ Lateral support spacing: per **T17** ____ [606.9]
- ☐ Anchor all roof & floor structures to masonry walls **F28** ____ [606.11]
- ☐ SDC A,B,C, & <30psf wind load–bolt embedment min 4in ____ [F606.11(1)]
- ☐ SDC C, D_0, D_1, & D_2 bolt embedment min 5in ____ [F606.11(1)&(2)]
- ☐ SDC C, D_0, D_1, & D_2 lateral ties around bolts & vert steel ____ [F606.11(1)&(2)]

Grout

- ☐ All cells w/ reinforcement must be filled **F28** ____ [606.13]
- ☐ Cleanouts req'd at bottom of each grouted cell for pours >4ft ____ [609.1.5.2]
- ☐ Grout continuous pour max lift 5ft ____ [609.1.4]
- ☐ Clean grout space–max 1/2in projections ____ [609.1.3]

Reinforcing for Masonry Units

- ☐ SDC D_0, D_1 & D_2: Min #3 vertical bars max 4ft o.c. & tied to footing reinforcement ____ [404.1.4]
- ☐ Vert. rebar ≤6% of grout space ____ [T609.1.2]
- ☐ Lap rebar splices 40× bar diameter ____ [F606.10(2)]
- ☐ Support/positioners: min 200 bar diameters (8ft #4 bar) ____ [609.4.1]
- ☐ Min cover: 2in to weather or soil ____ [606.12]

FIG. 28

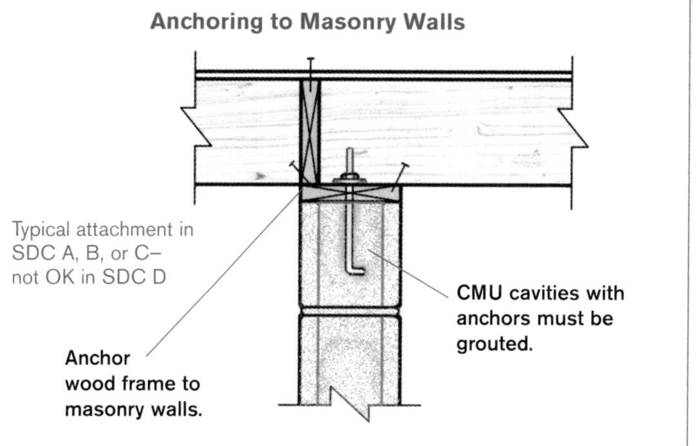
Anchoring to Masonry Walls
Typical attachment in SDC A, B, or C–not OK in SDC D
Anchor wood frame to masonry walls.
CMU cavities with anchors must be grouted.

TABLE 17 SPACING OF LATERAL SUPPORT FOR MASONRY WALLS [606.9]

Construction		Max. wall length to thickness[A] or wall height[B] to thickness
Bearing walls	Solid or solid grouted	20
	All other	18
Nonbearing walls	Exterior	18
	Interior	36

A. Thickness is nominal thickness perpendicular to face of wall, including sum of wythes.
B. An additional unsupported height of 6 ft. is permitted for gable walls.

BRACING & STRUCTURAL SHEATHING

Walls need to be braced to resist the lateral loads imposed by wind forces and earthquakes. The type and amount of bracing depend on the basic wind speed, the Seismic Design Category, and the building loads that the walls are supporting. In addition to the commonly used bracing rules shown here, there are other methods, such as continuous wood panel sheathing, in 602.10 of the IRC. We recommend these be reviewed with the building designer to determine the most appropriate bracing method for a particularly project.

Bracing Materials

- ☐ Amount & location of bracing per **T18** ______ [602.10.1]
- ☐ Braced panel construction methods: (**T18**) ______ [602.10.3]
  1. 1×4 let-in bracing from top to bottom plates @ 45°–60°angle from horiz.
  2. 5/8in diagonal wood sheathing on studs not >24in o.c.
  3. Structural sheathing min 5/16in for studs 16in o.c., min 3/8in for studs 24in o.c.
  4. 1/2in or 25/32in structural fiberboard sheathing installed vert or horiz on studs 16in o.c.
  5. Gypsum board min 1/2in thick on studs 24in o.c., nailed not screwed per **T21**
  6. Particleboard on studs 16in o.c.
  7. Portland cement plaster (hardcoat stucco) on studs 16in o.c.
  8. Hardboard panel siding min 7/16in thick

Bracing Installation

- ☐ Braced panels must begin within 12 1/2ft of end of braced wall line [602.10.1]
- ☐ All panel sheathing vert joints over common studs ______ [602.10.7][20]
- ☐ All panel sheathing horiz joints over common blocking min 1 1/2in thickness EXC ______ [602.10.7]
  Blocking not req'd in SDC A & B or detached dwellings in SDC C ______ [602.10.7X]
- ☐ Wood panels must have span rating on grade stamp **F41** ______ [604.1]
- ☐ Nail to schedule: min. 6in edge, 12in field ______ [T602.3(1&2)]
- ☐ Braced wall panels min width: 4ft exc let-in bracing & gypsum board [602.10.4]

TABLE 18 — WALL BRACING [TR602.10.1]

Seismic Design Category or Wind Speed	Condition	Type of Brace	Amounts of Bracing (25ft. o.c. & % of braced wall line)
A or B or ≤100 mph[A]	1 story	1–8	25 ft. o.c min. 16% for 2–8
	Top of 2 or 3 story		
	1st story of 2 story	1–8	25 ft. o.c., min. 16% for 3, 25% for 2, 4, 5, 6, 7, 8
	2nd story of 3 story		
	1st story of 3 story	2–8	25 ft. o.c., min. 25% for 3, 25% for 2, 4, 5, 6, 7, 8
C or <110 mph	1 story	1–8	25 ft. o.c., min. 16% for 3, 25% for 2, 4, 5, 6, 7, 8
	Top of 2 or 3 story		
	1st story of 2 story	2–8	25 ft. o.c., min. 30% for 3, 25% for 2, 4, 5, 6, 7, 8
	2nd story of 3 story		
	1st story of 3 story	2–8	25 ft. o.c., min. 45% for 3, 25% for 2, 4, 5, 6, 7, 8
D_0 & D_1 or <110 mph	1 story	2–8	25 ft. o.c., min. 20% for 3, 25% for 2, 4, 5, 6, 7, 8
	Top of 2 or 3 story		
	1st story of 2 story	2–8	25 ft. o.c., min. 45% for 3, 25% for 2, 4, 5, 6, 7, 8
	2nd story of 3 story		
	1st story of 3 story	2–8	25 ft. o.c., min. 60% for 3, 25% for 2, 4, 5, 6, 7, 8
D_2 or <110 mph	1 story	2–8	25 ft. o.c., min. 25% for 3, 25% for 2, 4, 5, 6, 7, 8
	Top of 2 or 3 story		
	1st story of 2 story	2–8	25 ft. o.c., min. 55% for 3, 25% for 2, 4, 5, 6, 7, 8
	Cripple walls[B]	3	25 ft. o.c., min. 75%

A. If wind speed >100mph & <110mph, use row for Seismic Design Category C.
B. See section on cripple walls on **p. 22**

Additional SDC D_0, D_1, & D_2 Requirements

- ☐ Spacing between braced wall lines: 1 room 35ft, rest max 25ft ____ [602.10.11.1]
- ☐ Braced wall panel at end of each braced wall line OR ____ [602.10.11.2]
  Structural sheathing within 8ft of corner & 24in-wide panels at corner OR ____ [602.10.11.2X1]
  Hold-downs each end of structural sheathed panels closest to corners ____ [602.10.11.2X2]

ALTERNATE BRACED WALL PANELS

Alternate braced wall panels are designed for higher strength and allow for narrower panels with greater height-to-width aspect ratios. These panels also include tie-downs (more commonly referred to as "hold-downs") at each end to secure the assembly. Though they can be assembled on site, manufactured panels are usually used for these purposes. A common application is the short wall at each side of a garage door opening.

- ☐ Alternate braced wall panels permitted to replace other bracing methods ____ [602.10.3X]
- ☐ Min construction 3⁄8 in structural sheathing nailed & blocked all edges ____ [602.10.6.1]
- ☐ 1st story of 2-story buildings: sheath both sides ____ [602.10.6.1]
- ☐ Anchor bolts at panel 1⁄4 points, hold-downs in end studs **F29** ____ [602.10.6.1]
- ☐ Hold-down bolts embedded & installed AMI **F29** ____ [602.10.6.1]
- ☐ Max height & min width: per **T19** ____ [602.10.6.1][21]
- ☐ Manufactured products OK w/ ICC report meeting performance of this section **F29** ____ [104.11]

FIG. 29

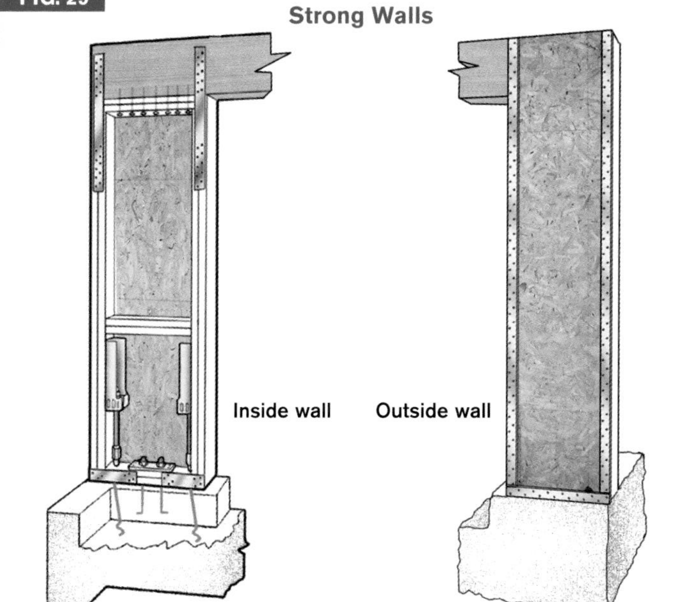

"Strong walls" are an example of a manufactured shear assembly designed for specific applications, such as the walls at the sides of a garage door. They can be a time saver in situations in which the aspect ratio (height to width) is limited & would otherwise req. steel moment frames or heavy site-constructed shear assemblies. They should bear an evaluation agency approval & be installed according to the exact specifications of the manufacturer.

TABLE 19 — MIN. WIDTHS & TIE-DOWN FORCES OF ALTERNATE BRACED WALL PANELS [T602.10.6]

Seismic Design Category & Wind-Speed	Tie-Down Force (lb)	Height of Braced Wall Panel				
		Sheathed Width				
		8 ft.	9 ft.	10 ft.	11 ft.	12 ft.
		2 ft. 4 in.	2 ft. 8 in.	2 ft. 8 in.	3 ft. 2 in.	3 ft. 6 in.
SDC ABC & wind <110 mph	1-story bldg	1,800	1,800	1,800	2,000	2,200
	1st of 2-story	3,000	3,000	3,000	3,300	3,600
SDC D_0, D_1, & D_2 & wind <110 mph		Sheathed Width				
		2 ft. 8 in.	2 ft. 8 in.	2 ft 8 in.	Ø	Ø
	1 story bldg	1,800	1,800	1,800	Not allowed–max. height is 10 ft.	
	1st of 2-story	3,000	3,000	3,000		

HOLD-DOWNS

Wind and seismic forces exert an overturning force against a building. Hold-downs in conjunction with wall bracing help resist these forces. Hold-downs must be installed in accordance with the manufacturer's instructions.

Hold Downs

- ☐ Alternate braced wall panels req hold-downs at each end stud _ [602.10.6.1]
- ☐ Bolts embedded & installed AMI ______ [602.10.6.1]
- ☐ Load path for lateral force must be provided between floor framing & braced walls panels above or below the floor framing ______ [502.2.1][22]
- ☐ Roof assemblies subject to wind uplift pressure ≥20psf req continuous load path to foundation ______ [802.11.1]

WIND UPLIFT

High-wind areas, such as much of the Gulf and eastern coasts, and Alaska, require reinforced framing connections to prevent building distortion and damage. In many ways, the requirements are similar to those in areas of high seismic activity. Fastening of roofing materials and siding also becomes a major consideration in high-wind areas. For regular-shaped buildings, SBCCI publishes a prescriptive code for hurricane-resistant construction, SSTD 10-99, which includes illustrations for ways to secure roof trusses and the building frame. Builders can use SSTD 10-99 and its prescriptive requirements for regular-shaped buildings–subject to approval by the local building official. In actual practice, most builders prefer to have engineering specifications provided for each project in a high-wind area. Uplift strength of different types of hardware is considered in the engineering drawings.

- ☐ Roof assemblies subject to uplift of ≥20psf req support by special hardware **F30** ______ [802.11.1]

FIG. 30

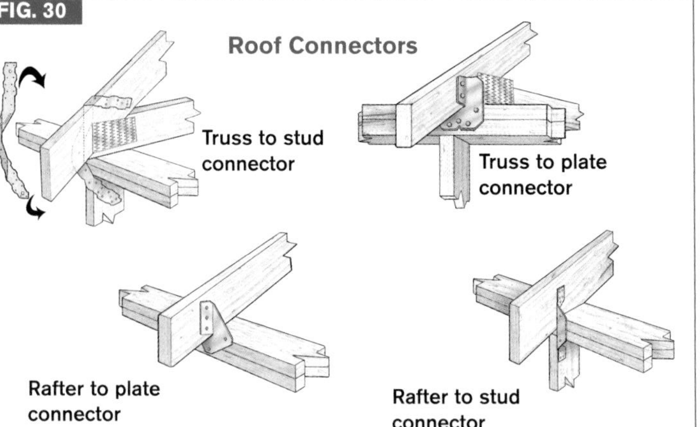

ATTICS

Water vapor migrates to attic spaces that are outside the building thermal envelope, and these spaces must be ventilated to prevent that vapor from condensing. Properly installed vapor retarders can reduce the amount of required ventilation. Insulation must be kept clear of ventilation openings. A new alternative allows the attic space to be part of the conditioned space

Access

- ☐ Attics >30in high & >30sq.ft req min 22in×30in access opening ____ [807.1]
- ☐ Min 30in unobstructed headroom at some point above opening ____ [807.1]
- ☐ Access opening must allow removal of mechanical equipment ____ [1305.1.3]

Ventilation

- ☐ Vent each enclosed attic & rafter bay ____ [806.1]
- ☐ Vent openings: req protective screen 1/8in to 1/4in mesh ____ [806.1]
- ☐ Total net-free vent opening area: min 1/150 of attic space area **F31** ____ [806.2]
- ☐ Reduction to 1/150 OK if 50%–80% of venting at least 3ft above eave or cornice vents & balance provided at eave, or cornice OR ____ [806.2]
  Warm-in-winter side of ceiling has vapor barrier not >1 perm ____ [806.2][23]
- ☐ Min 1in clearance between insulation & sheathing near vents **F31** ____ [806.3]

Insulation

- ☐ Installer must post certificate of R-value on materials not marked w/ R-value ____ [1101.4]
- ☐ Min value R-30, more in cold climates ____ [1102.1]
- ☐ Recessed lighting in building thermal envelope sealed to prevent air leakage: ____ [1102.4.3]
  - Non IC-rated lights in airtight insulated box w/ req'd clearances to light OR
  - IC-rated lights gasketed or w/ rated pressure differential & OK for insulation contact

Conditioned Attic Spaces

- ☐ OK if no interior vapor retarders & insulation on underside of roof deck [806.4][24]
- ☐ Must be installed to prevent dew point in roof deck ____ [806.4][24]

FIG. 31 Attic Ventilation

Ventilialation by ridge vent
or attic vent
Insulation held back to
allow 1in. air space.
Use baffles with loose
insulation.

WINDOWS & DOORS

Windows and doors must be selected to withstand the wind force that will act upon them. They must be secured to the building in accordance with manufacturer's recommendations to achieve their design strength. Windows must be flashed in accordance with the manufacturer's installation instructions to prevent leakage. Windows with a dropoff more than 6 ft. outside the window are a hazard to small children, and require protection.

General

- ☐ Exterior windows & sliding glass doors req 3rd-party testing & labeling [613.4]
- ☐ Windows & doors selected to resist design wind load of **T1** ____ [613.3]
- ☐ Windows & glass doors anchored to main structure AMI ____ [613.8.1]
- ☐ Windows flashed AMI **F37** ____ [613.8.1][25]
- ☐ Instructions for window flashing provided w/ each window ____ [613.8.1][26]
- ☐ Window sills >72in above grade: min 24in above finished interior floor EXC [613.2][27]
  Windows w/ guards or w/ all openings <4in ____ [613.2X][27]
- ☐ Safety glazing where req'd (see **p. 54 & 55**) **F70** ____ [308]

EXTERIOR WALL COVERS

Exterior wall systems are chosen for appearance, cost, function, and durability. The three basic methods of surface cladding are (1) the surface is the barrier to water entry (ex: wood panel siding), (2) a water-resistive membrane behind the surface protects the structure and drains water to the exterior (ex: stucco siding), and (3) a rain-curtain system in which the surface stops most water and an air space behind the surface drains the remaining water to flashings that extend to the exterior (ex: masonry veneer). All wall-cladding systems now require a water-resistive barrier behind the surface materials to protect the studs or sheathing.

General

- ☐ Exterior wall shall provide weather-resistant wall envelope ______ [703.1]
- ☐ Exterior wall shall prevent water accumulation & drain water to exterior ______ [703.1][28]
- ☐ Water-resistant barrier req'd over studs or sheathing of all exterior walls ______ [703.2][29]
- ☐ Type 1 felt or equivalent for barrier, lap 2in at horiz, lap 6in at vert ______ [703.2][30]
- ☐ Exterior sheathing must be dry before installing exterior cover ______ [701.2]
- ☐ Foam plastic sheathing board ≤½in thick OK as siding backer board ______ [314.2.5]
- ☐ All fasteners: corrosion resistive ______ [703.4]

Flashing

- ☐ Installed shingle fashion to prevent water reaching framing ______ [703.8][31]
- ☐ Windows flashed to exterior wall or to water-resistive membrane **F37** [703.8][32]
- ☐ Other req'd flashing locations: ______ [703.8]
  - Intersections of chimneys or other masonry w/ frame or stucco walls
  - Under & at ends of masonry, wood, or metal copings & sills **F35**
  - Continuously above all projecting wood trim
  - At attachments of exterior porches, decks, or stairs **F36,54**
  - At wall & roof intersections
  - At built-in gutters

Stucco–General

- ☐ Must comply w/ ASTM C 926 & ASTM C 1063 ______ [703.6][33]
- ☐ Lath fastener spacing: max 6in ______ [703.6.1]
- ☐ Not less than 3-coat system EXC **F32** ______ [703.6.2]
  2-coat OK over masonry or concrete or if concealed by other facing [703.6.2]
- ☐ Total thickness 3-coat work: ⅞in per ASTM C 926 ______ [703.6][34]
- ☐ Weep screed: req'd min 4in above earth 2in above paved areas **F34** [703.6.2.1]
- ☐ Lath must cover attachment flange of weep screed **F33,34** ______ [703.6.2.1]
- ☐ Water-resistive barrier at least equivalent to 2 layers Grade D paper EXC ______ [703.6.3][35]
  Rain-curtain (drain space between stucco & paper) 1 60-minute Grade D OK ______ [703.6.3X][36]
- ☐ Interval between coats **T20** ______ [T702.1(3)]
- ☐ No fasteners through top of horiz surfaces (copings, etc.) ______ [703.1]

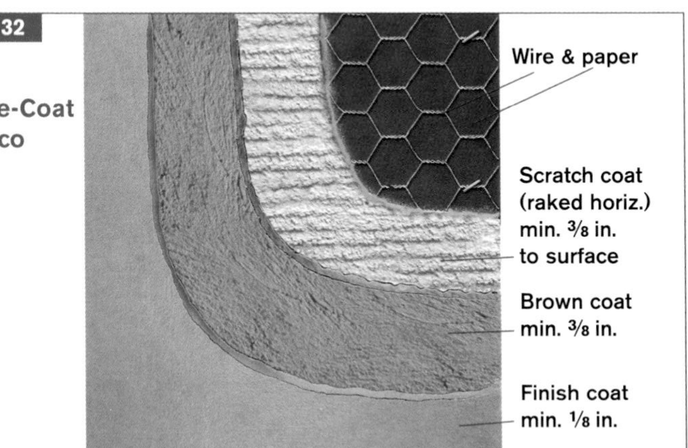

FIG. 32 Three-Coat Stucco

TABLE 20 PORTLAND CEMENT PLASTER (STUCCO) EXTERIOR WALL COVERING[A] [TR702.1(3)]

Coat	Portland Cement	Portland Cement Lime Plaster			Min. period moist coats	Min. interval before next coat
	Max. sand to cement volume ratio	Max. lime to cement volume ratio	Max. sand to cement + lime volume ratio	Approx. thickness during curing		
Scratch coat	4	3/4	4	3/8 in.	48 hr.	48 hr.
Brown coat	5	3/4	5	3/8 in. (3/4 in. min. total)	48 hr.	7 days
Finish coat	3	–	3	1/8 in.	n/a	n/a

A. ASTM C 926 allows different curing times depending on local climate.

FIG. 33 Weep Screed at Soffit

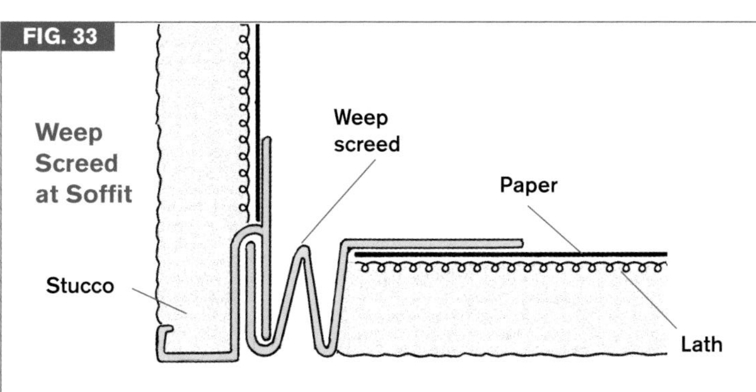

FIG. 34 Weep Screed

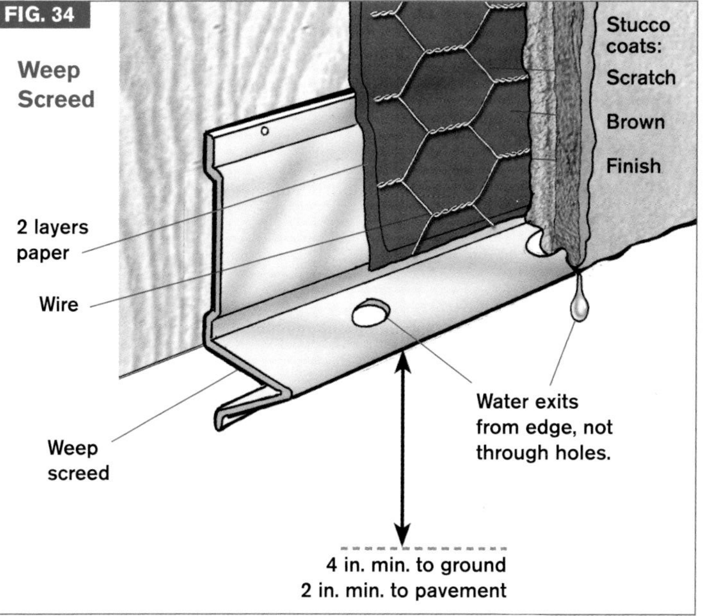

Exterior Insulation Finish Systems (EIFS)

- ☐ Must be installed AMI ____ [703.9]
- ☐ Must have water-resistive barrier behind exterior insulation ____ [703.9.1]
- ☐ Barrier Type 1 felt or equivalent, & must provide drainage to exterior [703.9.1]
- ☐ Terminate w/ flashing to exterior min 6in above finished ground level _ [703.9]
- ☐ No decorative trim face nailed through EIFS ____ [703.9]
- ☐ Lap barrier joints min 2in over lower layer, min 6in at vert joints ____ [703.9.1]

Masonry Veneer

- ☐ Max thickness in SDC A, B, & C 5in & 30ft height ____ [703.7X1][37]
- ☐ SDC C increase **T18** bracing by 50% for bottom of 2-story or bottom 2 floors of 3-story ____ [703.7X1, T703.7(1)]
- ☐ SDC D_0, D_1, & D_2 max thickness: 4in, max height: 20ft (30ft in 3-story SDC D_0) ____ [703.7X2, T703.7(2)][38]
- ☐ SDC D_0, D_1, & D_2 must have structural panel sheathing & increase bracing of **T18**, & must have hold-downs (exc 1-story in SDC D_0) ____ [703.7X2, T703.7(2)][38]

FIG. 35 Masonry Veneer

Metal ties max. 24 in. o.c.

Weep holes max. 33 in. o.c.

Min. 1in. air space

Sloped to drain

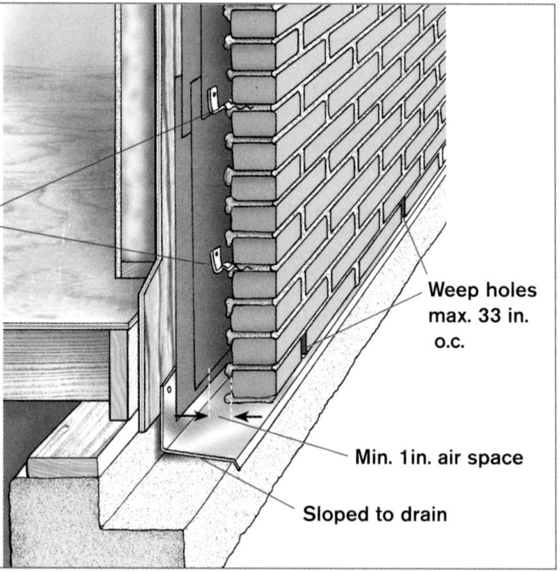

Masonry Veneer (cont.)

- ☐ Not OK to support additional loads on masonry veneer ____ [703.7.3]
- ☐ Min 3/4in gap between top of veneer & structural frame including roof rafters ____ [F703.7]
- ☐ 1in space between veneer & sheathing if metal ties used **F35** ____ [703.7.4]
- ☐ Approved water-repellant sheathing behind veneer or Type 1 felt ____ [F703.7]
- ☐ Air space can be filled if water-resistive barrer installed or approved water-resistive barrier-backed reinforcement ____ [703.7.4.3]
- ☐ Metal ties: max 24in horiz & vert spacing **F35** ____ [703.7.4.1]
- ☐ Ties in SDC D_1, & D_2 & townhouses in SDC C: max 2sq.ft of wall area per tie ____ [703.7.4.1X]
- ☐ Flashing to exterior req'd at all openings & horiz transitions ____ [703.8]
- ☐ Min 3/16in weepholes immediately above flashing max 33in o.c. **F35** [703.7.6]

Wood or Hardboard Panel or Lap Siding

- ☐ Panel siding vert joints over framing members or wood or structural sheathing ____ [703.3.1]
- ☐ Panel siding vert joints shiplapped or covered w/ batten ____ [703.3.1]
- ☐ Panel siding horiz joints lap 1in or shiplapped or Z-flashed ____ [703.3.1]
- ☐ Lap siding: min 1in lap or rabetted ____ [703.3.2]
- ☐ Ends of lap siding (vert joints): caulked, sealed, or flashed ____ [703.3.2]

Wood Shakes & Shingles

- ☐ Refer to Cedar Shake & Shingle Bureau for material grading ____ [703.5]
- ☐ May be single course or double course, over felt & sheathing, furring optional ____ [703.5.1]
- ☐ Max exposure 16in shingles: 7 1/2in single course, 12in if double course ____ [703.5.2]
- ☐ Fasteners: corrosion-resistant, not overdriven ____ [703.5.3]
- ☐ Fasteners must be concealed by course above ____ [703.5.3.1]

INTERIOR WALL SURFACES

Interior wall surfaces cannot be installed until rough inspections are complete, including framing, mechanical, plumbing, and electrical. Many jurisdictions then have a separate inspection for insulation and for drywall fastening before the joints are taped.

Gypsum Board

- ☐ Install only after all rough inspections complete ______ [109.1.2]
- ☐ Install after building is weathertight ______ [701.2]
- ☐ Install after exterior wall insulation (min R-13, more in cold climate) _[T1102.1]
- ☐ Interior gypboard not OK subject to weather or water ______ [702.3.5]
- ☐ Fastening per schedule **T21** ______ [T702.3.5]
- ☐ $5/8$in Type X on garage ceiling under habitable room req'd to be perpendicular to framing, min 1$7/8$in nails at 6in o.c. ______ [T702.3.5]
- ☐ OK to use adhesives + fasteners for foam plastics (ICF walls) AMI_ [702.3.4]
- ☐ Edges & ends over framing unless perpendicular to framing______ [702.3.5]
- ☐ Wood frame support: min 2× material or 1×2 furring strips over solid backing or framing ≤24in o.c.______ [702.3.2]
- ☐ Greenboard (water-resistant gypsum) not over vapor retarder in shower or tub ______ [702.3.8]
- ☐ Greenboard on ceilings–$1/2$in thick reqs 12in o.c. framing ______ [702.3.8]

TABLE 21 — $1/2$ IN. GYPSUM BOARD FASTENING SCHEDULE [T702.3.5] FOR APPLICATION W/O ADHESIVE

Location	Orientation to Framing	Max. Frame Spacing	Nails	Screws
Ceilings	Perpendicular	24 in.	7	12
Ceilings	Either	16 in.	7	12
Walls	Either	24 in.	8	12
Walls	Either	16 in.	8	16

Gypsum Board (cont.)

- ☐ Seal cut/exposed edges of greenboard______ [702.3.8]
- ☐ No greenboard in area of continuous high humidity or direct water exposure ______ [702.3.8.1][39]
- ☐ Cement board req'd as tile backer in tub & shower areas ______ [702.4.2][40]
- ☐ Nonabsorbant wall surface: min 72in above tub/shower floor______ [307.2]

FIG. 36 Door to Deck Flashing

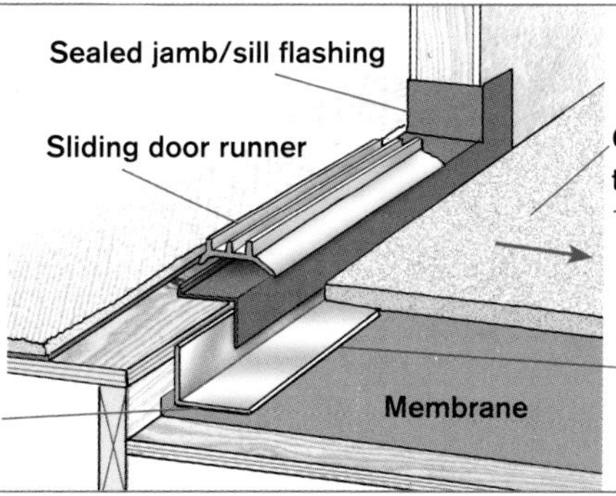

FIG. 37 Window Flashing

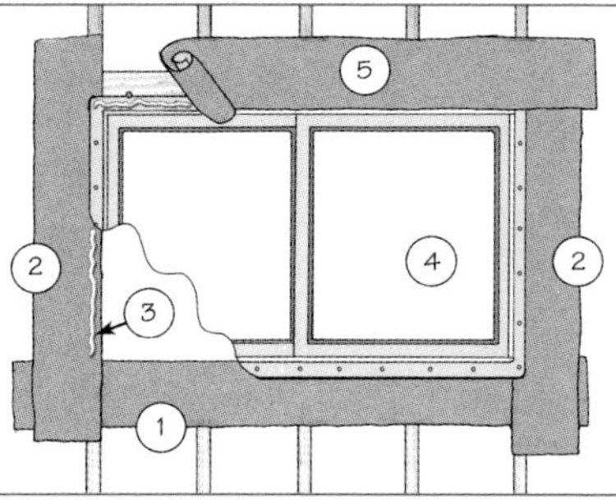

ROOF FRAMING

Roof framing systems must be sufficient to transfer live and snow loads and resist wind uplift. Sheathed roof assemblies also transfer lateral loads. In selecting trusses or ceiling joists, it is important to consider whether the attic spaces will be used as storage. The scope of the rules for wood-framed rafter details in the IRC is limited to roofs with a 3:12 or greater slope.

Rafters & Joists

- ☐ Ceiling joist spans per **T22** ______ [802.4]
- ☐ Use floor joist tables for attics w/ fixed stairway access **T13** ______ [502.3.1]
- ☐ Rafter spans per **T23** ______ [802.5]
- ☐ Rafter span can be measured from purlin support ______ [802.5.1]
- ☐ Purlin dimension no smaller than rafter ______ [802.5.1]
- ☐ Purlin braces: min 2×4, max 4ft o.c., max 45° from vert ______ [802.5.1]
- ☐ Cuts, notches, & holes in TJIs & other engineered lumber only AMI or by registered design professional ______ [802.7.2][41]
- ☐ Notching & boring similar to floor joists & girders – **F18** ______ [802.7]
- ☐ Ridges, hips, & valleys for roof slopes <3:12 must be supported as beams, i.e., not by opposition of the rafters alone ______ [802.3]
- ☐ Ridge board: min 1× & full depth of cut rafter ______ [802.3]
- ☐ Hip & valley rafters: min 2× & full depth of cut rafter ______ [802.3]
- ☐ Max allowable deflection ceiling joists: L/240* ______ [301.6]
- ☐ Max allowable deflection roof rafters: L/180* ______ [301.6]
- ☐ Ends of joists & rafters must have min 1½in bearing ______ [802.6]
- ☐ >2×10 req blocking at bearing points ______ [802.8]
- ☐ >2×12 req blocking, bridging, or 1×3 backer @ max 8ft intervals ______ [802.8.1]
- ☐ Rafter ties req'd if joists not tied to parallel rafters ______ [802.3.1][42]
- ☐ Rafter ties min 2×4 on lower ⅓ of every rafter pair ______ [802.3.1][43]
- ☐ Collar ties for wind uplift req'd in upper third, min 3-10d nails to each rafter ______ [802.3.1][44]

** L = length in inches.*

Rafters & Joists (cont.)

- ☐ Collar ties: min 1×4 max spacing 4ft o.c. ______ [802.3.1][45]
- ☐ Headers & trimmers >4ft must be doubled ______ [802.9]
- ☐ Header >6ft must be hung w/ hardware ______ [802.9]

TABLE 22 CEILING JOIST SPANS [T802.4(1)(2)]

Size DF#2	No Storage • 10psf Live Load			Limited Storage • 20psf Live Load		
	12 in. o.c.	16 in. o.c.	24 in. o.c.	12 in. o.c.	16 in. o.c.	24 in. o.c.
2×4	12 ft. 5 in.	11 ft. 3 in.	9 ft. 10 in.	9 ft. 10 in.	8 ft. 9 in.	7 ft. 2 in.
2×6	19 ft. 6 in.	12 ft. 10 in.	10 ft. 6 in.	15 ft. 6 in.	14 ft.1 in.	12 ft. 3 in.
2×8	25 ft. 8 in.	23 ft.	18 ft. 9 in.	18 ft. 9 in.	16 ft. 3 in.	13 ft. 3 in.
2×10	n/a	n/a	22 ft. 11 in.	22 ft. 11 in.	19 ft. 10 in.	16 ft. 3 in.

TABLE 23 ROOF RAFTER HORIZ. SPAN DISTANCE[A] [T802.5.1(1)(3)]

Size DF#2	No snow • 20psf live load			30 psf ground snow load		
	12 in.o.c.	16 in.o.c.	24 in.o.c.	12 in.o.c.	16 in.o.c.	24 in.o.c.
2×4	10 ft. 10 in.	9 ft. 10 in.	8 ft.	9 ft. 5 in.	8 ft. 2 in.	6 ft. 8 in.
2×6	16 ft. 7 in.	14 ft. 4 in.	11 ft. 9 in.	13 ft. 9 in.	11 ft. 11 in.	9 ft. 9 in.
2×8	21 ft.	18 ft. 2 in.	14 ft. 10 in.	17 ft. 5 in.	15 ft. 1 in.	12 ft. 4 in.
2×10	25 ft. 8 in.	22 ft. 3 in.	18 ft. 2 in.	21 ft. 4 in.	18 ft. 5 in.	15 ft. 1 in.
2×12	>26 ft.	25 ft. 9 in.	21 ft.	24 ft. 8 in.	21 ft. 5 in.	17 ft. 6 in.

A. Based on 10psf dead load, ceilings not attached to underside of rafters.

Roof Sheathing

- ☐ Spaced lumber sheathing: not allowed in SDC D_2 ______ [803.1]
- ☐ Wood structural panels must have grade stamp **F41** ______ [803.2.1]
- ☐ Spans for first # of grade stamp assume edge support **F41** _ [T503.2.1.1(1)]

ROOF & CEILING TRUSSES

The most common roof trusses have two-point bearing–they do not bear on the interior walls. The outside members of a truss are called chords, and the inside members are called the web (**F38**). Trusses must not be cut or altered from their original design and must be installed in accordance with the instructions included in the truss shipment. Because of possible seasonal truss movement, the attachment of interior partition walls to the trusses should be made only with hardware that allows vertical movement, as in **F39** and **F40**. During construction, trusses require temporary bracing. The Truss Plate Institute also publishes *BCSI 1-03–A Guide to Good Practice for Handling, Installing and Bracing Metal-plate-connected Wood Trusses.*

Trusses–General

- ☐ Design drawings must be submitted to BO & w/ truss shipment __ [802.10.1]
- ☐ Trusses must conform to accepted engineering practice ________ [802.10.2]
- ☐ Applicability limits for 70psf snow: 36ft×60ft, 2 stories, slope 3:12 to 12:12 ________ [802.10.2.1][46]
- ☐ Bracing per drawings; if none shown, then follow BCSI 1-03 ___ [802.10.3][47]
- ☐ No alteration of trusses w/o approval of registered design professional ________ [802.10.4]
- ☐ No additional loads w/o verification of truss loading per drawings_ [802.10.4]
- ☐ Truss connections to wall plates uplift resistance min 175 lb._____ [802.10.5]

FIG. 38 Typical Fink Truss

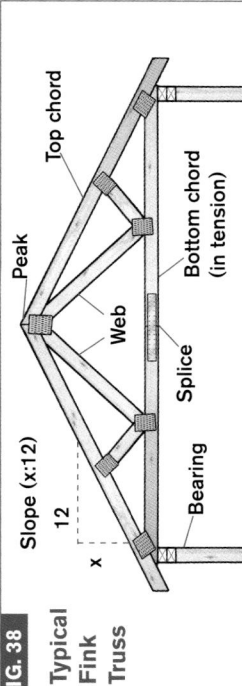

Slope (x:12)
12
x
Peak
Top chord
Web
Splice
Bearing
Bottom chord (in tension)

FIG. 39 FIG. 40

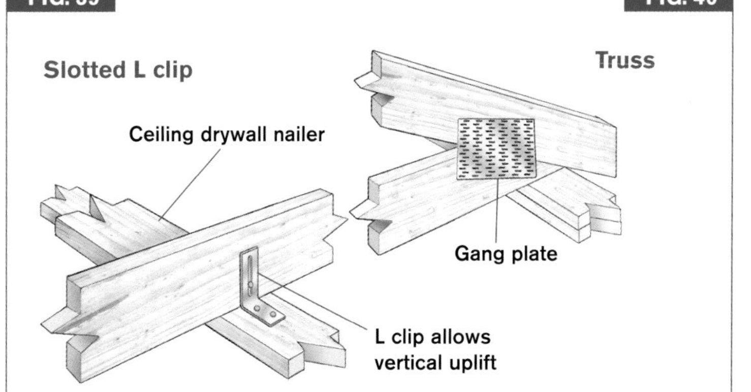

FIG. 41 Wood Structural Panel Grade Mark

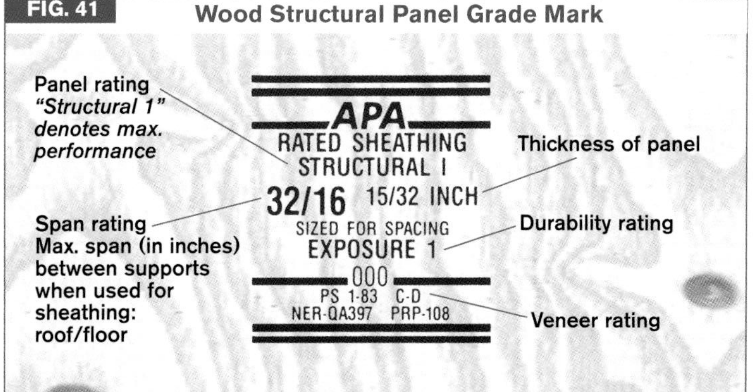

ROOFS

Roof systems are categorized by their fire rating, class A, B, C, and non rated. Local or state laws may specify certain minimum ratings, so it is important to check with the local jurisdiction when selecting a roofing or reroofing system. Other local factors include the need for ice barriers and wind or hail-resistant materials. The IRC provides specific prescriptive rules for roof installations, and more detailed instructions are often specified by the manufacturer or the ASTM standards or the NRCA recommendations.

Roof Surfaces–General

- ☐ Class A, B, or C req'd per local laws or if <3ft of PL ________ [902.1]
- ☐ Slate, tile, & metal considered Class A fire resistant ________ [902.1]
- ☐ Fire-retardant-treated wood roofs req test agency label each bundle _ [902.2]
- ☐ Follow approved manufacturer's installation instructions______ [903.1,904.1]
- ☐ Flashing req'd at all penetrations & intersections ________ [903.2.1]
- ☐ Penetrations >30in wide in sloped roofs req cricket **F46,47,71** _ [905.2.8.3][48]
- ☐ Drains at each low point of roof unless designed to run over edges __ [903.4]
- ☐ Overflow drains req'd w/ inlet 2in above low point of roof________ [903.4.1]
- ☐ Overflow drains not allowed to be connected to roof drain lines ___ [903.4.1]
- ☐ Hail exposure must be determined before material selection ________ [903.5]
- ☐ Roofing materials must bear identification & test agency labels ______ [904.4]

Asphalt Shingles

- ☐ Min slope: 2:12 (double underlayment if <4:12) ________ [905.2.2]
- ☐ Fasteners: min 3/4in into sheathing (or through if sheathing <3/4in) **F43** [905.2.5]
- ☐ Min 4 fasteners per strip shingle (3ft wide) **F42** ________ [905.2.6]
- ☐ Special fastening AMI on steep slopes: (>20:12) or in high-wind areas ________ [905.2.6]
- ☐ Ice-dam barrier req'd to 24in horiz inside building exterior wall__ [905.2.7.1][49]
- ☐ Drip flashing below underlayment, rake flashing above underlayment **F45** ________ [905.2.8.1]
- ☐ Closed valleys req min 36in wide valley lining underlayment______ [905.2.8.2]

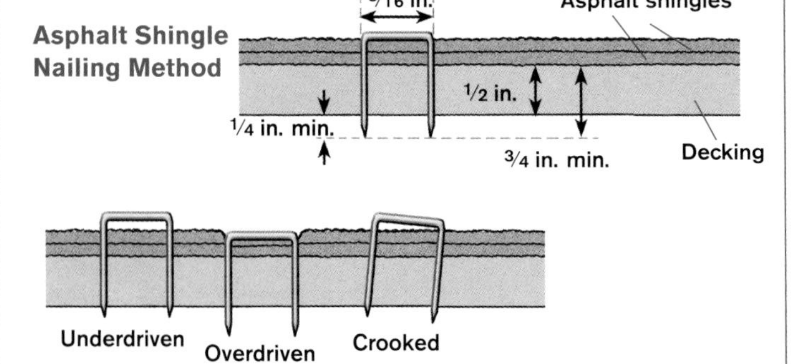

Clay & Concrete Tile

- ☐ Clay/concrete tile min slope: $2^1/_2$:12, double underlayment if ≤4:12 [905.3.2]
- ☐ Double underlayment if slope <4:12 ______ [905.3.3.1]

(Note: Manufacturer's requirements are more stringent, requiring built-up roofing, or other membrane systems on roofs below 3:12 slope.)

- ☐ Tile must comply w/ ASTM standards ______ [905.3.4,5]
- ☐ Tile application AMI & must be appropriate for climate, roof slope, underlayment system, & type of tile ______ [905.3.7]
- ☐ Valley flashing min 11in from center line **F44** ______ [905.3.8]

(Note: The Tile Roofing Institute publishes installation manuals for cold & snow regions, moderate climate regions, & Florida.)

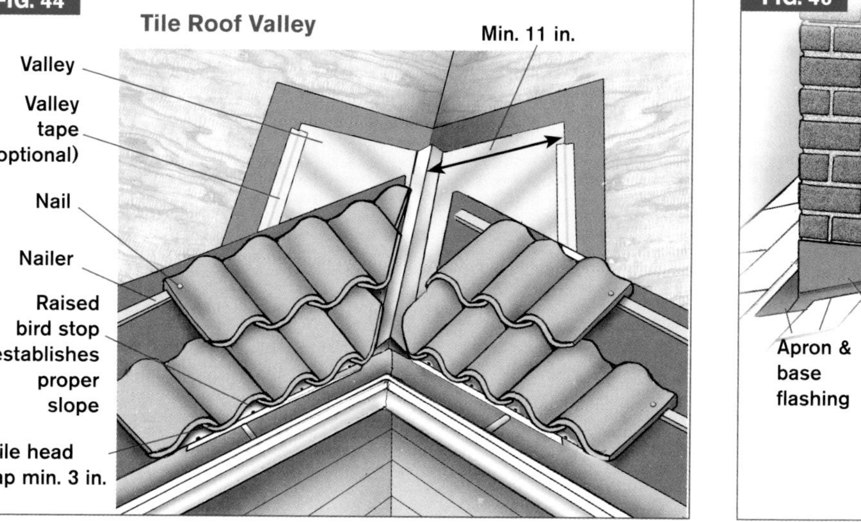

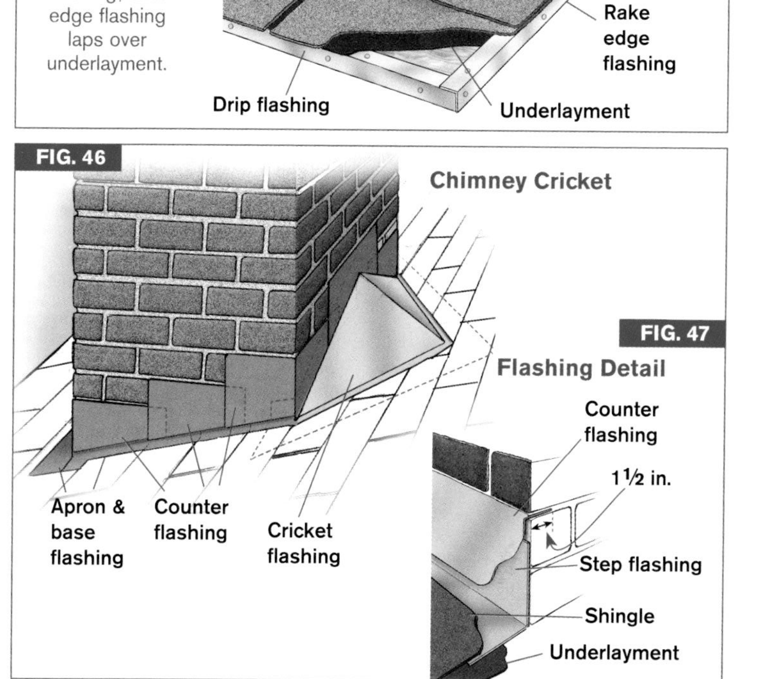

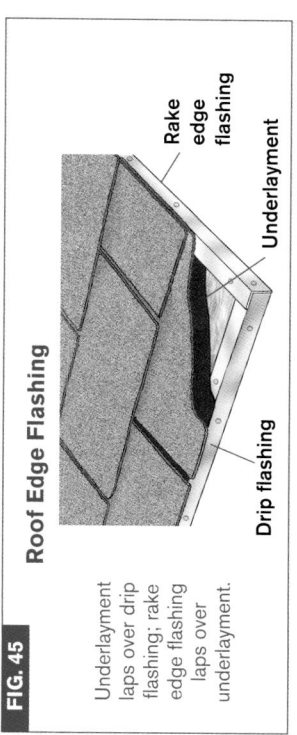

Wood Shingles

- ☐ 16in shingles min slope: 3:12 [905.7.2]
- ☐ Ice-dam barrier req'd to 24in horiz inside building exterior wall [905.7.3.1][49]
- ☐ Sidelaps: min 1 1/2in, no 2 keyways aligned in any 3 adjacent courses **F48** [905.7.5]
- ☐ Exposure per **T24** [905.7.5]
- ☐ Spacing between shingles: min 1/4in, max 3/8in **F48** [905.7.5]
- ☐ 2 fasteners per shingle, penetration into sheathing: min 1/2in, fasteners: ≤3/4in from edge & ≤1in above exposure line **F48** [905.7.5]

FIG. 48 Wood Shingle Application

Shingle keyway spacing, min. 1/4 in., max. 3/8 in.
Adjacent course joints should be offset min. 1 1/2 in.
2 fasteners per shingle &: Into sheathing–min. 1/2in. From edge–≤3/4in. Above exposure line: ≤1in.
Exposure line

Wood Shakes

- ☐ Min slope: 3:12 [905.8.2]
- ☐ Ice-dam barrier req'd to 24in horiz inside building exterior wall [905.8.3.1][49]
- ☐ Sidelaps: min 1 1/2in, spacing: min 1/8in–5/8in (1/4in–3/8in for PT shakes) **F49** [905.8.6]
- ☐ Exposure per **T25** [905.8.6]
- ☐ 2 fasteners per shake: ≤2in above exposure line & 1in from edge **F49** [905.8.6]
- ☐ Max exposure standard #1 shake: 10in if slope ≥4:12 [905.8.6]
- ☐ No felt exposed to sunlight **F49** [905.8.7]

For further information on cedar shingle & shake installations, refer to www.cedarbureau.org

FIG. 49 Wood Shakes Application

Felt not exposed to sunlight
Shake keyway spacing: min. 1/8in.–5/8in. Pressure-treated shakes: 1/4in.–3/8in.
Adjacent course joints offset min. 1 1/2in.
2 fasteners per shake &: From edge–1in. Above exposure line–≤2in.
Exposure line

TABLE 24 WOOD SHINGLE WEATHER EXPOSURE [T905.7.5]

Shingle Length (in.)	Grade & Label Color	Exposure (in.)	
		3:12 Slope to <4:12	4:12 Slope or Steeper
16	No. 1–blue	3 3/4	5
	No. 2–red	3 1/2	4
	No. 3A–black	3	3 1/2
18	No. 1–blue	4 1/4	5 1/2
	No. 2–red	4	4 1/2
	No. 3A–black	3 1/2	4
24	No. 1–blue	5 3/4	7 1/2
	No. 2–red	5 1/2	6 1/2
	No. 3A–black	5	5 1/2

TABLE 25 WOOD SHAKE WEATHER EXPOSURE [T905.8.6]

Material	Length (in.)	Grade	Exposure[A] (in.)
Naturally durable wood (cedar)	18	No. 1	7 1/2
	24	No. 1	10
PT taper-sawn Southern Pine[B]	18	No. 1	7 1/2
	24	No. 1	10
	18	No. 2	5 1/2
	24	No. 2	7 1/2
Taper-sawn naturally durable wood (cedar)	18	No. 1	7 1/2
	24	No. 1	10
	18	No. 2	5 1/2
	24	No. 2	7 1/2

A. Assumes a 4:12 or greater slope.
B. These shakes are graded by the Forest Products Laboratory of Texas. Other shakes are graded by the Cedar Shake & Shingle Bureau.

FIG. 50

Three-Ply Built-Up Roof

Inter-ply bitumen must be installed in a continuous firmly bonded film w/ no voids between the plies of material. Approximately 25 lb. of asphalt per square is req'd. The temperature must be maintained at the proper range for the specific type of asphalt.

Low-Slope Built-Up Roofing (BUR)

- ☐ BUR (asphalt & gravel): min slope 1/4:12 ______ [905.9.1]
- ☐ Comply with ASTM standards & install AMI **F50** ______ [905.9.2,3]

NRCA Recommendations & ASTM Requirements:

- Store rolls on end not sides to prevent deformation
- Water-based materials must be protected from freezing before installation
- Protect insulation from moisture
- Do not install roofing while rain, snow, or ice are present
- Use cant strips to limit bends to 45° at horiz to vert intersections
- Sample temperature (typical 350°F to 425°F Type I asphalt)
- Aggregate dry & clean to adhere to hot bitumen

Other Low-Slope Roofing

- ☐ Mineral-surfaced roll roofing only on solidly sheathed roofs ______ [905.5.1]
- ☐ Mineral-surfaced roll roofing min slope: 1:12 ______ [905.5.2]
- ☐ EPDM (ethylene propylene diene monomer) roofing min slope: 1/4:12 [905.12.1]
- ☐ EPDM must comply w/ ASTM standards & installed AMI ______ [905.12.2,3]
- ☐ PVC roofing min slope: 1/4:12 ______ [905.13.1]
- ☐ PVC roofing must comply w/ ASTM standards & installed AMI _ [905.13.2,3]
- ☐ Spray foam roofing min slope: 1/4:12 ______ [905.14.1]
- ☐ Spray foam roofing complies w/ ASTM standards & installed AMI [905.14.2,3]
- ☐ Spray foam roofing coated within 72 hr. of application ______ [905.14.3]

MASONRY FIREPLACES & CHIMNEYS

Fireplace and chimney construction must safely separate the combustible building structure from heat sources. Chimneys must safely convey dangerous combustion products to the exterior. Also, environmental pollution must be considered and many communities restrict the use of solid-fuel. Refer to *Code Check HVAC* for further information on wood-burning stoves.

Masonry Fireplace & Chimney Construction

- ☐ Footing: min 12in thick & 6in beyond all sides **F1** [1001.2,1003.2]
- ☐ Footing: min 12in below grade [1001.2,1003.2]
- ☐ Chimney walls: min 4in thick [1003.10]
- ☐ SDC D_0, D_1, & D_2 reinforcement:
  - Min 4-#4 vert reinforcing bars for chimneys ≤40in wide **F51** [1001.3.1, 1003.3.1]
  - 2 more bars if >40in or for each additional flue [1001.3.1, 1003.3.1]
  - Min 1/4in horiz ties each 18in [1001.3.2, 1003.3.2]
  - Anchor at each floor or ceiling >6ft above grade unless chimney completely enclosed by building exterior **F51** [1001.4, 1003.4]
- ☐ Grout to enclose rebar & not bond to flue liner [1001.3.1,1003.3.1]
- ☐ Chimneys entirely outside exterior walls min 1in to combustible framing [1003.18]
- ☐ All other chimneys min 2in clearance to combustible framing **F51,62** [1001.11,1003.18]
- ☐ Exterior siding & trim OK within 12in of flue interior [1001.11X3,1003.18X3]
- ☐ Fireblocking req'd between chimneys & ceiling **F62** [1001.12,1003.19]

Flue & Termination

- ☐ Size flue for proper draft per fireplace opening size & chimney height [1003.15]
- ☐ Termination: min 3ft above roof **F52** [1003.9]
- ☐ 2ft above any part of building within 10ft **F52** [1003.9]
- ☐ Spark arrester: min area 4× flue opening [1003.9.1]
- ☐ Spark arrester mesh: 3/8in min, 1/2in max [1003.9.1]

Flue & Termination (cont.)

- ☐ Spark arrester must be accessible & removable for flue cleaning [1003.9.1]
- ☐ Crickets req'd if chimney >30in wide **F46** [1003.20]

FIG. 51

Chimney Reinforcement

#4 bars

Min. of 4 #4 vert. bars for chimneys up to 40 in. wide Wider chimneys req. at least 2 additional bars.

2×4 ties must cross min. 4 joists.

2×4 tie

Chimney anchoring strap detail

Anchor straps must hook around the outer bars of the chimney & fasten to the framing w/ min. of 2 bolts: min. ½ in. dia.

FIG. 52

Min. Chimney Height

Dirty chimneys are a fire hazard due to the accumulation of combustible creosote. Consult a certified chimney sweep annually for cleaning & inspections.

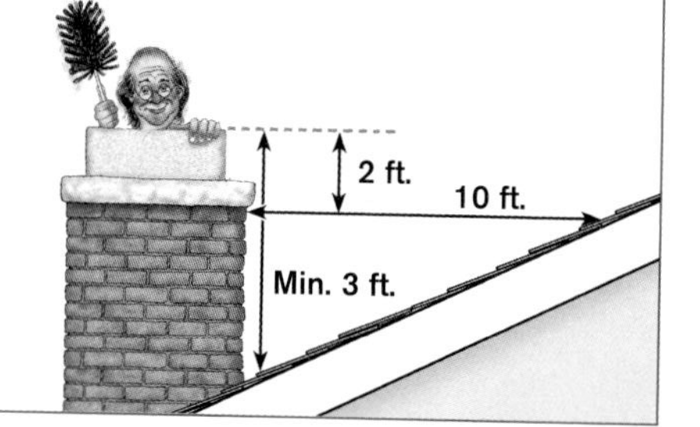

Masonry Fireplaces

- ☐ Firebox walls: min 8in if including 2in liner, otherwise 10in solid masonry [1001.5]
- ☐ Min depth 20in, throat min 8in above fireplace opening, min 4in deep EXC [1001.6]
  Rumford fireplaces 12in depth OK if at least 1/3 of fireplace width [1001.6X]
- ☐ Damper req'd [1001.7.1]
- ☐ Damper: min 8in above opening [1001.7.1]
- ☐ Hearth & extension req reinforcing to carry their weight & imposed loads [1001.9]
- ☐ All combustible material removed from under hearth & extension [1001.9]
- ☐ Hearth slab min thickness: 4in [1001.9.1]
- ☐ Hearth extension min thickness: 2in [1001.9.2]
- ☐ Hearth extension depth: min 16in front, 8in side if opening <6sq.ft **F53** [1001.10]
- ☐ Hearth extension depth: min 20in front, 12in side if opening ≥6sq.ft **F53** [1001.10]
- ☐ No combustible material within 6in of opening **F53** [1001.11X4]
- ☐ Combustible material <12in of opening limited to max 1/8in projection for each inch of clearance **F53** [1001.11X4]
- ☐ All combustion air from outside [1006.1]
- ☐ Exterior air intake no higher elevation than firebox, & screened [1006.2]
- ☐ No exposed corbelling inside smoke chamber [1001.8]
- ☐ Lintel must extend min 4in each side of fireplace opening [1001.7]

FACTORY-BUILT FIREPLACES & CHIMNEYS

Manufactured fireplaces and chimneys are very common in modern construction. They are tested and listed by independent laboratories, and must be installed in exact accordance with the terms of their listing per UL 127.

Factory-Built Fireplaces & Chimneys

- ☐ Must be L&L & installed AMI [1004.1,1005.1]
- ☐ Must conform to UL 127 [1004.1,1005.4]
- ☐ Noncombustible hearth extension AMI & distinguishable from floor [1004.2]
- ☐ Decorative shrouds only if specifically L&L & AMI [1004.3, 1005.2]
- ☐ Outside combustion air ducts L&L for fireplace & installed AMI [1006.1.1]

Factory-Built Fireplaces & Chimneys (cont.)

- ☐ Firestop spacer AMI where passing through ceiling [1005.4]
- ☐ Insulation shield AMI if loose-fill attic insulation [1005.4]

FIG. 53

Masonry-Fireplace Clearances

No combustible material is allowed within 6 in. of the fireplace opening. 1 in. clearance is needed for each 1/8 in. of projection of combustible material within 12 in. of the fireplace opening. For example, a 3/4 in. projection needs 6 in. of clearance.

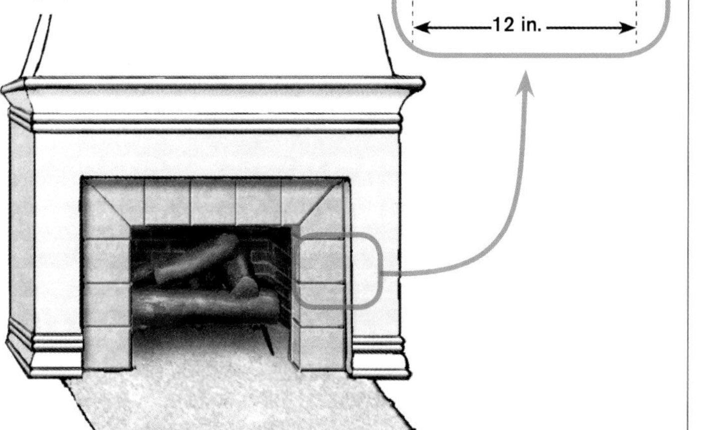

EGRESS

A means of egress (emergency escape) is an essential safety issue at the heart of the codes. Properly sized doors must open from the inside without use of a key or tool to ensure an escape route to the exterior. A second means of egress to the exterior is needed for each bedroom and for basements with habitable space. The egress path must be stable. If it includes a balcony or deck, the deck must be securely attached to the building. Stairs with enclosed construction beneath them must have drywall surfaces protecting the area under the stairs.

Doors

- ☐ Min 1 exit door req'd direct to exterior–not through garage ______ [311.4.1]
- ☐ Req'd exit min 3ft wide × 6ft 8in high & hinged on side ______ [311.4.2]
- ☐ All doors req keyless operation from interior ______ [311.4.4]
- ☐ Hallway: min 36in wide ______ [311.3]

FIG. 54

Deck Ledger Flashing

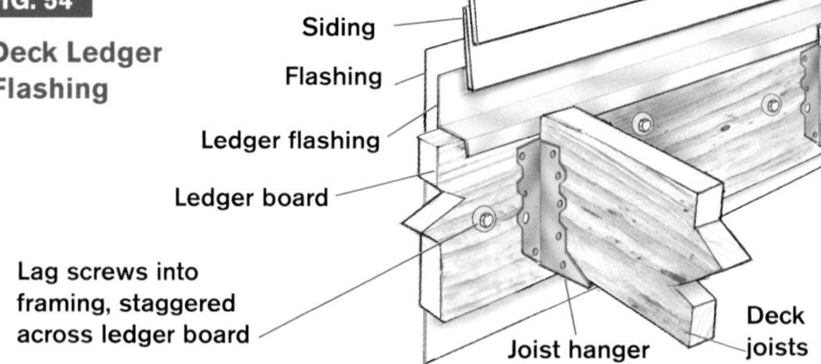

Thresholds & Landings at Doors

- ☐ Max threshold height above floor at exterior doors: $1\frac{1}{2}$in ______ [311.4.3]
- ☐ Floor or landing req'd each side of exterior doors EXC ______ [311.4.3]
  Stairs w/ 2 risers OK outside in-swing exterior doors, not req'd exits [311.4.3X1]
- ☐ Exterior landings OK $7\frac{3}{4}$in below top of threshold if door swings in [311.4.3X2]
- ☐ Landings may have slope up to 2% ______ [311.4.3][50]
- ☐ Landing min width same as door, min depth in direction of travel 36in [311.4.3]
- ☐ Screen doors may swing over landings that are lower than floors [311.4.3X1,2]
- ☐ Decks supported by bldg structure (not by veneer) OR ______ [502.2.2,703.7.3]
  Independent support OK exc when part of req'd exit system ______ [502.2.2]
- ☐ Req'd exit balconies req positive anchor to structure for vert & lateral loads **F54** ______ [311.2.1]
- ☐ Attachment cannot use toenails or nails subject to withdrawal ______ [311.2.1,502.2.2]
- ☐ Canilevered decks req design to resist uplift from deck live load ______ [502.2.2]

Guards & Guardrails

- ☐ Req'd for any walk off >30in above floor or grade **F55** ______ [312.1]
- ☐ Screened porches req guards if walk off >30in ______ [312.1]
- ☐ Min guard height: 36in, 34in from nosing w/ handrail on open side of stair **F55** [312.1]
- ☐ Max opening size: <4in EXC **F55** ______ [312.2]
- ☐ 6in opening OK at tread/riser/rail triangle ______ [312.2X1]
- ☐ Openings in rail at open side of stairs: <$4\frac{3}{8}$in ______ [312.2X2]
- ☐ Open risers on stairs must have <4in opening between treads ______ [311.5.3.3]
- ☐ Strength at top must resist 200lb point load any direction ______ [301.5]
- ☐ In-fill components must resist 50lb load spread over 1sq.ft area ______ [301.5]

TABLE 26 — BEDROOM WINDOW EGRESS: MIN. HEIGHT & WIDTH REQS. TO MEET REQ'D. 5.7 SQ. FT. OPENING SIZE (IN.)

Width	20	20½	21	21½	22	22½	23	23½	24	24½	25	25½	26	26½	27	27½	28	28½	29	29½	30	30½	31	31½	32	32½	33	33½	34
Height	41	40	39½	38½	37½	36½	35½	35	34½	33½	33	32½	31	31	30½	30	29½	29	28½	28	27½	27	26½	26½	25½	25½	25	24½	24

TABLE 27	BEDROOM WINDOW EGRESS FOR 5.0 SQ. FT. OPENING: DIRECT GRADE LEVEL ACCESS ONLY (IN.)																			
Width	20	20½	21	21½	22	22½	23	23½	24	24½	25	25½	26	26½	27	27½	28	28½	29	29½	30
Height	36	35	34½	33½	33	32	31½	31	30	29½	29	28½	28	27½	27	26½	26	25½	25	24½	24

FIG. 55

Guardrails

FIG. 56 **Bedroom Window Egress**

When the req'd second exit in a bedroom is a window, it must open to a size that provides an escape route for residents & an entrance for a firefighter wearing an oxygen pack. The sill must not be higher than 44 in. from the finished floor. The openable portion of the window must be at least 24 in. high & 20 in wide, w/ a net area at least 5.7 sq. ft., per **T26**. If the window is at direct grade level, the overall size may be reduced to 5.0 sq. ft. per **T27**.

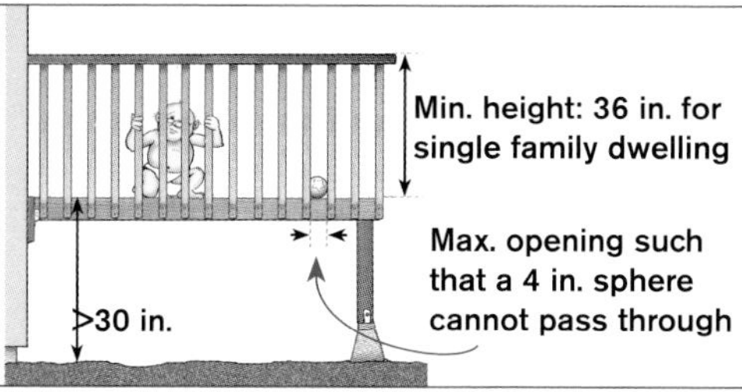

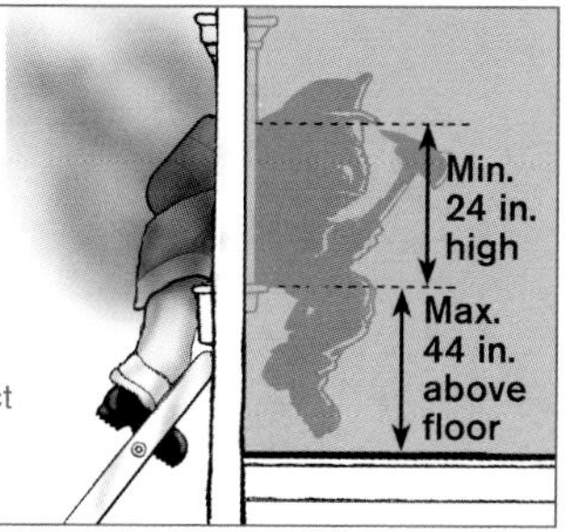

Emergency Escape & Rescue Openings

- ☐ Basements & sleeping rooms req emergency escape & rescue opening EXC ______ [310.1][51]
  Basements ≤200sq.ft for mech equipment ______ [310.1X][51]
- ☐ Basement sleeping rooms req individual emergency openings ______ [310.1][52]
- ☐ Basement areas adjoining sleeping rooms do not req emergency openings ______ [310.1][52]
- ☐ Escape window OK under deck if escape path not <36in high under deck ______ [310.5][53]
- ☐ 5.7sq.ft min clear opening EXC **T26** ______ [310.1.1]
- ☐ 5.0sq.ft OK if direct grade level access **T27** ______ [310.1.1X]
- ☐ Window sill height: 44in max above finished floor **F56** ______ [310.1]
- ☐ Min 20in width & min 24in height & per **T26,27** ______ [310.1.2,3]
- ☐ No windows closer than 3ft to PL ______ [302.1]
- ☐ Security bars must have release hardware operable from inside w/o use of key, tool, special knowledge, or greater force than to open window _ [310.4][54]

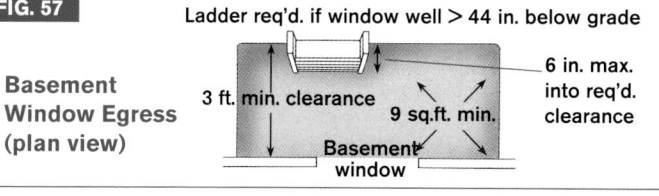

Basement Window Wells

- ☐ Window wells req'd when threshold below adjacent ground ______ [310.1]
- ☐ Window well min 9sq.ft & 36in min horiz dimension **F57** ______ [310.2]
- ☐ Wells >44in below grade req permanent ladder **F57** ______ [310.2.1]
- ☐ Ladder may project 6in into req'd well space **F57** ______ [310.2X]
- ☐ Bulkhead enclosure doors must provide min req'd opening when open [310.3]
- ☐ Bulkhead stairs <8ft vert no landing OK if hinged bulkhead door fully covers stair ______ [311.5.8.2]

FIG. 57

Basement Window Egress (plan view)

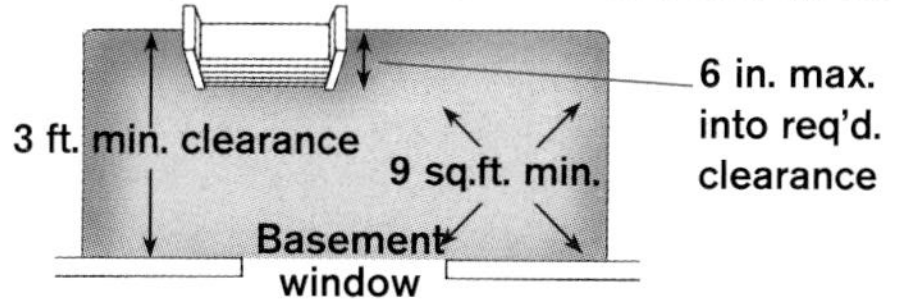

STAIRS

Stairs are the most frequent location of injury accidents in the home. A consistent rise and run is important for safety. The tallest riser cannot exceed the shortest by more than 3/8in., and the deepest tread cannot exceed the smallest by more than 3/8in. Proper handrails and landings are important safety items .

Stairways–General

- ☐ Min stairway width above handrail 36in ____ [311.5.1]
- ☐ Min stair width above rail 36in ____ [311.5.1]
- ☐ Min width below rail: 31 3/8in clear if 1 handrail, 27in if 2 handrails ____ [311.5.1]
- ☐ Headroom: min 6ft 8in (spiral 6ft 6in) **F58** ____ [311.5.2, 311.5.8.1]
- ☐ Max riser height: 7 3/4in **F61** ____ [311.5.3.1]
- ☐ Min tread depth: 10in **F61** ____ [311.5.3.2]
- ☐ Tallest riser not >3/8in than shortest ____ [311.5.3.1]
- ☐ Greatest tread depth not >3/8in more than smallest ____ [311.5.3.2]
- ☐ Nosing: min 3/4in max 1 1/4in req'd on stairs w/ solid risers **F61** ____ [311.5.3.3]
- ☐ Beveling of nosing: max 1/2in ____ [311.5.3.3]
- ☐ Curvature radius of tread leading edge: max 9/16in **F61** ____ [311.5.3.3]
- ☐ Nosing not req'd on stairs w/ treads ≥11in ____ [311.5.3.3X1]
- ☐ Walking surface of treads & landings no >2% slope ____ [311.5.5]
- ☐ Illumination req'd for stairs & landings ____ [303.6, 311.5.7]
- ☐ Exterior stair lighting at top landing, control inside dwelling ____ [303.6, 303.6.1]
- ☐ Interior stair req's switch at each floor level if 6 or more risers ____ [303.6.1]
- ☐ Winding stairs: tread min 6in at inner edge **F60** ____ [311.5.3.2]
- ☐ Winding stairs: min 10in tread depth within 12in of inside **F60** ____ [311.5.3.2]
- ☐ Enclosed accessible space below stairs reqs 1/2in gypsum board on surfaces ____ [311.2.2]

Landings at Stairways

- ☐ Landing req'd at top & bottom of stairs EXC **F58** ____ [311.5.4]
  Door at interior stair (including garage) may open away from top step [311.5.4.X]

Landings at Stairways (cont.)

- ☐ Landing: min width same as door, min depth in direction of travel 36in [311.5.4]
- ☐ Landing not req'd at stairs w/ 2 risers outside exterior door that swings in & is not req'd exit ____ [311.4.3X1]
- ☐ Max 12ft vert stairway rise between landings ____ [311.5.4]

Handrails

- ☐ Grippable rail req'd if 4 or more risers ____ [311.5.6]
- ☐ Handrail height top of nosing to top of rail min 34in, max 38in **F59** [311.5.6.1]
- ☐ Grips must be 1 1/4in–2in circular cross section **F59** ____ [311.5.6.3]
- ☐ Max projection into stairway: 4 1/2in **F59** ____ [311.5.1]
- ☐ Ends shall return to wall or newel post or volute **F58** ____ [311.5.6.2]
- ☐ Min 1 1/2in space between handrail & wall **F59** ____ [311.5.6.2]
- ☐ Rail: sufficient strength to resist 200lb point load any direction ____ [301.5]

Spiral Stairways

- ☐ Min width: 26 in, all treads identical, min headroom 6ft 6in ____ [311.5.8.1]
- ☐ Min tread depth: 7 1/2in at 12in from center post ____ [311.5.8.1]]

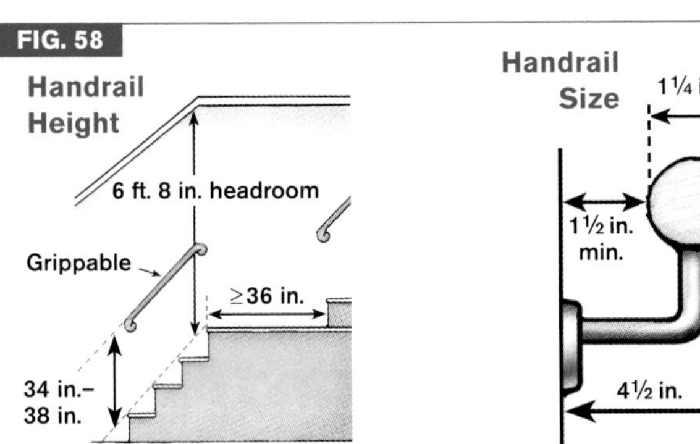

FIG. 60

Winding Stairs

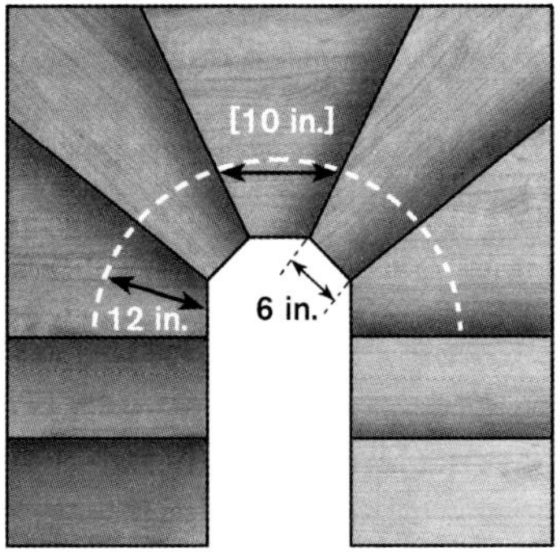

FIG. 61

Stair Profile

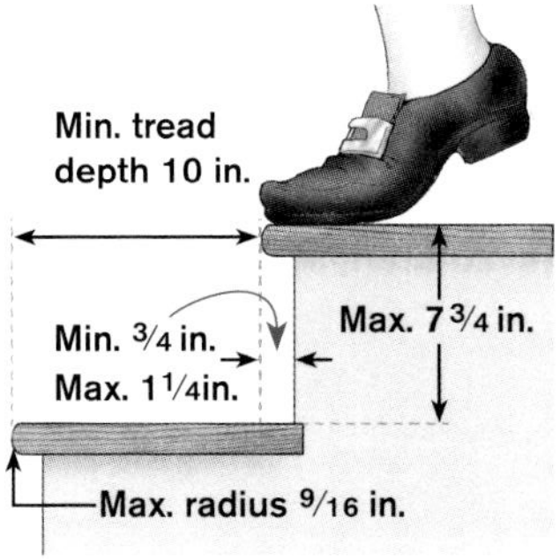

LIGHT & VENTILATION

Openable windows, doors, louvers, and skylights provide natural ventilation of habitable spaces. Mechanical ventilation can be substituted for these openings. Skylights, windows, and doors provide natural light for habitable rooms.

Light

- ☐ In habitable rooms, natural light min 8% floor area ________ [303.1]
- ☐ Artificial light OK for rooms where emergency egress not req'd & supplied w/ mechanical venting ________ [303.1X2]
- ☐ Bathrooms: min 3sq ft glazing or artificial light & mechanical venting [303.3X]

Ventilation

- ☐ Natural: ≥4% of floor area OR ________ [303.1]
  Mechanical: 0.35 air changes/hr in room or 15cfm/occupant for house [303.1X1]
- ☐ Bathroom window min 1 1/2sq.ft openable OR ________ [303.3]
  Artificial light & 50cfm fan ________ [303.3X]
- ☐ No bath exhaust to any part of attic, must go outdoors ___ [1501.1,1506.2][55]
- ☐ 2 rooms are considered 1 for light & ventilation if opening between them min 50% of common wall *and* ≥10% floor area of interior room *and* at least 25sq.ft ________ [303.2]

POOL SAFETY

Drowning is one of the leading causes of death or injury to young children in areas where swimming pools are common. Appendix G of the IRC provides guidelines for barriers and protection against entrapment.

Barriers

- ☐ Applies to all pools or spas >24in deep ________ [AG102.1]
- ☐ Barrier fence req'd min 48in high ________ [AG105.2]
- ☐ Bottom of fence max 2in above grade, 4in above concrete ______ [AG 105.2]
- ☐ Max opening must prevent passage of 4in sphere
- ☐ Fence cannot be climbable ________ [AG 105.2]
- ☐ Gate lockable, self-closing, open away from pool ________ [AG 105.2]
- ☐ If latch <54in high, must be poolside & min 3in from top ________ [AG 105.2]
- ☐ No openings >1/2in within 18in of gate latch ________ [AG 105.2]
- ☐ Doors & screen doors w/ direct pool access req alarm audible throughout house within 7 seconds of opening for 30 seconds_ [AG 105.2][56]
- ☐ Alarm control: min 54in high, automatically resetting ________ [AG 105.2]
- ☐ Safety covers acceptable alternative to barriers ________ [AG105.5]

Entrapment Protection

- ☐ Suction outlets anti-vortex type or min 18in×23in drain grate _____ [AG106.2]
- ☐ Min 2 suction outlets, min separation 3ft ________ [AG106.4]

FIRE PROTECTION

Fireblocking slows the spread of a fire in concealed building cavities thereby giving the occupants more time for escape and firefighters more protection against building collapse. In multi-family dwellings, larger open spaces are divided by draftstops to prevent the spread of flame. The garage is an area with the greatest fire load, so surfaces common to the house require materials that slow the spread of a fire. Buildings with different occupancies, such as a two-family dwellings or townhouses, require rated fire separations.

Fireblocking: Required Locations

- ☐ Wall fireblocking req'd vertically at ceiling & floor levels ____________ [602.8]
- ☐ Wall fireblocking req'd horizontally at max 10ft intervals ____________ [602.8]
- ☐ Intersections of concealed vert & horiz spaces (soffits, drop ceilings) **F63,66,67** ____________ [602.8]
- ☐ Concealed spaces between stair stringers at top & bottom of stairs __ [602.8]
- ☐ Openings around vents, pipes, ducts, cables, & wires at ceiling & floor level **F64** ____________ [602.8][57]
- ☐ Spaces between chimneys & floors/ceilings **F62** ________ [602.8, 1003.19]
- ☐ Space between factory-built fireplace chimneys & ceiling opening __[1005.4]

Fireblocking Materials & Methods

- ☐ Approved materials include 2× lumber, 2 thickness of 1×, gypsum board, cement board, glass fiber battens or approved materials **F62,64,67** _ [602.8.1]
- ☐ Unfaced fiberglass insulation as wall cavity fireblock min 16in vert [602.8.1.1]
- ☐ Approved caulking materials for cables, wires, ducts, & pipes **F64** ___ [602.8]
- ☐ Metal firestop spacers OK as fireblocking at vents & manufactured chimneys ____________ [602.8,1005.4]
- ☐ Masonry chimney fireblocking self-supporting or laid on metal or metal lath **F62** ____________ [1003.19][58]

Draftstopping

- ☐ Req'd at floor/ceiling assemblies >1,000 sq.ft (open-web trusses or ceiling suspended under framing) ____________ [509.12]

Draftstopping (cont.)

- ☐ Draftstopping should divide space into approximately equal areas___[509.12]
- ☐ Materials: min 1/2in gypboard or 3/8in. particleboard, parallel to framing[509.12.1]

FIG. 62

Chimney Fire Block

Fire blocking, such as compressed fiberglass must be self-supporting or supported on metal lath.

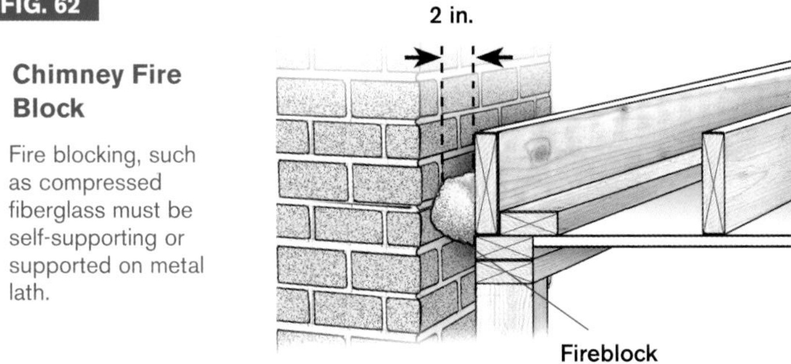

FIG. 63

Coved or Dropped Ceiling

FIG. 64

Fire Blocking Openings around Pipes

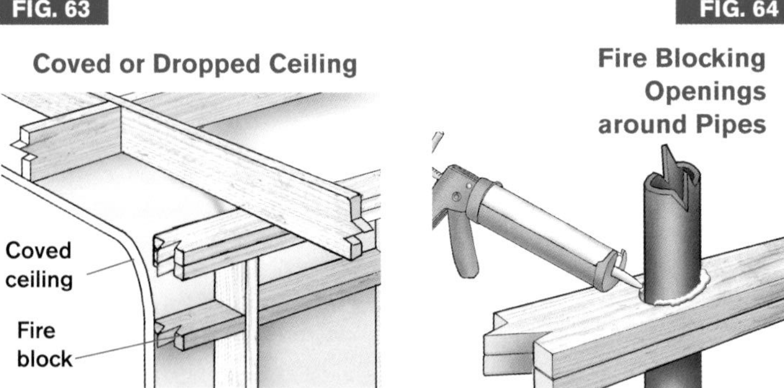

Garages

- ☐ Min 1/2in gypboard on garage side of walls common to house ____ [309.2]
- ☐ Min 5/8in Type X gypboard on garage ceiling common to house ____ [309.2]
- ☐ Min 1/2in gypboard on walls that support ceilings common to house __ [309.2]
- ☐ <3ft from house req 1/2in gypboard on walls parallel to house ____ [309.2][59]
- ☐ No openings from garage to room used for sleeping purposes ____ [309.1]
- ☐ Door to house rated 20-minute or 1 3/8in solid core ____ [309.1]
- ☐ Ducts through wall or ceiling common to house min 26-gauge steel [309.1.1]
- ☐ Seal duct/pipe penetrations of drywall w/ approved firestop caulking ____ [309.1.2][60]
- ☐ No duct openings in garage ____ [309.1.1]
- ☐ Floor noncombustible w/ slope to vehicle door or a drain ____ [309.3]
- ☐ Carport means open on at least 2 sides ____ [309.4]
- ☐ Carport floor must be noncombustible or asphalt ____ [309.4]

Two-Family Dwellings

- ☐ 2-family reqs 1hr construction at common walls & floor/ceiling EXC __ [317.1]
  1/2hr OK if automatic sprinkler system present ____ [317.1X1]
- ☐ 1hr rated floor/ceiling assemblies must extend to exterior walls ____ [317.1]
- ☐ Supporting construction for 1hr floor/ceiling assembly: 1hr rating ___ [317.1.1]
- ☐ 1hr rated wall assemblies extend to underside of roof sheathing EXC [317.1]
  Need not extend through attic if ceilings 5/8 Type X gypboard & 1/2in gypboard on frame supporting the ceiling & attic is draftstopped ____ [317.1X2]

Townhouses

- ☐ Each unit considered separate bldg for separation & openings rules (**p.1**) [317.2]
- ☐ Townhouses req 2hr construction at common walls **F65** ____ [317.2X]
- ☐ No plumbing, or mechanical equipment in common-wall cavity ____ [317.2X]
- ☐ Fire-resistance rated walls continuous from foundation to underside of roof [317.2.1]
- ☐ Parapets min 30in high req'd above separation walls OR ____ [317.2.2]
  Class C or better roofing & noncombustible or protected decking _ [317.2.2X]

FIG. 65

Rated Two-Hour Wall

Typical separation wall in townhouse (Based on *Gypsum Association Manual WP3910*)

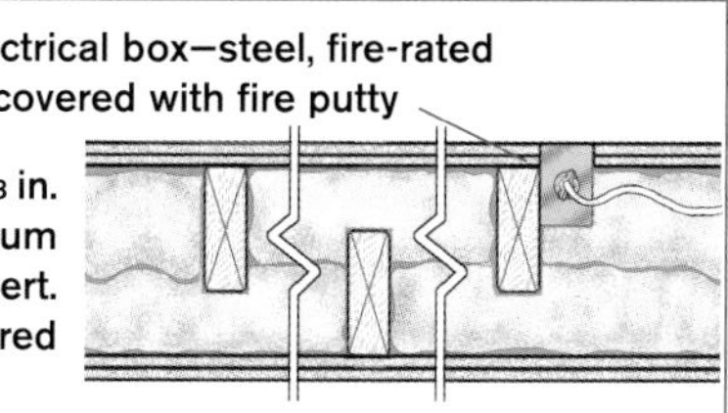

Firewall Penetrations

- ☐ Through-penetrations req approved penetration firestop system EXC [317.3.1]
  Metal pipes, tubes, or conduits in concrete or masonry walls can be grouted ____ [317.3.1X]
- ☐ Membrane penetrations (1 side only) for electrical boxes allowed if: [317.3.2X]
  - Boxes are steel or fire-resistance rated plastic **F65** ____ [317.3.2X1&2]
  - 16sq.in max size box, max 100sq.in in 100sq.ft ____ [317.3.2X1]
  - No back-to-back boxes (24in horiz separation) OR ____ [317.3.2X1]
  - Putty pad system in accordance w/ listing ____ [317.3.2X1]

FIG. 66

Air Flow through Soffit

Air communicates through soffit to attic or ceiling space above

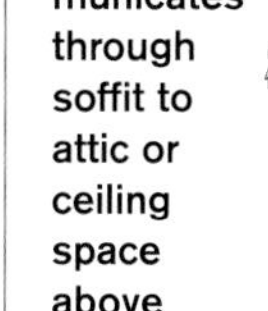

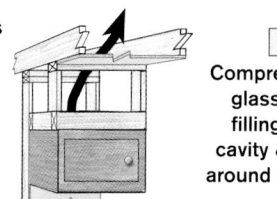

FIG. 67

Soffit Fireblocking Options

Fireblocking achieved by 1, 2, or 3

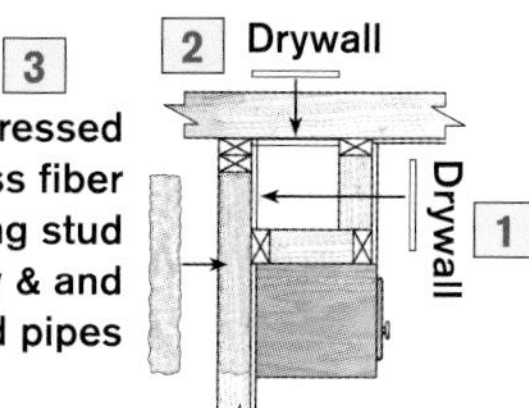

SMOKE ALARMS

General Requirements

- ☐ NFPA 72 fire alarms OK if equivalent to smoke alarms ____ [313.1][61]
- ☐ Req'd in each sleeping room & adjoining area **F68** ____ [313.2]
- ☐ Min 1 detector each story including basements **F68** ____ [313.2]
- ☐ Interconnect so activation of one alarm sets off all alarms ____ [313.2]
- ☐ Compliance req'd for work requiring permit EXC ____ [313.2.1]
  Work on exterior (porches, re-siding, re-roofing, etc.) ____ [313.2.1X2][62]
- ☐ Power from bldg wiring plus battery backup ____ [313.3]
- ☐ Battery power alone OK for remodel where no finishes removed ____ [313.2.1X1]

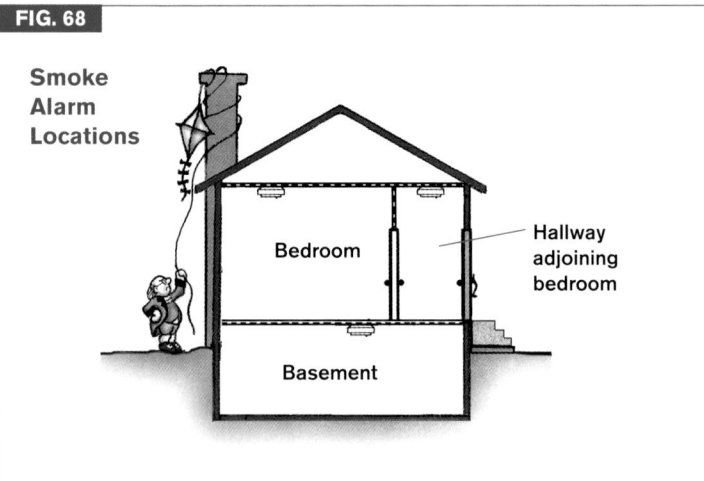

FIG. 68

Smoke Alarm Locations

Bedroom
Hallway adjoining bedroom
Basement

SAFETY GLASS

Safety glass can be laminated, tempered, or an approved plastic. Proper use of safety glass is critical in areas that are subject to human impact. Typical causes of accidents are failure to see the glass, intentional breakage, slips, and falls. The code rules are designed to minimize the hazards of these situations.

Identification

- ☐ Tempered glass reqs permanent etched label EXC **F69** ____ [308.1]
  Label permitted other than manufacturer's designation ____ [308.1][63]
- ☐ Tempered spandrel glass may have removable paper label ____ [308.1X2]
- ☐ Glazing ≤1ft in multipane may bear "16-CFR 1201" label **F69** ____ [308.1.1]

FIG. 69

Tempered Glass Label

The 16 CFR 1201 standard reqs the glass thickness to be labeled.
It is not a requirement of the IRC.

Hazardous Locations

- ☐ Safety glass must meet 16 CFR 1201 Class II impact standard (400ft lb) exc lites <9sq.ft in doors & in sidelights can be Class I (150ft lb) ____ [308.3]
- ☐ All swinging doors exc jalousies ____ [308.4]
- ☐ Glass panes in doors if 3in sphere could pass through glass **F70** ____ [308.4]
- ☐ Fixed & sliding panels of sliding doors & bifold closet doors ____ [308.4]
- ☐ Storm doors ____ [308.4]
- ☐ All tub & shower doors & enclosures EXC ____ [308.4]
  Glass block is considered masonry not glazing ____ [610]
- ☐ Glazing in bldg walls <60in above tub or shower standing surface ____ [308.4]
- ☐ Sidelites w/ vert edge of glazing >24in of door edge & lowest edge of glass <60in above floor EXC **F70** ____ [308.4]
  Lites that are perpendicular to door & on latch side OR
  Lites adjacent to a door to a closet <3ft deep ____ [308.4X4][64]
- ☐ All glass in railings ____ [308.4]
- ☐ All glass in walls or fences enclosing swimming pools or hot tubs w/ lower edge of glass <60in above walking surface & <60in horiz of water's edge ____ [308.4]
- ☐ Glass within 36in horiz of stairs or 60in from bottom tread if any part of glass <60in above walking surface EXC ____ [308.4]
  Guardrail between stair & glass >18in from railing OR ____ [308.4X9]
  Solid wall from walking surface to 34in–36in above ____ [308.4X9][65]
- ☐ Decorative glass (leaded, etched, or beveled) exempt in doors, sidelites, & picture windows ____ [308.4X2]
- ☐ Closet door mirrors mounted to continuous backing need not be safety glass if they meet impact requirement ____ [308.4X8]

Hazardous Locations (cont.)

- ☐ Windows w/ a walk-through hazard meeting all the following: **F70** ____ [308.4]
  - Exposed area of glazing >9sq ft. +
  - Bottom edge <18in above floor or ground +
  - Top edge >36in above floor or ground +
  - Within 36in horiz of walking surface EXC
  - Min 1 1/2in high protective guard installed 36in (±2in) above floor & capable of 50lb./linear foot w/o contacting glass ____ [308.4X5]

FIG. 70

If all conditions (A+B+C+D) exist, then window must be safety glass. Safety glazing would also be req'd if within 2ft. of the door frame.

SKYLIGHTS

Most residential skylights are factory-made unit skylights. These are designed to meet requirements of the Codes and to provide ease of installation. Many are self-curbing and self-flashing, and the manufacturer's instructions must be followed to provide a weather-tight installation. Skylights can also provide the required light and ventilation discussed on **p.26**.

General

- ☐ Skylights may be tempered, heat-strengthened, laminated, wired, or approved rigid plastics [308.6.2]
- ☐ Tempered or heat-strengthened glass reqs screen below glass EXC [308.6.3]
  - Panes <16sq.ft & ≤12ft above walking surface OR [308.6.5]
  - Greenhouses <20 ft above grade [308.6.6]
- ☐ Min 4in curb if roof slope < 3/12 **F71** [308.6.8]

FIG. 71

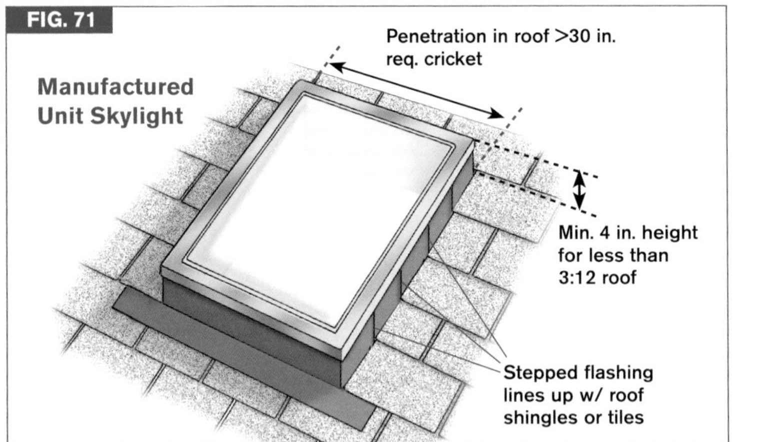

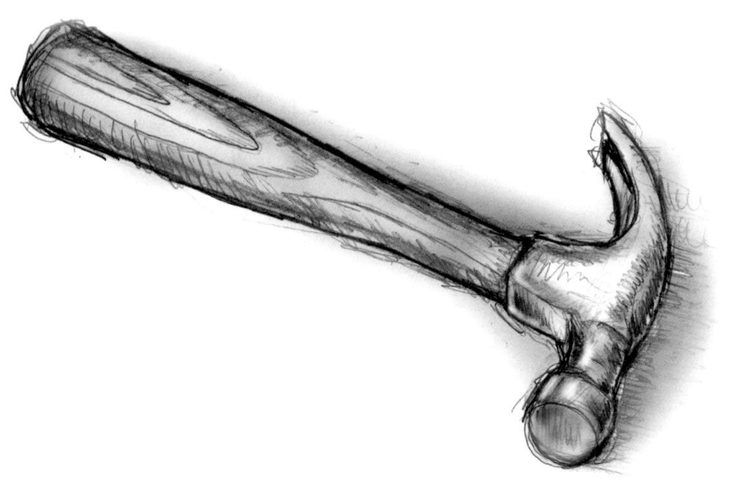

GLOSSARY

A

Admixture: Material other than water, aggregate, or hydraulic cement used as an ingredient of concrete and added to concrete before or during its mixing to modify its properties.

Aggregate: Granular material, such as sand, gravel, crushed stone, and iron blast-furnace slag, used with a cementing medium to form a hydraulic cement concrete or mortar.

Air-entrained: Concrete containing microscopic air cells to relieve internal pressure on the concrete from expansion of water when it freezes. Typical air-entrained concrete contains 5%–8% of air by volume

Aspect ratio: For buildings with unbalanced basements, the ratio of length to width. For structural sheathing or reinforced CMU shear walls, the ratio of height to width of the braced wall.

Bond beam: A horizontal grouted element within masonry embedded with reinforcement.

Building thermal envelope: Basement walls, exterior walls, floor, roof, and any other part of the building that encloses conditioned spaces.

Concrete: Mixture of Portland cement or any other hydraulic cement, fine aggregate, coarse aggregate, and water, with or without admixtures. (ACI)

- **Type 1** is normal Portland cement and is used in most residential construction.

- **Type 2** is used for structures in water or soil containing moderate amounts of sulfate or when heat buildup is a concern.

- **Type 3** is used when high early strength is desired. It is not common in single-family residential construction.

- **Type 4** is low-heat Portland cement and is used when the amount and rate of heat generation must be kept to a minimum. It is not common in single-family residential construction.

- **Type 5** is sulfate-resistant Portland cement and is used when the water or soil is high in alkali.

Corrosion-resistant: For items such as screening, galvanized coatings and construction of copper. For fasteners, materials must be hot-dipped galvanized, stainless steel, silicon bronze, or copper. Corrosion-resistant hangers and hardware are recommended when connecting to pressure-preservative treated wood.

Cripple wall: Wood-framed wall extending from the foundation to joists below the first floor. Found in the underfloor area.

Dampproofing: A coating intended to protect against the passage of water vapor through walls or other building elements. It is a lesser degree of protection than waterproofing.

Dead load: The weight of all materials of the building and fixed equipment.

Design required: Conditions requiring a set of plans and specifications created by a registered (licensed) design professional and beyond the prescriptive requirements of the IRC.

GLOSSARY

Diaphragm: A horizontal or nearly horizontal system acting to transmit lateral forces to the vertical resisting elements (e.g., roof sheathing).

Grade: The finished ground level adjoining the building at all exterior walls.

Habitable space: Space in a building for living, sleeping, eating, or cooking. Bathrooms, bathroom closets, halls storage, or utility areas are not considered habitable space.

Ice-dam barrier: A material consisting of either self-adhering modified bitumen, or two layers of underlayment cemented together, and extending horizontally to a point at least 2 feet inside the exterior wall line of the building. Ice-dam barriers prevent water from melted snow from backing up under the roofing surface.

Insulating sheathing: Insulating board having a minimum rating of R-2.

King stud: Wood framing member that extends unbroken from the bottom plate to the top plate next to a headered opening.

Live loads: Loads produced by use and occupancy of the building and not including wind, snow, rain, earthquake, flood, or dead loads.

Monolithic: Concrete cast in one continuous operation with no joints, such as a footing and floor slab, or a footing and foundation stem wall.

Perm: The unit of measurement of water vapor transmission through a material, based on the number of grains of water vapor at a given pressure differential. Vapor retarders are rated in perms.

Plain concrete: Structural concrete with no reinforcement or with less reinforcement than the minimum amount specified for reinforced concrete.

PSI (pounds per square inch): Common method of measuring concrete strength. Most common concretes come in 2,000, 2,500, 3,000, 3,500 and 4,000 psi strengths.

Seismic design category: Classification assigned to buildings based on location and severity of earthquake ground motion expected at the site.

Shear wall: A term for walls designed to resist racking and distortion from wind or seismic forces.

Story: That portion of a building that is between the upper surface of one floor and below the upper surface of the next floor above or the roof.

Story above grade: That part of the building which is more than 6 feet above grade for more than 50% of the total building perimeter or more than 12 feet above ground at any point.

Slump test: A method of field measuring of the stiffness of fresh concrete, usually performed with a cone shaped device. More water in the mix means higher slump and weaker concrete. Both the type of concrete ordered and psi will affect slump.

Townhouse: Single-family dwelling unit constructed in groups of three or more attached units in which each unit extends from foundation to roof and with open space on at least two sides.

Trimmer stud: The stud that supports the header. Also called jamb stud. In horizontal applications, the header is fastened to the trimmer stud.

Waterproofing: Materials that protect walls or other building elements from the passage of moisture as either vapor or liquid under hydrostatic pressure.

Wood structural panel: A panel manufactured from veneers (plywood) or wood strands (OSB) and bonded with waterproof synthetic resins. Wood structural panels must bear a grade stamp (see **F41 p. 41**) and are used in floors, roof diaphragms, and shear walls

SIGNIFICANT CODE CHANGES TO THE 2006 IRC

1. Unrated walls were allowed 3ft from property line in 2003.
2. There were no restrictions on percentage of openings in 2003.
3. Sheds & accessory buildings over 200 sq. ft req'd permits in 2003.
4. Sidewalks & driveways now all exempt, formerly only if <30in above grade.
5. Swale now acceptable where lot lines or physical barriers prevent 6in in 10ft.
6. Interior footings of monolithic slabs were req'd to be 18in deep from the top in 2003.
7. Reduction to one bolt for 2ft offset braced wall section is new.
8. New bolting exemption for offsets up to 12in in braced wall lines.
9. Ext revision of tables for concrete & CMU basements to allow taller walls.
10. New requirement for design of retaining walls that are not supporting buildings.
11. New safety factor requirement in retaining wall design.
12. New prescriptive requirements for lateral restraint of basement walls.
13. New requirement for support of reinforcing mesh in slabs during pour.
14. The 2003 IRC allowed a reduction to $^1/_{1500}$ if a vapor barrier was present.
15. The rules for unvented crawlspaces are new.
16. The 2003 IRC limited steel framing to bldgs 36ft wide.
17. The 2003 IRC limited the story height of steel bldgs to 10ft.
18. Manufacturer's instructions w/o further engineering are acceptable for notches & alterations of TJIs.
19. Joints in top plates are now specifically allowed to not occur over studs.
20. Vert joints of structural sheathing must land on a common stud. 2003 IRC would have allowed the joints on separate studs next to each other.
21. Table for aspect ratios & uplift force of alternate braced wall panels is new.
22. New requirement to provide load path for lateral loads transferred from one floor to another.
23. The 2006 code clarifies that the vapor barrier goes on the warm-in-winter side, not simply the "warm" side, which would be the top in the summertime.
24. The provisions for conditioned attic spaces are new.
25. The requirement for flashing AMI is now explicit.
26. The requirement for installation instructions delivered w/ the windows is new.
27. Child-safe windows now req when >6ft above grade.
28. The requirement to drain accumulated water to the exterior is new.
29. The term *water-resistant barrier* has replaced "weather-resistant."
30. The 2003 IRC only req'd Type 1 felt or equivalent when called for by Table 703.4.
31. The specific purpose & installation method of wall flashing has been clarified.
32. Window flashing must now be installed shingle fashion. The term self-flashing window has been deleted.
33. Stucco must now comply w/ these two ASTM standards, which include extensive specific requirements beyond the wording in the codes.
34. The 2003 IRC referred to Table 702.1(1) for plaster thickness & the 200 refers to ASTM C 926. For three-coat work, the net result is the same.

35. The min standard for the water-resistive barrier has been increased.
36. Rain-curtain systems (drainable air space behind surface) were not addressed in 2003.
37. The rules & exceptions for veneer height & bracing are now in table form.
38. Requirements for bracing behind veneer are now more stringent & there are specific differences for SDC D_0.
39. The limitations on greenboard have been expanded beyond just tubs & showers.
40. Cement board is now explicitly recommended for tub & shower walls.
41. Manufacturer's instructions w/o further engineering are acceptable for notches & alterations of TJIs.
42. Rafter ties are now defined as being in the lower third of the attic space.
43. Rafter ties must now be at least 2×4s & must be installed at every rafter not tied to ceiling joists.
44. Collar ties are now defined as being in the upper third & are now required.
45. The minimum size of collar ties is the same as the former minimum size for rafter ties.
46. The applicability limits for trusses are new.
47. BCSI 1-03 is the new bracing guide for trusses, replacing HIB-91.
48. The rule for crickets now applies to any roof penetration wider than 30in., including items such as skylights & chimneys.
49. The requirement for an ice barrier now depends on the local history. In the 2003 IRC, it depended on the average daily January temperature.
50. The rule for slope of landings is new.
51. In 2003, only basements w/ habitable space req'd escape & rescue openings. Now the only exemption is for basements that are only mech eqpmt rooms & <200sq.ft.
52. The requirement for individual escape & rescue openings in basement bedrooms is new. These openings are sufficient to serve the non-bedroom basement areas.
53. The escape & rescue clearance under a deck is now defined.
54. "Special knowledge" was added to the list.
55. Bath exhaust must terminate on the building ext.
56. The 7-second interval & 30-second duration requirements are new.
57. Cables openings through plates must now be fire blocked.
58. The 2003 IRC req'd chimney fireblocking to be at least 1in in depth.
59. The requirement for protection in garages closer than 3ft is new.
60. The requirement for penetration firestop sealant is new.
61. Central station fire alarm systems have different standards than smoke alarms.
62. The exemption for work only on the exterior is new.
63. Installers can apply the permanent label to tempered glass without including the thickness. A designation of thickness is required by CFR 1201 but not by the ICC.
64. Perp sidelites on wall opposite from that toward which door swings open are exempt.
65. Glazing above the handrail is exempt when the surface below is solid.

Pipe Wrench

Code ✓Check® Plumbing Third Edition

BY DOUGLAS HANSEN, REDWOOD KARDON, AND MICHAEL CASEY

Code Check Plumbing is based on the 2006 International Residential Code and the 2006 Uniform Plumbing Code–the two most widely used residential plumbing codes. For areas that using older code editions, this book has a feature that highlights recent code changes and summarizes them on pp. 120–121. Before beginning a building project, we recommend checking with your local building department to determine which codes are used in your area.

KEY TO USING CODE CHECK PLUMBING

Code references are followed by two bracketed numbers, ex:

☐ Code reference T1,F1 ____________________ [123.4] {123.4}

Code numbers on left, in brackets, ex: [123.4], refer to IRC codes.

Code numbers on right, in braces, ex: {123.4}, refer to UPC codes.

T1 refers to Table 1. F1 refers to Figure 1.

[manu] = Typically required by manufacturer's installation instructions.

[2407.15X] = An X after a code number refers to an exception in code.

EXC = When placed at end of text line signals an exception in following line.

OR = When placed at end of text line signals an alternative in following line.

n/a] = not addressed by the IRC. {n/a} = not addressed by the UPC.

[Ø] = Prohibited by the IRC. {Ø} = Prohibited by the UPC.

[brackets] or {braces} in a text line restrict that text to the [IRC] or {UPC} codes.

Codes ending in numbers separated by commas refer to multiple code sections, ex: [1802.1,2407.3] = IRC sections 1802.1 and 2407.3. A colored code citation with a superscript number indicates a change in the code.

ex: [3005.2.9][9] refers to a code change in the IRC, listed as #9 in the code change summary on p. 120.

Example (from p. 68)

☐ Sleeve pipes through footings, fndns
{& concrete floors}{does not apply to bored holes} ___[2603.5] {313.10.1}[1]

The IRC and UPC require a sleeve around any pipe passing through footings and foundations. In addition, the UPC requires a sleeve for pipes passing through concrete floors. The braces indicate that the rule for floors is only in the UPC, not the IRC. The different color for the UPC citation tells us it is a code change, and the superscript 1 tells us the change is #1 in the list on the p.120.

Example (from p. 86)

☐ Vent connection at interconnection of fixt drains OR ___ [3107.2] {905.6}
[Downstream of interconnection] F23,34 __________ [3107.2] {Ø}

In the IRC, a vent may connect where fixture drains are joined or downstream from where they connect. In the UPC, they must connect where they are joined. These practices are shown in Figures 23 and 34.

ABBREVIATIONS

AAV(s) = air admittance valve(s)
ABS = acrylonitrile-butadiene-styrene drain (black plastic drain pipe)
ACH = air changes per hour
AHJ = authority having jurisdiction (agent of building official)
appl(s) = appliance(s)
AWG = American wire gauge
B (vent) = double-walled gas appliance flue
BA = bathroom
bldg(s) = building(s)
BO = building official
BR = bedroom
BT = bathtub
Btu(s) = British thermal unit(s)
CI = cast iron
CO = cleanout
CO_2 = carbon dioxide
CPVC = chlorinated PVC
CSST = corrugated stainless-steel tubing (for gas)
cu = cubic
Cu = copper
CW = clothes washer
CWV = combination waste & vent
DFU = drainage fixture unit
dia = diameter
DWV = drain, waste, & vent
eqpmt = equipment
ex. = example
exc = except
ext = exterior
Fe = iron or steel pipe
fixt(s) = fixture(s)
FLR = flood level rim
fndn(s) = foundation(s)
ft = foot, feet
FVIR = flammable vapor ignition resistant
gal = gallons
GPM = gallons per minute
horiz = horizontal
in = inch(es)
k = 1,000
KS = kitchen sink
lav(s) = bathroom lavatory sink(s)
L&L = listed & labeled
LT = laundry tub
manu(s) = manufacturer(s)
max = maximum
min = minimum
O.C. = on center
PB = polybutylene
PE = polyethylene
PEX = cross-linked polyethylene water tubing
PEX-AL-PEX = PEX-aluminum-PEX tubing
PMI = according to manufacturer's instructions
PRV = pressure-relief valve
psi = pounds per square inch
psig = pounds per square inch gauge
PVC = polyvinyl chloride
req(s) = requirements
req's, req'd = requires or required
SDC = seismic design category
sq = square
temp = temperature
TJI® = manufactured I-joists
TPRV = temperature & pressure relief valve
vert = vertical
w/, w/o = with, without
WC(s) = water closet(s), (toilet[s])
WH(s) = water heater(s)

CODE CHECK PLUMBING CODES

Code Check Plumbing 3rd Edition references these two codes:

2006 IRC (*International Residential Code*–published by ICC–the International Code Council) and

2006 UPC (*Uniform Plumbing Code*–published by IAPMO–the International Association of Plumbing and Mechanical Officials)

If your area is using an earlier edition of one of these codes, take special note of the highlighted code changes and use the list on pp. 120–121 to aid in proper application of the appropriate rules.

In 1752, Benjamin Franklin brought the first bathtub to the United States. After designing a more comfortable model, he took it with him on his many travels to Europe.

CODE CHECK PLUMBING CONTENTS

FIG. 1

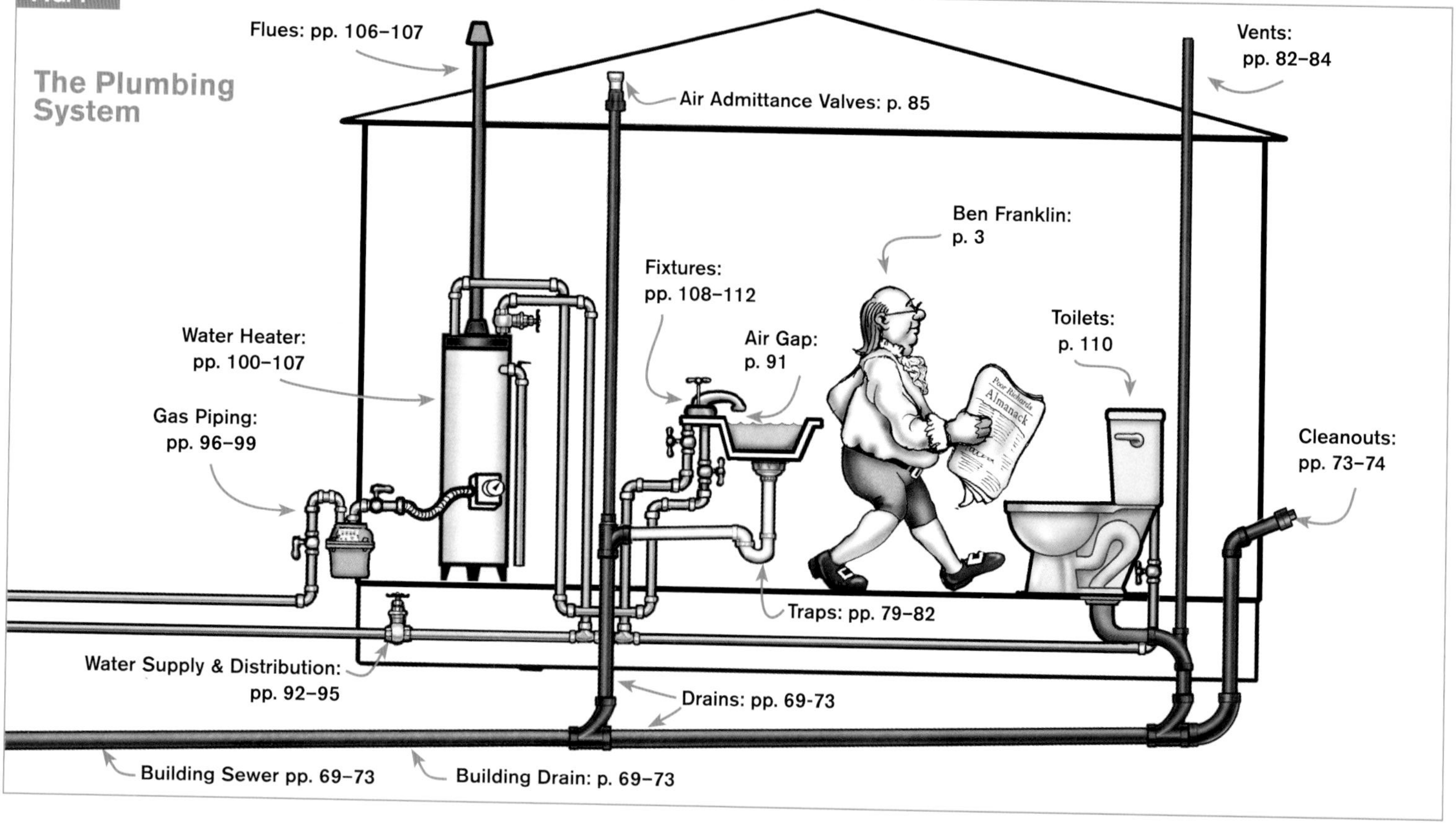

The Plumbing System

Flues: pp. 106–107
Air Admittance Valves: p. 85
Vents: pp. 82–84
Water Heater: pp. 100–107
Gas Piping: pp. 96–99
Fixtures: pp. 108–112
Air Gap: p. 91
Ben Franklin: p. 3
Poor Richard's Almanack
Toilets: p. 110
Cleanouts: pp. 73–74
Traps: pp. 79–82
Water Supply & Distribution: pp. 92–95
Drains: pp. 69-73
Building Sewer pp. 69–73
Building Drain: p. 69–73

TABLE 1 **MAXIMUM PIPE SUPPORT SPACING** [T2424.1 & T2605.1] & {T3-2}

Pipe Material	Horiz. Spacing [IRC]	Horiz. Spacing {UPC}	Vert. Spacing [IRC]	Vert. Spacing {UPC}
ABS/PVC DWV	4 ft.	4 ft. Allow for expansion every 30 ft. Support at each horiz. branch connection	10 ft. Mid-story guides if ≤2 in. dia.	Base & each story Provide mid-story guide Provide for expansion every 30 ft.
Threaded steel water pipe	12 ft.	≤3/4 in. dia.–10 ft. ≥1 in. dia.–12 ft.	15 ft.	25 ft. & every other story
CI (no hub)	5 ft. 10 ft. OK when 10 ft. lengths of pipe are used	Every other joint unless over 4 ft. Within 18 in. of joints Horiz.. bracing max. 40 ft. All horiz. branch connections No hangers on couplings	Base & each story 15 ft.	Base & each story 15 ft.
Cu water tubing	≤1 1/4 in. dia.–6 ft. ≥1 1/2 in. dia.–10 ft.	≤1 1/2 in. dia.–6 ft. ≥2 in. dia.–10 ft.	10 ft.	10 ft. & each story
CPVC	≤1 in. dia.–3 ft. ≥1 in. dia.–4 ft.	≤1 in. dia.–3 ft. ≥1 1/4 in. dia.–4 ft.	10 ft. Mid-story guides if ≤2 in. dia.	Each story Mid-story guides
Flexible plastic (PEX)	32 in.	32 in.	10 ft. Mid-story guides if ≤2 in. dia.	Base & each story Provide mid-story guides
Threaded steel or brass gas pipe	≤1/2 in. dia.–6 ft. ≥3/4 in. & 1 in. dia.–8 ft. ≥1 1/4 in. dia.–10 ft.	≤1/2 in. dia.–6 ft. ≥3/4 in. & 1 in. dia.–8 ft. ≥1 1/4 in. dia.–10 ft.	≤1/2 in. dia.–6 ft. ≥3/4 in. & 1 in. dia.–8 ft. ≥1 1/4 in. dia.–each story	≤1/2 in. dia.–6 ft. ≥3/4 in. & 1 in. dia.–8 ft. ≥1 1/4 in. dia.–each story

TRENCHING & PIPE PROTECTION

Pipes installed in the soil must be properly supported for their entire length. Improperly supported piping can be stressed and may prematurely fail. Smooth, self-compacting backfill, such as sand or small gravel, helps eliminate sags that could cause water to be trapped and lead to blockage. All pipes must be protected from sharp rocks or other debris when backfill is placed. Piping encased in concrete must be protected.

Breakage & Strain Relief	IRC	UPC
☐ Wrap embedded piping to prevent strain & corrosion	[2603.3]	{313.1,2}
☐ Provide for expansion & contraction	[2606.3]	{313.2}
☐ Sleeve pipes through footings, fndns {& concrete floors}{does not apply to bored holes}	[2603.5]	{313.10.1}[1]
☐ Sleeve pipe passing through fndn	[2603.5]	{313.10.1}
☐ Sleeve size min 2 pipe sizes larger than pipe passing through fndn	[2603.5]	{n/a}
☐ Seal spaces between pipes & sleeves	[2603.4]	{313.10.3}

Piping Support & Depth in Trench	IRC	UPC
☐ Pipe supported on firm bed for entire length F2	[2604.1]	{314.3}
☐ No rocks supporting or touching pipes	[2604.1]	{315.4}
☐ No rocks or debris in first 12in backfill over pipe	[2604.3]	{315.4}
☐ Trenches not to undermine footings within 45° F3	[2604.4]	{313.3}
☐ Water pipe min 12in deep & 6in {12in} below frost line	[2603.6]	{609.1}

Piping in Common Trench	IRC	UPC
☐ Water & sewer OK in same trench if sewer materials approved for use within bldg F5	[2904.4.2]	{609.2}
☐ Water min 5ft from sewer if sewer not approved for use within bldg F4 OR	[2904.4.2]	{609.2}
Water service on shelf 12in above sewer F6 OR	[2904.4.2]	{609.2}

Piping in Common Trench (cont.)	IRC	UPC
Water pipe crossing sewer: min 12in above OR	[2904.4.2]	{609.2}
Sleeved for 5ft each way from centerline of sewer	[2904.4.2X]	{n/a}

FIG. 2 Pipes in Trench

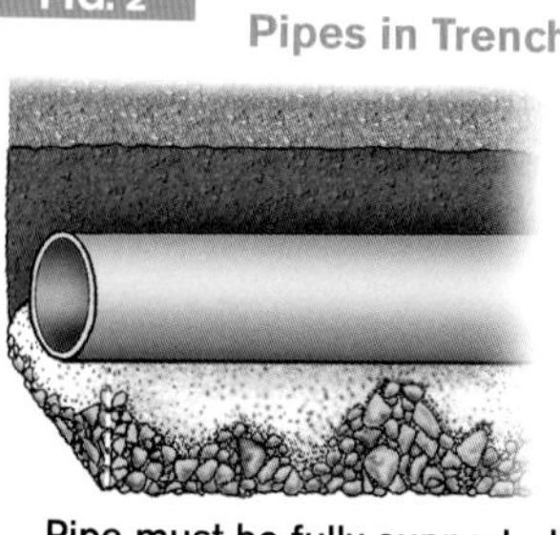

Pipe must be fully supported

FIG. 3 Pipes Near Footing

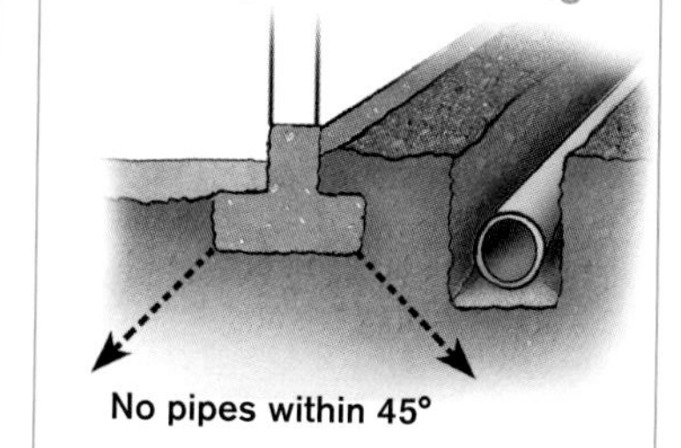

No pipes within 45°

FIG. 4 Separate Trenches

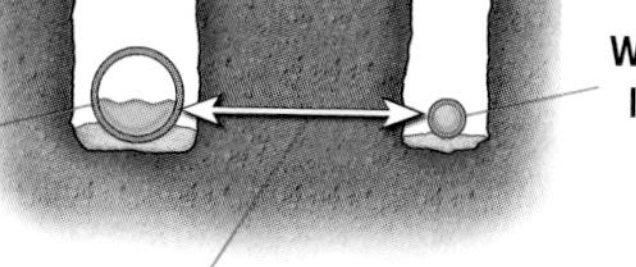

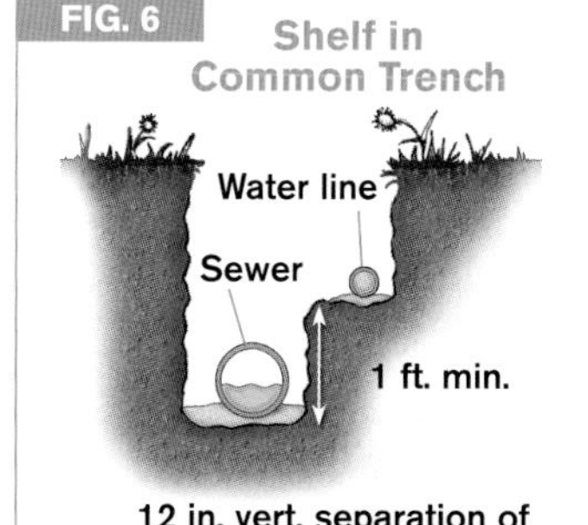

Sewer material approved within the bldg.

12 in. vert. separation of water & sewer if sewer not approved within the bldg.

DRAINAGE

Drain and waste pipes carry contaminated water out of the building and discharge it to the building sewer. They must have sufficient slope and be free of constriction or obstructions. Pipe sizes are based on the number of drainage fixture units each pipe must serve. Changes of direction in pipes must be done with fittings that will not cause an obstruction in flow. The change from vertical to horizontal requires the greatest sweep, because the liquid and solids in the vertical pipe have greater velocity than in the horizontal pipe.

General

	IRC	UPC
☐ Materials must be listed and/or approved T2	[2608.4]	{301.1.1}
☐ All pipes & fittings marked by manu	[2608.1]	{301.1.2}
☐ Size per DFU loads {& vert pipe length} T3–5	[3005.4]	{703.1}
☐ Min slope ¼in/ft EXC	[3005.3]	{708.0}
⅛in/ft OK for 3in or larger pipe	[3005.3]	{Ø}
☐ {⅛in/ft OK for 4in pipe if structurally necessary & BO approves}	[n/a]	{708.0}
☐ No drilling or tapping (ex. saddle fitting) connections	[3003.2]	{311.2}

General (cont.)

	IRC	UPC
☐ No reductions in direction of flow EXC	[3002.3.1, 3005.1.7]	{316.4.1}
3in by 4in WC bend OK F7	[3005.1.7]	{316.4.2
☐ ABS & PVC cannot be directly glued together EXC	[3003.2(5)]	{316.1.6}
ABS & PVC bldg drain to bldg sewer OK w/ listed transition solvent	[Ø]	{316.1.6}[2]
☐ No metal straps directly contacting ABS or PVC F8	[2605.1]	{301.1.1}
☐ Proper fittings for changes of direction F9,10	[3005.1]	{706.1}
☐ Horiz branches join at least 10 pipe dia downstream of base of stack	[3005.5][3]	{n/a}

TABLE 2 — DRAINAGE MATERIALS [3002.1] & {701.1}

Material	Aboveground inside bldg.[4]		Underground inside bldg.[4]		Underground outside bldg. (Sewer)	
	IRC	UPC	IRC	UPC	IRC	UPC
Cu DWV	OK	OK	OK	OK	Ø[5]	OK
Cu type M	OK	OK	OK	Ø	Ø[6]	Ø
Cu type L or K	OK	OK	OK	Ø	OK	Ø
CI no hub	OK	OK	OK	OK	OK	OK
PVC/ABS schedule 40	OK	OK[A]	OK	OK	OK	OK
Galvanized steel	OK[B]	OK[B]	Ø	Ø	Ø	Ø
Vitrified clay	Ø	Ø	Ø	OK	OK	OK[C]
PVC/DR PS 35, 50, 100, 140, 200	Ø	Ø	OK[7]	Ø	OK	Ø

A. Must be firestopped when passing through rated walls, floors, or assemblies.
B. Galvanized steel to be kept min. 6 in. aboveground.
C. Vitrified clay to be min. 12 in. belowground.

FIG. 7 Closet Bend Reductions

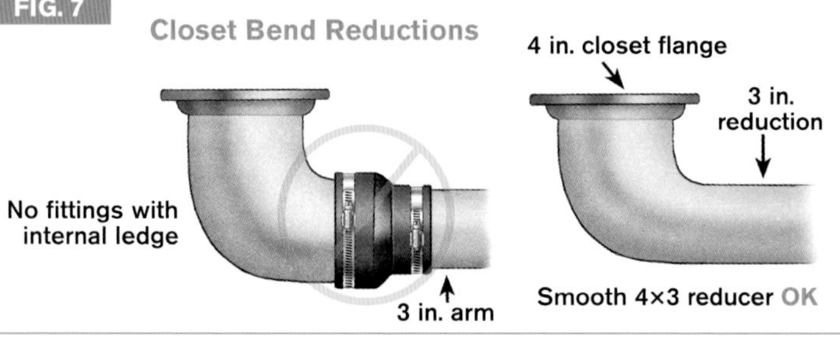

FIG. 8 Plastic Pipe Support & Spacing

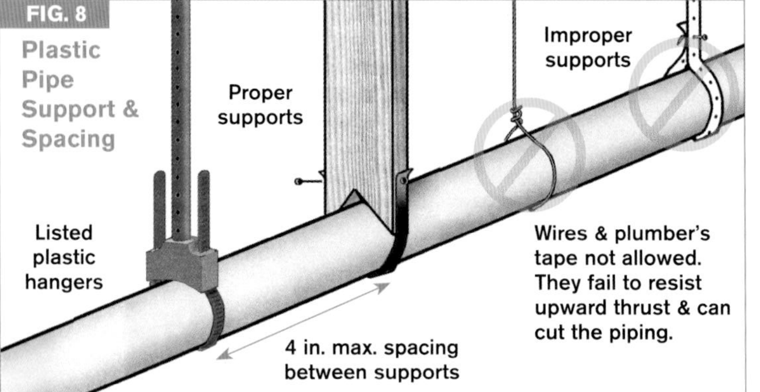

TABLE 3 DFU LOAD & TRAP SIZE [3004.1] & [3201.7] {T7-3}

Fixt.	Fixt. Units		Min. Trap Size (in.)	
	IRC	UPC	IRC	UPC
Bar sink	1	1	1 1/2	1 1/2
Bath/shower	2	2	1 1/2	1 1/2
Full BA group[A]	6	n/a	n/a	n/a
Bidet	1	1	1 1/4	1 1/4
CW standpipe	2	3	2	2
LT	2	2	1 1/2	1 1/2
Laundry group	3	n/a	n/a	n/a
KS	2	2	1 1/2	1 1/2[B]
DW	2	2	1 1/2	1 1/2[B]
KS w/ DW	2	2	1 1/2	1 1/2[B]
KS w/ disposer	2	2	1 1/2	1 1/2[B]
Lav.	1	1	1 1/4	1 1/4
Shower stall	2	2	1 1/2[C 8]	2
WC >1.6 gal./flush	4	4	n/a	n/a
WC < 1.6 gal./flush	3	3	n/a	n/a
Floor drain	0	0	2	2

A. The IRC recognizes a BA group designation that includes the WC, lav, bidet, and BT or shower located together on the same floor. The designation is important because it allows for horiz. wet venting. For bath groups with 1.6 gal./flush toilets: subtract 1 fixt. unit.
B. For these fixts., the UPC req's a 2 in. drain after the trap arm. The IRC allows the same size drain as the trap.
C. Applies only to showers w/ a total flow rate through showerheads and bodysprays totaling 5.7 GPM. For showers >5.7 GPM the min. trap size is 2 in.[8]

TABLE 4	MAXIMUM DFUs ON BUILDING DRAIN & SEWER [3005.4.2] & {T7-5}		
Pipe Size (in.)	Slope per foot		
	1/8 in.	1/4 in.	1/2 in.
1 1/2	Ø	Ø[A]	Ø[A]
2	Ø	21 {8}[B]	27 {8}[B]
3[B]	36 {Ø}	42 {35}	50 {35}
4	180 {172}[C]	216 {216}	250 {216}

A. 1 1/2 in. pipe is limited to branches serving not more than 2 waste fixts. or 1 fixt. only if pumped discharge or garbage disposer.
B. Min. size of drain serving a WC: 3 in.
C. 1/8 in./ft. slope allowed only when approved by AHJ.

TABLE 5	MAXIMUM DFUs ON BRANCHES & STACKS [3005.4.1] & {T7-5}			
Pipes size (in.)	IRC DFUs		UPC DFUs	
	Horizontal	Vertical	Horizontal	Vertical
1 1/4[A]	1	1	1	1
1 1/2	3	4	1	2[B]
2	6	10	8	16
2 1/2	12	20	14	32
3	20	48	35	48
4	160	240	216[C]	256

A. 1 1/4 in. pipes are limited to the discharge of a single fixt.
B. No sinks, urinals, or DWs (lav. OK).
C. If <1/4 in./ft., refer to T4.

HIEROGLYPHIC HUMOR

FIG. 9 DWV Fittings

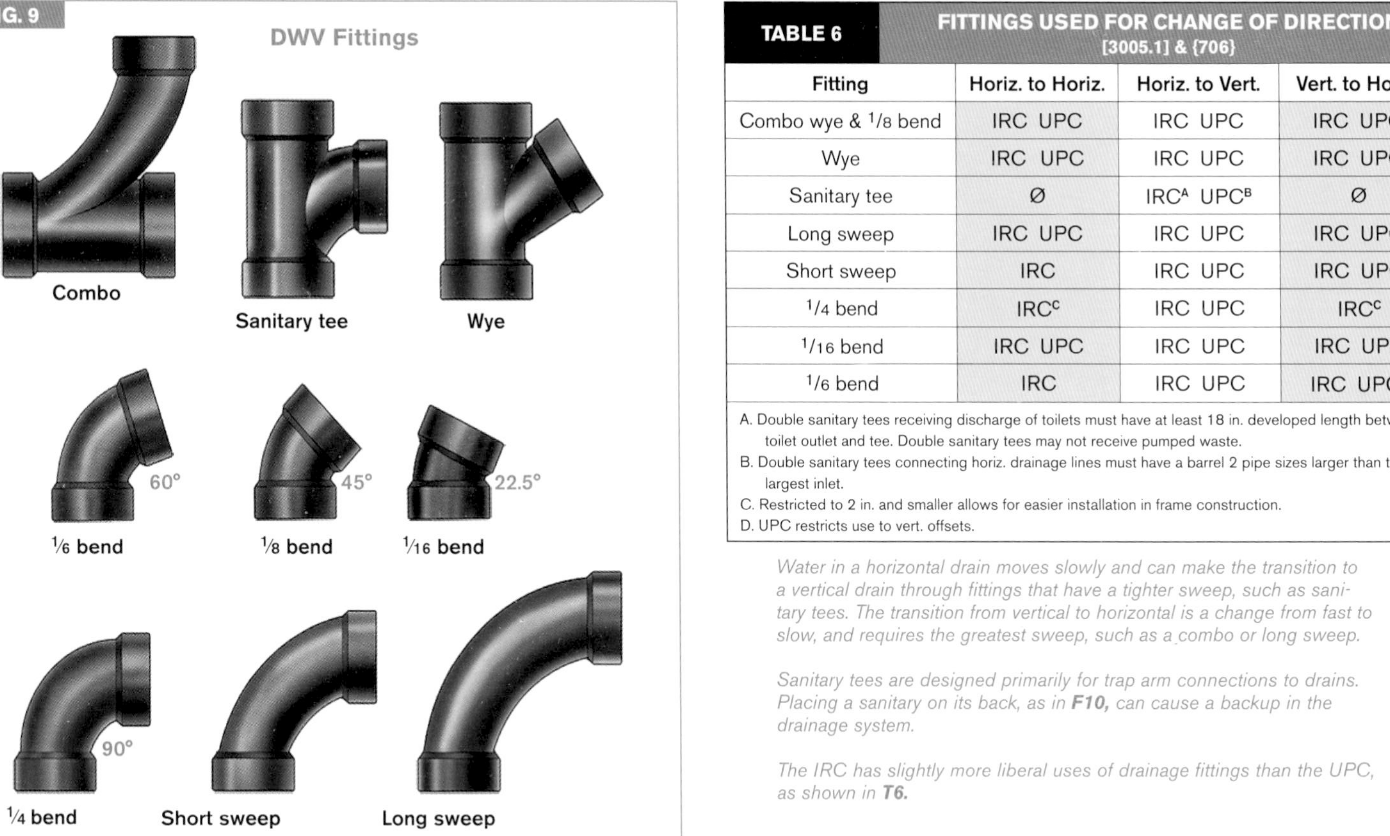

TABLE 6 FITTINGS USED FOR CHANGE OF DIRECTION [3005.1] & {706}

Fitting	Horiz. to Horiz.	Horiz. to Vert.	Vert. to Horiz.
Combo wye & 1/8 bend	IRC UPC	IRC UPC	IRC UPC
Wye	IRC UPC	IRC UPC	IRC UPC
Sanitary tee	Ø	IRC[A] UPC[B]	Ø
Long sweep	IRC UPC	IRC UPC	IRC UPC
Short sweep	IRC	IRC UPC	IRC UPC
1/4 bend	IRC[C]	IRC UPC	IRC[C]
1/16 bend	IRC UPC	IRC UPC	IRC UPC
1/6 bend	IRC	IRC UPC	IRC UPC[D]

A. Double sanitary tees receiving discharge of toilets must have at least 18 in. developed length between toilet outlet and tee. Double sanitary tees may not receive pumped waste.
B. Double sanitary tees connecting horiz. drainage lines must have a barrel 2 pipe sizes larger than the largest inlet.
C. Restricted to 2 in. and smaller allows for easier installation in frame construction.
D. UPC restricts use to vert. offsets.

Water in a horizontal drain moves slowly and can make the transition to a vertical drain through fittings that have a tighter sweep, such as sanitary tees. The transition from vertical to horizontal is a change from fast to slow, and requires the greatest sweep, such as a combo or long sweep.

*Sanitary tees are designed primarily for trap arm connections to drains. Placing a sanitary on its back, as in **F10,** can cause a backup in the drainage system.*

*The IRC has slightly more liberal uses of drainage fittings than the UPC, as shown in **T6.***

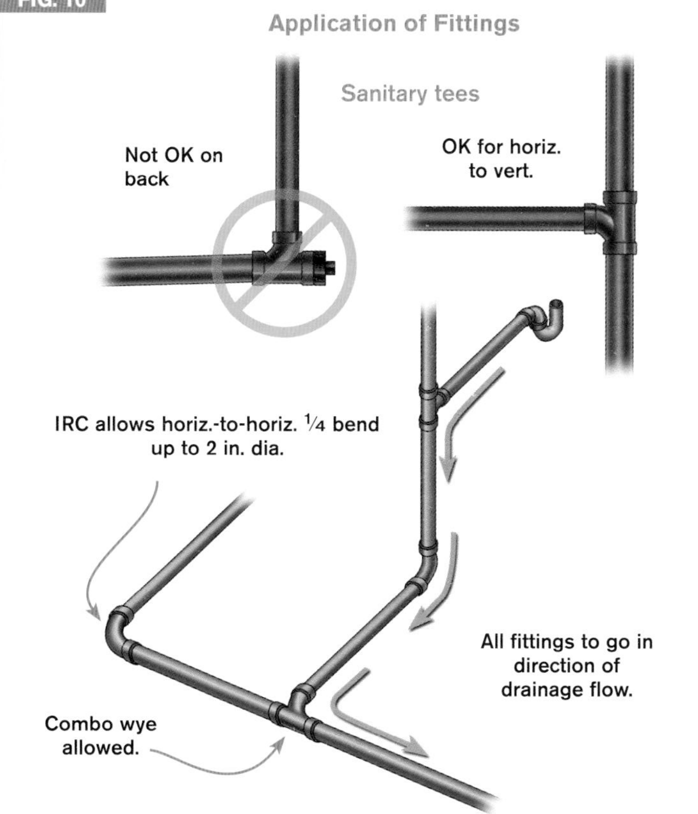

CLEANOUTS

Drain cleanouts at accessible locations provide access for snaking obstructions from a pipe or viewing the pipe interior with a sewer camera.

IRC CO Requirements — IRC

- ☐ Same size as drain pipes up to 4in EXC ____ [3005.2.9][9]
  CO in stack can be one size smaller than drain ____ [3005.2.9][10]
- ☐ Removable traps OK as CO for drains up to 1 size larger than trap ____ [3005.2.9X1, 10]
- ☐ Req'd at base of stack or outside within 3ft of bldg wall F14 ____ [3005.2.6]
- ☐ Req'd near junction of bldg drain & bldg sewer EXC ____ [3005.2.7]
  Not req'd if CO on 3in soil stack located within 10ft of bldg sewer [3005.2.7][11]
- ☐ 2-way CO OK at bldg drain & bldg sewer ____ [3005.2.7][12]
- ☐ Pipes <3in req 12in clearance; ≥3in: req 18in ____ [3005.2.5]
- ☐ Req'd each 100ft of straight horiz run ____ [3005.2.2]
- ☐ Req'd in horiz drains each change of direction >45° exc only 1 CO req'd per 40ft of run F12 ____ [3005.2.4]
- ☐ CO openings: not OK for new fixts unless new CO installed ____ [3005.2.11]

UPC CO Requirements — UPC

- ☐ Size per T7 ____ {707.11}
- ☐ Req'd at upper terminal of horiz runs on first floor OR 2-way CO near junction of bldg drain & bldg sewer F11 ____ {707.4}
- ☐ Additional CO req'd horiz runs aggregate >135° F12 ____ {707.5}
- ☐ Not req'd for horiz runs <5ft (exc sinks) ____ {707.4X1}
- ☐ Not req'd if slope >4in/ft ____ {707.4X2}
- ☐ Takeoff above flow line exc wye branch or end of line ____ {707.6}
- ☐ Pipes ≤2in req 12in clearance; >2in req 18in clearance T7 ____ {707.10}
- ☐ Underfloor CO not >20ft from access door ____ {707.10}
- ☐ Passageway to underfloor CO min 30in horiz by 18in vert ____ {707.10}

FIG. 11

Horizontal Distances & Cleanout Locations

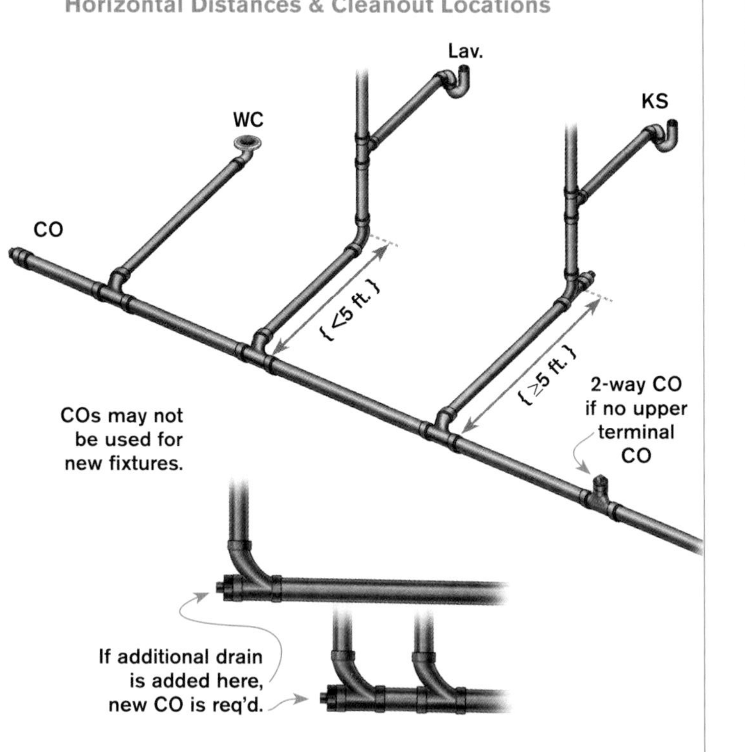

TABLE 7	CLEANOUTS [3005.2] & {707}	
Pipe Size (in.)	UPC CO Size (in.)	UPC Clearance (in.)
1 1/2	1 1/2	12
2	1 1/2	12
2 1/2	2 1/2	18
3	2 1/2	18
4 or larger	3 1/2	18

IRC 3005.2.9 req's COs to be the same size as the pipe they serve up to 4in.[9]
UPC 707.1 req's CO fittings w/ raised square heads or countersunk slots.

FIG. 12

Bends & Clearances

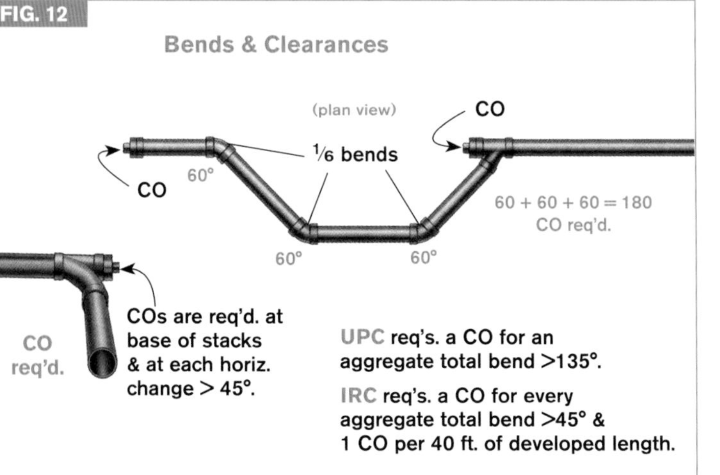

UPC req's. a CO for an aggregate total bend >135°.

IRC req's. a CO for every aggregate total bend >45° & 1 CO per 40 ft. of developed length.

FIXTURES BELOW MANHOLE COVER

The drain lines serving fixtures located below the manhole cover but above the sewer will drain by gravity to the sewer. During a time when the street sewer is blocked, the drains could back up to those fixtures. To prevent this backup and sewer contamination from reaching the fixtures in the house, backwater valves (a form of check valve) must be installed.

Backwater Valves	IRC	UPC
☐ Backwater valve req'd for fixts below next upstream manhole cover F13	[3008.1]	{710.1}
☐ Fixts above manhole cover: elevation not allowed to discharge through backwater valve F13	[3008.1]	{710.1}
☐ Backwater valves req'd to be accessible for service	[3008.1]	{710.6}}

*Fixtures above the sewer, but below the flood weir of the next upstream manhole are required to be protected by an accessible backwater valve (a type of check valve) as shown in **F13**. The valve protects the building from sewer contamination in the event of a street main backup. The UPC also requires a backwater valve for fixtures below the manhole cover of a private sewer system.*

FIG. 13

Fixtures Downstream of Manhole

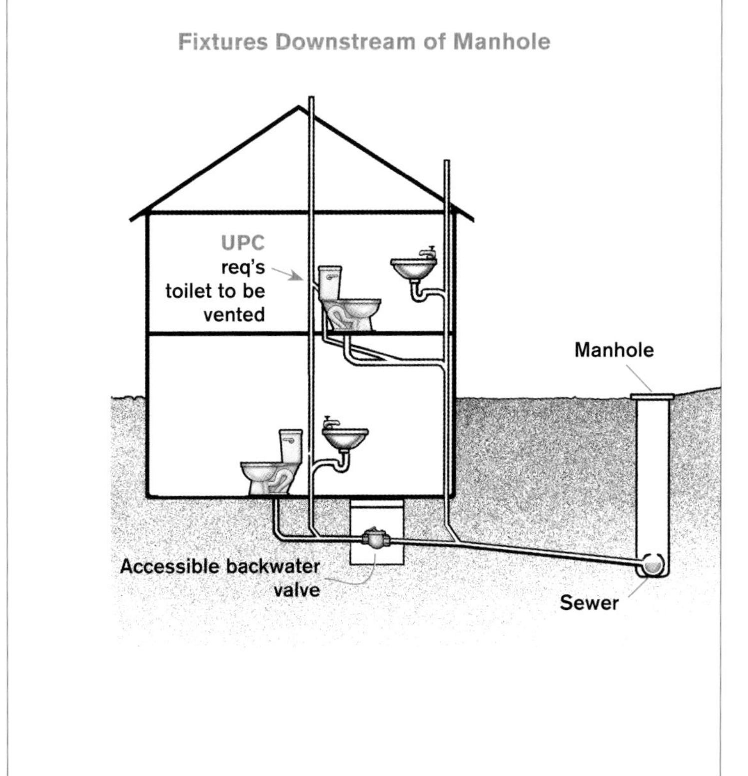

WASTE STACKS & VENTS

A waste stack provides a convenient way of discharging clustered fixtures on different floor levels. Toilets and urinals may not discharge into the waste stack. The waste stack must be undiminished in size to its vent terminal and may not have offsets. The UPC does not allow waste stacks and vents.

General Requirements	IRC	UPC
☐ Waste stack must be vert w/ no offsets	[3109.2]	{Ø}
☐ Vent offsets min 6in above highest FLR **F14,26**	[3109.3][13]	{Ø}
☐ No toilets or urinals allowed on waste stack	[3109.2]	{Ø}
☐ Waste & vent stack same size for entire length **F14**	[3109.3]	{Ø}
☐ Size waste stack per total DFUs discharging into it **T8**	[3109.4]	{Ø}

TABLE 8 WASTE STACK VENT SIZE [T3109.4]

Stack size (in.)	Total Discharge into One Branch Interval (DFUs)	Total Discharge for Stack (DFUs)
$1^1/_2$	1	2
2	2	4
$2^1/_2$	no limit	8
3	no limit	24
4	no limit	50

FIG. 14

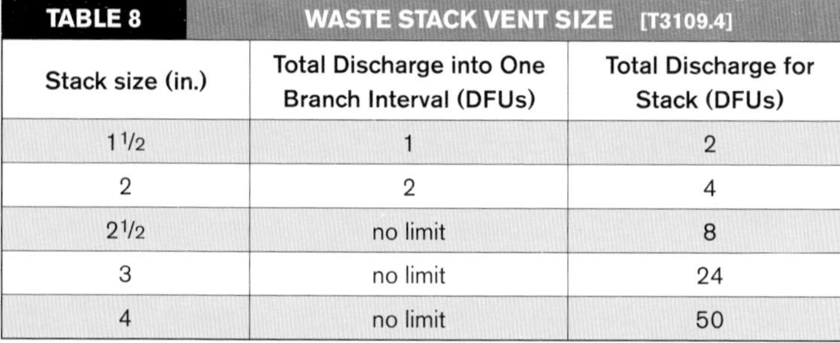

SUMPS & SEWAGE EJECTORS

Fixtures located below the level of the building drain and sewer will not drain into the sewer by gravity. Their contents must be lifted by a specially designed sump and pump, as shown in **F15 & F16**. The discharge from the pump requires a backwater valve (to prevent short-cycling) and a fullway valve (to allow cleaning and servicing) on the discharge side of the pump. A 2 in. pipe is typical on the discharge side of a sewage ejector.

Fixtures below Sewer	**IRC**	**UPC**
☐ Fixts to drain by gravity where practical **F16** ______	[3007.2.1]	{709.0}
☐ Sump discharge must be lifted above gravity drain **F16**	[3007.2]	{710.2}
☐ Connect to horiz gravity drain at top through wye **F16** ___	[n/a]	{710.4}
☐ Min vent from sump 1¼in {1½in} **F15** ________	[T3113.4.1]	{710.10}
☐ Sumps shall be fitted w/ watertight removable covers _	[3007.2]	{710.10}
☐ Lowest inlet min 2in higher than pump start level **F16** ____	[n/a]	{710.9}
☐ Min ejector capacity [1.9ft per second = 14.2GPM] {20GPM} ________	[3007.1]	{710.3.1}
☐ Backwater valve req'd on ejector discharge pipe **F15,16**	[3007.1]	{710.4}
☐ Gate valve req'd on discharge side of check valve **F15,16**	[3007.1]	{710.4}
☐ Min 2in discharge piping **F15** EXC ________	[3007.1]	{710.3.2}
1¼in discharge OK w/ grinder ejectors ________	[3007.1]	{Ø}
☐ Gravity drains receiving discharge from ejector sized at 1½ {2} DFU for each GPM of pump ________	[T3007.1]	{710.5}

FIG. 15 Sump Pump

Vent min. 1¼ in. {1½ in.}
Discharge pipe min. 2 in. dia.
Gate valve
Backwater or check valve
Discharge pipe

FIG. 16

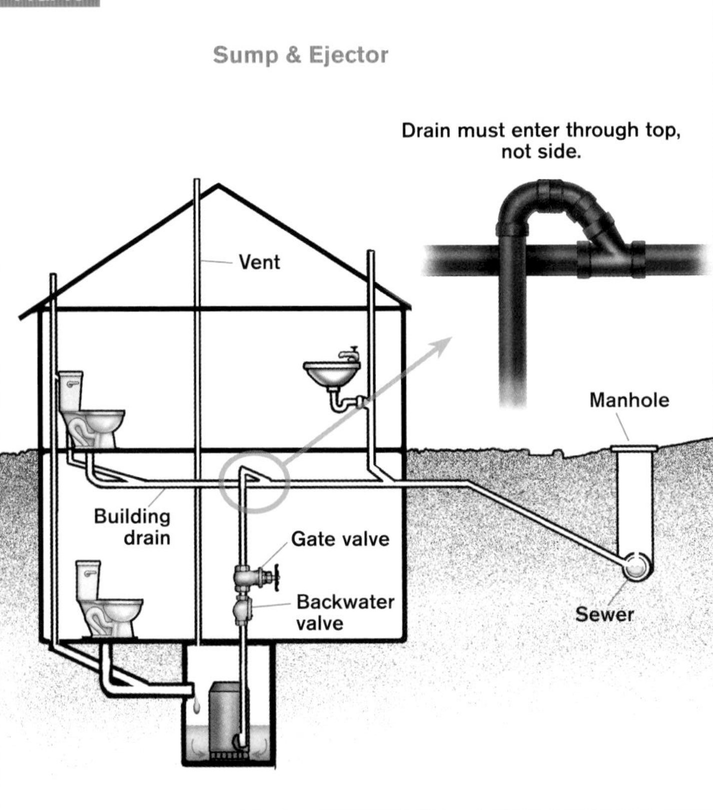

The discharge on the pump requires a backwater valve (or check valve) and full-flow shutoff valve (or gate valve) after the check valve as seen in ***F16****. The shutoff turns off the waste in the line, thereby allowing replacement of the check valve. The discharge for the pump should be 2 in. minimum diameter. Pipes > 2 in. for discharge could prematurely wear out the pump due to increased head pressure. The discharge pipe should lift as vertically as possible to a point above the gravity drains and enter the gravity drain from the top. Sump enclosures must be accessible for routine cleaning. The sump cover should have a tight seal around electrical connections to prevent escape of sewer gases.*

TRAPS & TAILPIECES

Traps prevent sewer gases, vermin, and other contaminants from entering the dwelling. The trap seal must be a sufficient depth (2 in.) to maintain a seal and not so deep (4 in. maximum) as to become blocked with sludge or create a siphoning effect. Trap arms (fixture drains) must be vented, otherwise the negative pressure created by water moving down the pipe will cause air to be sucked through the trap seal. The maintenance of proper trap seals is the underlying principle behind the code rules for drainage, traps, and venting.

Traps, General	IRC	UPC
☐ Fixts req separate trap as close as possible to fixt	[3201.6]	{1001.1,4}
☐ No separate traps for integral-trapped fixts (toilets)	[3201.6]	{1001.1}
☐ Trap seal 2in min–4in max depth F17	[3201.2]	{1005.0}
☐ Set traps level & protect from freezing	[3201.3]	{1005.0}
☐ No S traps, bell traps, crown-vented traps, drum traps, or traps w/ moving parts F24	[3201.5]	{1004.0}
☐ Tubular brass traps min 20gauge {17gauge}	[3201.1]	{1003.1}
☐ Size per table T3	[3201.7]	{1003.3}
☐ Double traps (in series) prohibited	[3201.6]	{1004.0}

Fixture Tailpieces	IRC	UPC
☐ Fixt tailpiece max 24in vert 30in horiz EXC F20	[3201.6]	{1001.4}
CW standpipes 18–42in {18–30in} F70	[2706.2]	{804.1}
☐ 1 center trap may serve up to 3 sinks, LTs, or lavs of same type in same room, max 30in apart F18	[3201.6]	{1001.2}
☐ Trap shall not be >1 pipe size larger than tailpiece	[n/a]	{1003.3}
☐ All fittings must be drainage type F9,10	[3002.3]	{701.2}
☐ Directional fittings req'd for continuous wastes from disposer or DW (ex: wyes, combos, or tees w/ baffles) F19	[2707.1]	{404.4}

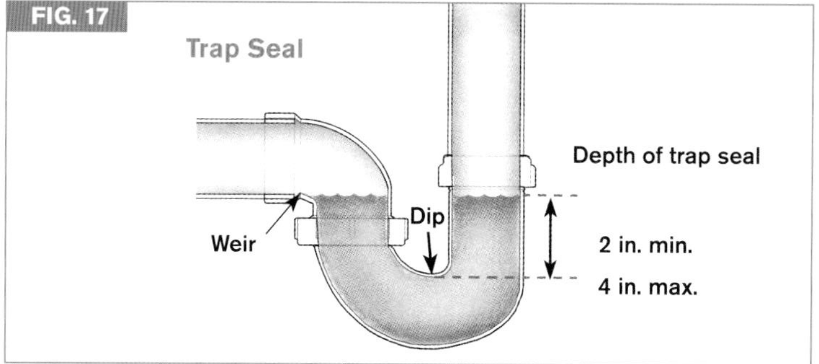

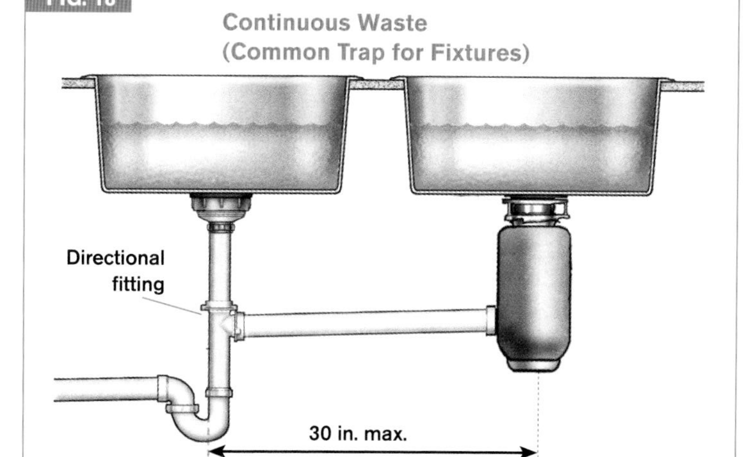

FIG. 19

Directional Fittings for DW & Disposer Discharge

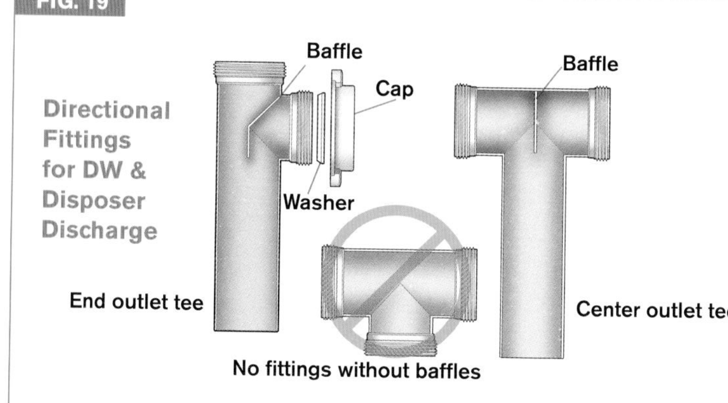

FIG. 20

Fixture Tailpieces

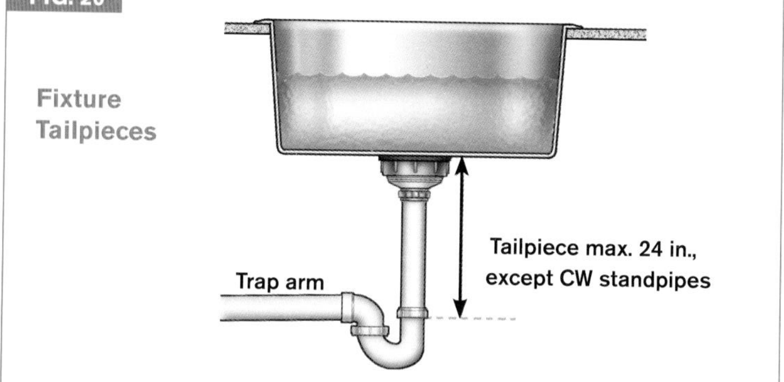

Trap Arms

	IRC	UPC
☐ Trap size no larger than trap arm	[3201.7]	{1003.3}
☐ Trap weir to vent distance min 2× trap arm dia F28	[3105.4]	{1002.2}
☐ Trap arm length & slope F21 per T9 EXC	[3105.2]	{1002.2}
Trap arm length from WC [unlimited] {6ft}	[3105.1X]	{T10-1}
☐ Trap arms <3in dia min slope $^1/_4$in/ft	[3002.3.1]	{708.0}
☐ Total fall of trap arm max 1 pipe dia	[3105.2]	{T10-1}
☐ Only 1 trap permitted on trap arm	[n/a]	{1001.1}
☐ 2 traps allowed on 1 trap arm if vent sized correctly F23	[3107.1]	{Ø}
☐ Size vent per T10	[3107.3]	{Ø}
☐ Tubular traps req listed accessible trap adapter F22	[3003.1, 2704.1]	{1003.2}
☐ Max 1 slip joint allowed on outlet side of trap F25	[3201.1]	{1003.2}
☐ CO req'd if direction change >45°{90°}	[3005.2.4]	{1002.3}
☐ Slip joints must be accessible min 12in by12in opening F27	[2704.1, 3201.1]	{405.2}

TABLE 9 MAXIMUM TRAP ARM DISTANCE [T3105.1] & {T10-1}

Trap Arm (in.)	[IRC] Distance Trap to Vent	{UPC} Distance Trap to Vent
$1^1/_4$	5 ft.	2 ft. 6 in.
$1^1/_2$	6 ft.	3 ft. 6 in.
2	8 ft.	5 ft.
3	12 ft.	6 ft.
4 or larger	16 ft.	10 ft.

Trap arm length from WC [unlimited] {6 ft.}.

FIG. 21

Trap Arms & Vents

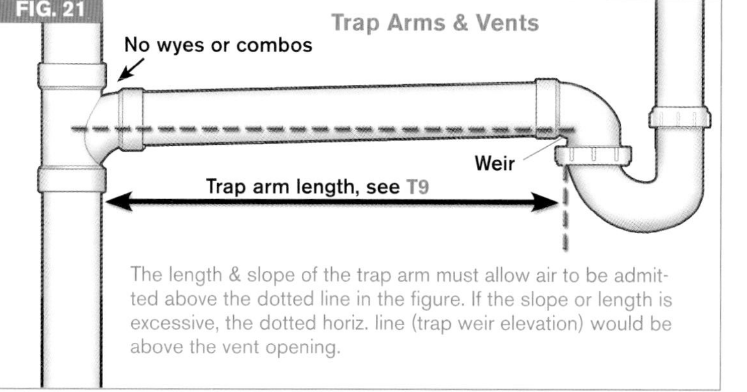

The length & slope of the trap arm must allow air to be admitted above the dotted line in the figure. If the slope or length is excessive, the dotted horiz. line (trap weir elevation) would be above the vent opening.

FIG. 22

Trap Adapter for Tubing Traps

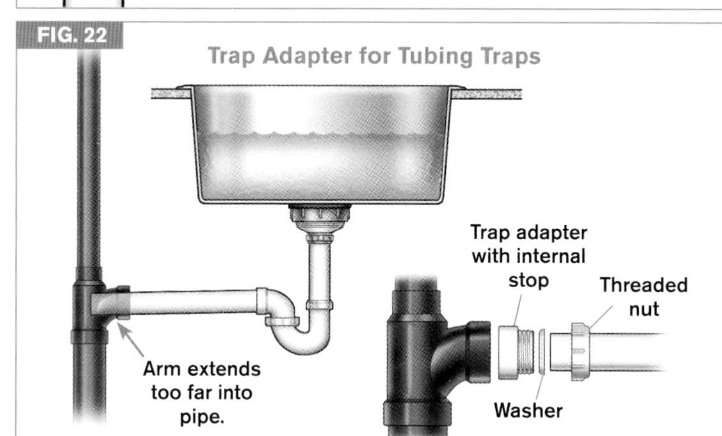

FIG. 23

Common Trap Arm IRC

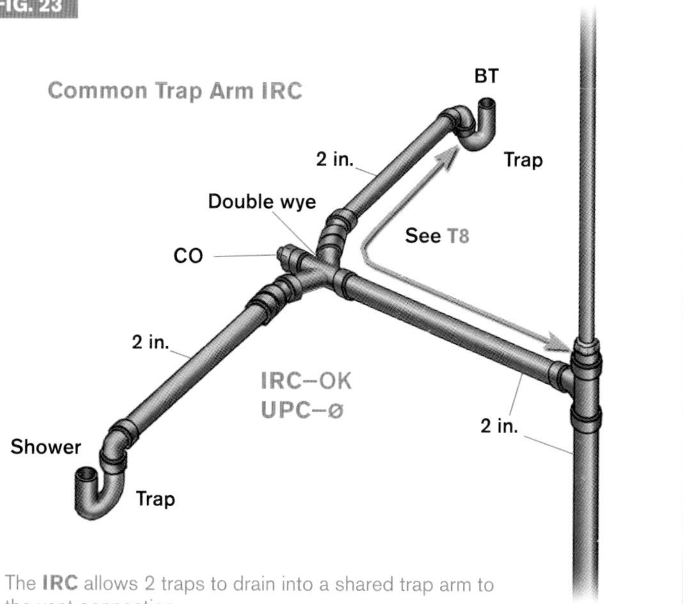

BT
Trap
2 in.
Double wye
See T8
CO
2 in.
IRC—OK
UPC—ø
2 in.
Shower
Trap

The **IRC** allows 2 traps to drain into a shared trap arm to the vent connection.

The **UPC** specifically prohibits this practice & all traps must enter the vent through individual trap arms.

FIG. 24

S Trap

Water filling the downstream vert. portion of the S trap will cause siphoning & loss of trap seals. Trap seals must be maintained to prevent sewer gases & vermin from entering the dwelling. The combination waste & vent (**F38**) is not an S trap because it has a horiz. arm & an increased size in the vert. drain.

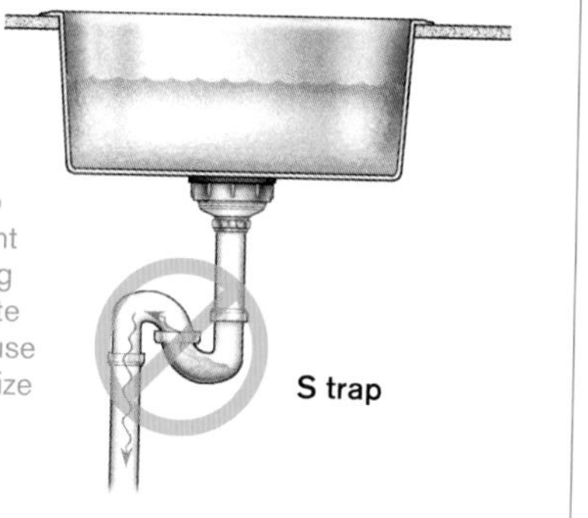

FIG. 25

Slip Joints & Access

Must provide a 12-in. by 12-in. min. access to concealed slip joints for replacement or repair. The access can be through the wall or the ceiling.

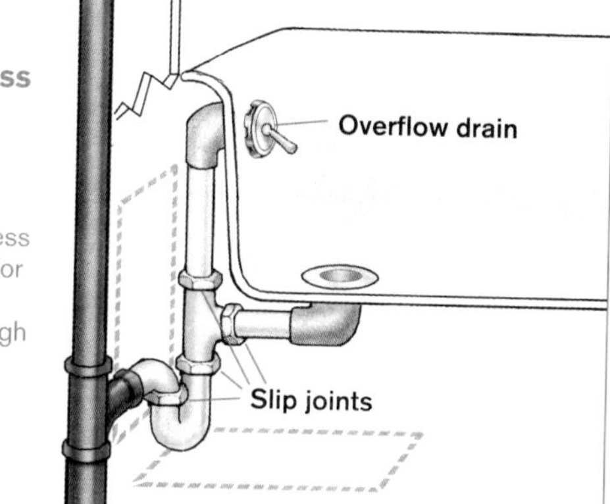

VENTS

Vents protect traps against atmospheric pressure differences that could cause them to siphon or to blow out, leaving the occupants unprotected against the potential contaminants downstream from the trap. **The IRC and the UPC have very different approaches to venting.**

General

	IRC	UPC
☐ All fixt traps req venting	[3101.2.1]	{901.0}
☐ Vent system not to be used for any other purposes	[3101.3]	{n/a}
☐ Vent to ext req'd for bldg drain or its extension **F30**	[3102.1][14]	{904.1}
☐ Req'd ext vent must be dry vent	[3102.2][15]	{904.1}
☐ Req'd ext vent not upstream from ejector or backwater valve or other device that may obstruct air flow	[3102.2]	{904.1}
☐ No flat dry vents (take off above horiz centerline) **F27**	[3104.3]	{905.2}
☐ Slope vents toward soil or waste piping **F29**	[3104.2]	{905.1}
☐ No vent opening below trap weir exc toilets	[3105.2][16]	{905.5}
☐ Vent min 2 pipe dia from trap (no crown vents) **F28**	[3105.3]	{1002.2}
☐ Horiz (<45° to horiz) dry vents min 6in above FLR **F26**	[3104.4]	{905.3}
☐ Horiz (<45° to horiz) branch vents min 6in above FLR **F26**	[3104.5]	{905.3}
☐ Vent piping <6in above FLR req's drainage type fittings **F26**	[3104.2]	{905.3}

Size

	IRC	UPC
☐ Min size $1\,^{1}/_{4}$in	[3113.1]	{904.1}
☐ Vents min $^{1}/_{2}$ size of drain served	[3113.1]	{904.1}
☐ Size per # of DFUs served & length of vent **T10**	[n/a]	{904.1}
☐ Increase 1 pipe size if developed vent length >40ft	[3113.1]	{n/a}
☐ Increase 1 pipe size if $>^{1}/_{3}$ of vent is horiz	[n/a]	{904.2X}
☐ Waste stack vent size must be = to waste stack **F14**	[3109.3]	{n/a}
☐ Total area of vents ≥ to bldg sewer **T11**	[n/a]	{904.1}

TABLE 10	VENT SIZING [3113]^A & {T7-5}					
Pipe Size (in.)	1 1/4	1 1/2	2	2 1/2	3	4
UPC max. DFUs	{1}	{8}^B	{24}	{48}	{84}	{256}
UPC max. Length^C (ft.)	{45}	{60}	{120}	{180}	{212}	{300}

A. IRC vents must be at least half the size of the drain served & min. 1 1/4 in.
Ex: A 3 in. horiz. branch may serve 20 DFUs & needs a 1 1/2 in. vent.
B. No WCs.
C. Vents are measured from drain connection to termination; the total developed length includes both the horiz. & the vert. sections. IRC vents have no max. length & are increased 1 pipe size if >40ft.
UPC vents must be increased one size if >1/3 of the length is horiz.

TABLE 11	AGGREGATE VENT AREA FILL-IN TABLE {904.1}		
Vent Size (in.)	Area (sq. in.)	# Vents	Net Vent Area
1 1/4	1.23		
1 1/2	1.77		
2	3.14		
3	7.07		
4	12.57		
TOTAL			

Example:
Req'd 4 in. bldg. drain = 12.57 sq. in.
Three 2 in. vents = 9.42 sq. in.
One 14 in. vent = 1.23 sq. in.
1.23 + 9.42 = 10.65 sq. in.
Thus, more venting would be req'd.

Sidewall Vent Termination

	IRC	UPC
☐ Min 10ft horiz from property line	[3103.6]	{Ø}
☐ Min 10ft above highest grade within 10ft	[3103.6]	{Ø}
☐ May not terminate under vented soffit	[3103.6]	{Ø}
☐ Sidewall vent req's screen or other bird & rodent protection	[3103.6]	{n/a}

FIG. 26 Horizontal Offsets

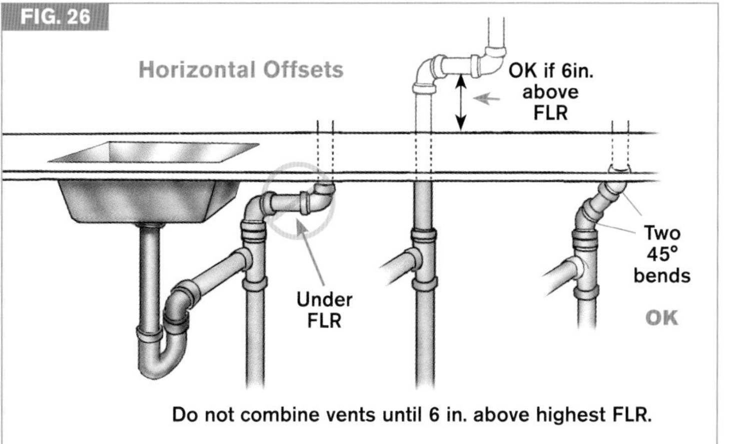

Do not combine vents until 6 in. above highest FLR.

FIG. 27 Vent above Center Line

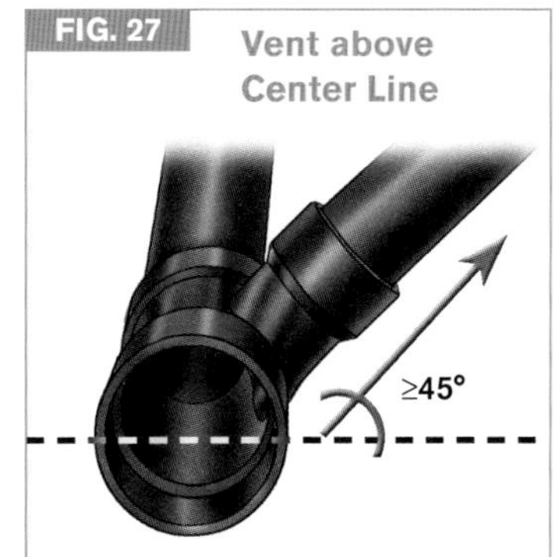

FIG. 28 Vent Too Close to Trap

Improper application of sanitary tee, cannot be placed on back

Vent Termination

	IRC	UPC
☐ Main vent {all vents} must terminate outdoors {above roof} F68	[3102.1][17]	{906.1}
☐ Vents through roof min 6in above roof	[3103.1][18]	{906.1,3}
☐ Min 12in horiz from any vert surface	[n/a]	{906.1}
☐ Min 7ft above roof used as deck {if within 10ft horiz}	[3103.1]	{906.3}
☐ Clearance above bldg openings 2ft {3ft} if ≤10ft horiz OR	[3103.5]	{906.2}
☐ Min 4ft below any window or opening within 10ft horiz	[3103.5]	{Ø}
☐ Vent terminations min 3ft distance from property line	[n/a]	{906.2}
☐ Provide flashing for roof penetrations	[3103.3]	{906.5}
☐ No flagpoles, aerials, or similar items secured to vent pipes [unless pipe anchored in approved manner]	[3103.4]	{906.3}

Snow or Frost Closure

	IRC	UPC
☐ Min 3in {2in} vent size through roof [or wall]	[3103.2]	{906.7}
☐ Transition to larger dia: min 1ft below roof [or inside wall]	[3103.2]	{906.7}
☐ Min 6in above snow line {10in above roof}	[3103.1]	{906.7}

FIG. 29

Branch Vents

Two or more individual vents may combine to form a branch vent, thereby reducing the number of penetrations through the roof. Horiz. piping must still have 1/4 in. per foot of slope, & fittings must be oriented in the direction of water flow. Refer to T10 for sizing of the branch vent.

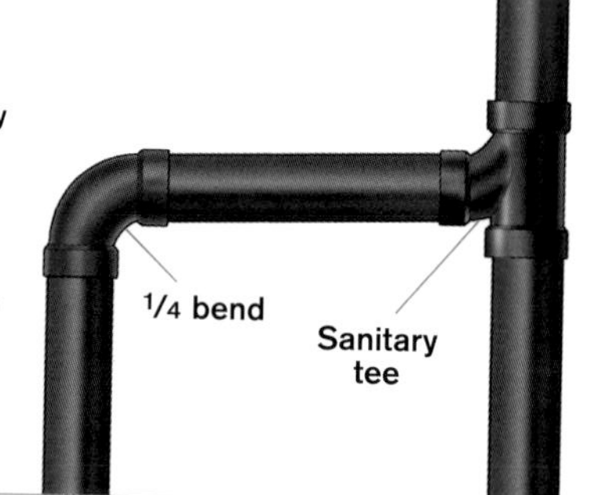

FIG. 30

Main Vent IRC

The **IRC** & **UPC** have very different approaches to venting. In the **UPC**, *all* vents must terminate through the roof. AAVs are permitted under the **UPC** only with authorization of the AHJ.

In the **IRC**, AAVs are permitted & only 1 vent needs to terminate on the exterior. The **IRC** also allows side wall vent terminations.

Main vent sized per bldg. drain

Increase by 1 pipe size if vent >40 ft.

AIR ADMITTANCE VALVES

The air admittance valve is a relatively modern innovation that allows a vent pipe to perform its function without the necessity of bringing the pipe outside the building. AAVs are not the same as the older style mechanical vents that are no longer approved. They operate by gravity, as shown in **F32**, and have no metal or rubber parts that could corrode or deform. The UPC does not explicitly approve AAVs, except under the provisions for Alternate Materials and Methods in 301.2. If the UPC is the code in your area, be sure to check with the local building department before installing AAVs.

Air Admittance Valves	IRC	UPC
☐ Install after DWV leak test	[3114.2]	{local}
☐ AAVs permitted at individual & branch vents **F31**	[3114.3]	{local}
☐ Individual fixt AAV min 4in above fixt drain **F31**	[3114.4]	{local}
☐ Stack-type AAV min 6in above FLR of highest fixt	[3114.4]	{local}
☐ AAV within same max distance as conventional vent **T10**	[3114.4]	{local}
☐ AAVs terminating in attic min 6in above insulation **F1**	[3114.4]	{local}
☐ AAVs must be accessible	[3114.5]	{local}
☐ Space containing AAV must be ventilated	[3114.5]	{local}
☐ Min 1 vent {all vents} to outdoors for bldg drain	[3114.7]	{906.1}

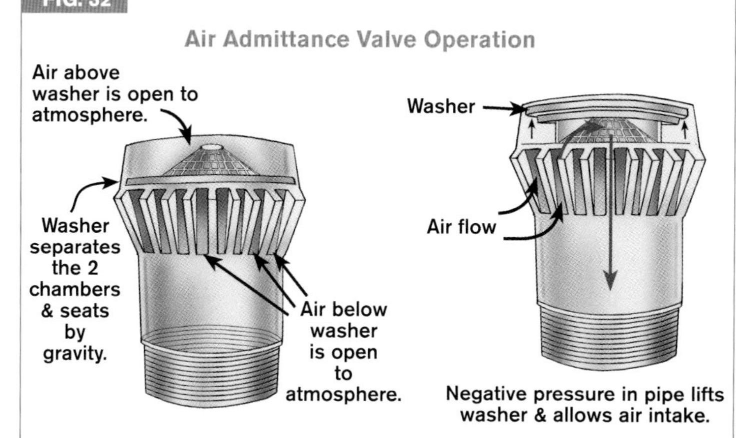
FIG. 31
Air Admittance Valve
AAVs must remain accessible.
When AAVs are placed in attics, they must be at least 6 in. above insulation.
Flood level rim
Air admittance valve
4 in. min.
FIG. 32
Air Admittance Valve Operation
Air above washer is open to atmosphere.
Washer separates the 2 chambers & seats by gravity.
Air below washer is open to atmosphere.
Washer
Air flow
Negative pressure in pipe lifts washer & allows air intake.

SPECIAL VENTING SYSTEMS

The IRC offers a number of options for island sinks, including AAVs, combination waste and vent, and the loop vents shown in F33 and F34. The UPC allows only the method shown in F33, unless specific approval is obtained from the AHJ.

Island Sinks	IRC	UPC
☐ Island venting limited to sinks (disposal OK) & lavs	[3112.1]	{909.0}
☐ Island vented w/ drainage pattern fittings only F33,34	[3112.3]	{909.0}
☐ Island vent above fixt drain outlet {as high as possible} before offsetting horiz & down	[3112.2]	{909.0}
☐ Lowest part of island vent shall connect full size to a vert drain or top half of a horiz drain	[3112.3]	{Ø}
☐ COs req'd in island fixt vents & drains	[3112.3]	{909.0}
☐ Connect island vent downstream of fixt drain	[n/a]	{909.0}
☐ Foot vent req'd through wye branch off below-floor vent F33	[n/a]	{909.0}
☐ CO req'd in vert section of foot vent F33	[n/a]	{909.0}
☐ No upstream fixt on drain serving island	[n/a]	{909.0}

Common Venting

☐ Common-vented fixt must be on the same floor level	[3107.1]	{905.6}
☐ Max 2 fixt to {vert} common vent F35	[3107.1]	{905.6}
☐ Vent connection at interconnection of fixt drains OR [Downstream of interconnection] F23,34	[3107.2] [3107.2]	{905.6} {Ø}
☐ Must connect at approved double fitting F37	[n/a]	{905.6}
☐ Fixt at different levels (same floor) Vert drain between fixts is a wet vent & must be properly sized	[3107.3, T3107.3]	{n/a}
☐ Size common wet vent per DFUs of upper drain T12	[3107.3]	{n/a}
☐ Toilet cannot be upper fixt	[3107.3]	{908.1}

FIG. 33

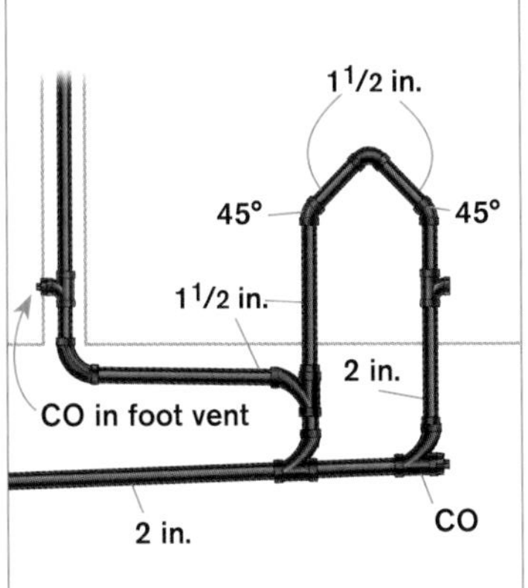

FIG. 34

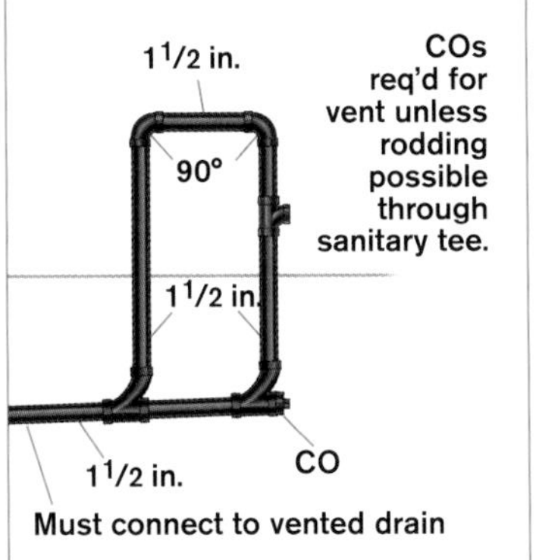

TABLE 12 SIZING COMMON VENTS [T3107.3]

Pipe Size (in.)	Max. Discharge from Upper Fixture Drain (Fixture Units)
$1\frac{1}{2}$	1
2	4
$2\frac{1}{2}$ to 3	6

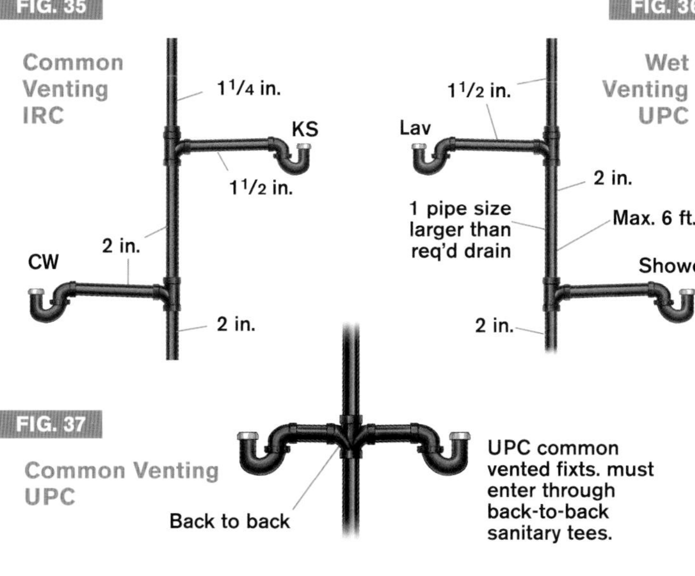

Vertical Wet Venting

	IRC	UPC
☐ Limited to trap arm of 1- & 2-unit fixt	[n/a]	{908.1}
☐ All wet vented fixtures to be on same story	[3108.4]	{908.1}
☐ 6ft max (vert only) developed length of wet vent F36	[n/a]	{908.1}
☐ Size wet vent per DFUs of upper drains T13	[3108.3]	{n/a}
☐ Wet vent 1 pipe size larger than the req'd waste–min 2in	[n/a]	{908.2}
☐ 4 fixts max on wet vent	[n/a]	{908.1}

TABLE 13 — MAX. DFU LOADS FOR WET VENTS & CWV
[T3108.3 & 3111.3] & {908.2 & 908.4.3}

Pipe Size (in.)	Wet Vents		CWV [IRC]	
	IRC	UPC	To Branch	To Bldg. Drain
$1\frac{1}{2}$	1	Ø	Ø	Ø
2	4	4	3	4
$2\frac{1}{2}$	6	8[A]	6	26
3	12	8[B]	12	31
4	32	8[B]	20	50

A. For horiz. went vents, max. load is 4 DFUs on $2\frac{1}{2}$ in. pipe.
B. More than 8 possible for horiz. wet vents.

Horizontal Wet Venting

	IRC	UPC
☐ Horiz wet venting OK for any combination of fixt within 1 or 2 BA groups on same floor F39	[3108.1]	{908.4.1}[19]
☐ Only fixts within BA groups allowed on wet vent. Other vented fixts may connect to the drain downstream of wet vent	[3108.1]	{908.4.1}[20]
☐ Dry vent connection to wet vent must be individual or common vent connected to lav, bidet, BT, or shower	[3108.2][21]	{908.4.2}[22]
☐ Side inlet 1/4 bends OK as a wet vent connection if not serving a WC	[3005.1.2]	{n/a}
☐ Max trap arm length measured from trap weir to wet vent	[3108.5][23]	{908.4.1}
☐ Size wet vent per DFU & T13	[n/a]	{910.1}

Combination Waste & Vent IRC F38

	IRC	UPC
☐ UPC CWV req's specific approval from AHJ	[n/a]	{910.1}
☐ Only sinks (no disposal), lavs, standpipes, & floor drains	[3111.1]	{Ø}
☐ One vert pipe (max 8ft length) allowed between fixt drain & horiz CWV pipe	[3111.2][24]	{Ø}
☐ Max slope of CWV piping $^{1}/_{2}$in/ft	[3111.2.1]	{n/a}
☐ CWV shall connect to a vented horiz branch OR	[3111.2.2]	{910.3}
Connect vent to CWV pipe	[3111.2.2]	{910.3}
☐ Vent connected to CWV pipe must rise vert 6in min above fixt FLR before horiz offset	[3111.2.2]	{n/a}
☐ Fixt drain (trap arm) max length per T10; vert portion of CWV is considered beginning of vent for purposes of T10	[3111.2.4]	{n/a}

FIG. 38

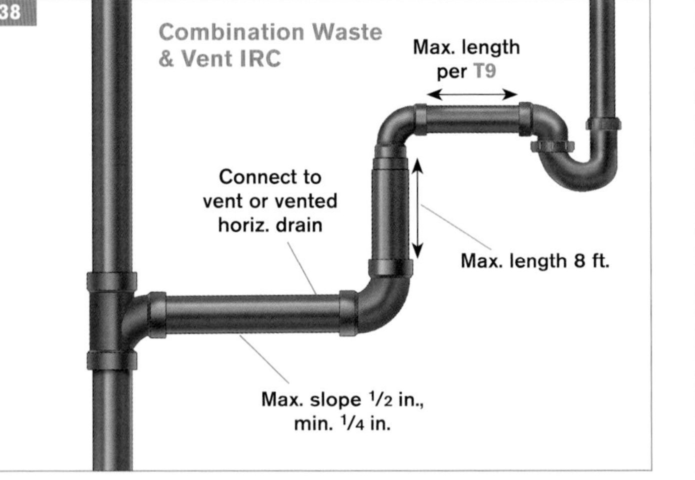

FIG. 39

Horizontal Wet Venting

Lav.

Dotted line represents UPC wet vent.

2 in.

Lav.

WC

BT

IRC does not require vents for toilet

Wet vent

Shower

FIG. 40

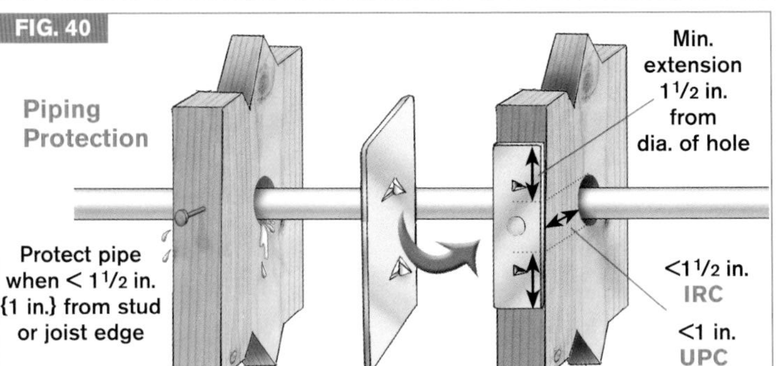

WATER SUPPLY & DISTRIBUTION

Water supply piping must provide an adequate flow of clean potable water and be free of any cross-connections that would introduce contaminants. Piping systems must be protected against damage and movement. Modern plumbing systems often use plastic pipe and tubing; and in some systems, the branch piping originates from a central manifold, rather than a traditional series or main line with branches.

General

	IRC	UPC
☐ Use only approved materials T14	[2904.4,5]	{604.1}
☐ Proper installation & support req'd T14	[2605.1]	{314.0}
☐ Min water service 3/4in	[2903.7]	{610.8}
☐ Pressure tank & pump req'd if supplied pressure <40psi {15psi}	[2903.3]	{608.1}
☐ Regulator req'd if pressure >80psi F44	[2903.3.1]	{608.2}
☐ Steel-plate protection for plastic & Cu pipes in notches or bored holes if closer than 1 1/2in{1in} of face of framing F40	[2603.2.1]	{313.9}
☐ Protection min 2in above sole plates & below top plates	[2603.2.1]	{n/a}
☐ Protection min 1 1/2in beyond outside of pipe	[n/a]	{313.9}[25]
☐ Water hammer arrestors req'd near quick-close valves (DW or CW)	[2903.5]	{609.10}
☐ No pipes previously used for other than potable water	[2608.5]	{604.7}

Pex Tubing

	IRC	UPC
☐ Bend radius PMI T17, F49	[manu]	{manu}
☐ No PEX in first 18in connected to WH F49	[manu]	{604.11.2}
☐ 3/8in tubing limited to 60ft developed length	[2903.8.2][26]	{manu}
☐ If WH fed from end of cold-water manifold; manifold must be 1 size larger than WH feed F47	[2903.8.2]	{manu}

TABLE 14 — WATER PIPE MATERIALS [T2904.5] & {604.1}

Material	IRC		UPC	
	Service	Distribution	Service	Distribution
ABS plastic pipe	OK	Ø	Ø	Ø
Cu type K,L,M	OK	OK	OK[A]	OK[A]
CPVC	OK	OK	OK	OK
Ductile iron	OK	Ø	OK[B]	OK[B]
Galvanized steel	OK[C]	OK[C]	OK[B]	OK[B]
PB plastic pipe	OK	OK	Ø	Ø
PE plastic tubing	OK	Ø	OK	Ø
PEX plastic tubing	OK	OK	OK	OK
PEX-AL-PEX pipe	OK	OK	Ø	Ø
PVC plastic pipe	OK	Ø	OK	Ø

A. UPC allows type M inside a bldg. only if aboveground {604.2x}.
B. UPC ferrous piping in or under a concrete slab floor must be machine wrapped & joints field-wrapped to equivalent protection {609.3.1}.
C. IRC does not allow steel pipe under slabs [2904.5.1].

Joints & Connections

	IRC	UPC
☐ Cu to Fe req's brass [or dielectric] fitting	[2904.17.1]	{316.2.1}
☐ Cu joints in or under concrete slab on grade within bldg req brazed wrought-Cu fittings	[2904.14]	{609.3.2}
☐ Slip joints: only at exposed fixt supply	[2904.6]	{606.2.3}
☐ Unions req'd ≤12in of WH	[2801.3]	{609.5}
☐ Unions req'd ≤12in of softeners, filters, regulators, etc.	[n/a]	{609.5}

Prohibited Joints

	IRC	UPC
☐ Joints between different types of plastic req adapter	[2904.17.2]	{316.2.3}
☐ No female threaded PVC fittings	[2904.17.2]	{606.2.2}

Required Valves

	IRC	UPC
☐ Accessible main valves req'd [at entrance to structure]	[2903.9.1]	{605.2}
☐ Main & WH valve must be full-bore type F41–43	[2903.9.1]	{605.2}
☐ Main valve must be on discharge side of water meter	[n/a]	{605.2}
☐ Main valve must have bleed orifice or separate drain	[2903.9.1]	{n/a}
☐ Valves req'd at fixt supply pipe exc tub & shower	[2903.9.3]	{605.5}
☐ Valves only at distribution manifold OK if labeled	[2903.8.6]	{605.5}
☐ Valve w/ drain req'd at hose faucets subject to freezing	[2903.10]	{313.6}
☐ Valves req'd on cold-water supply at each WH	[2903.9.2]	{605.2}
☐ All shutoffs req access	[2903.9.3]	{605.6}

FIG. 41 FIG. 42 FIG. 43

Gate Valve

Ball Valve

Globe Valve

Full-bore valves *req'd* for mains & water heater

Throttling valves *not approved* for mains & water heater

PRESSURE REGULATORS

Excessive pressure increases the chance of leaks and scalding, and puts a strain on fixtures and fittings. When supply pressure exceeds 80psi, a regulator is required. A screen in the regulator assembly can help prevent it from clogging with sediment. Regulators without an integral bypass feature create a closed system downstream from the regulator. As the water heater recovers its heat, the pressure rises in the closed system downstream of the regulator. To prevent excessive pressure, expansion tanks are required and usually placed in the cold-water line just above the water heater.

General

	IRC	UPC
☐ Req'd when water pressure at bldg exceeds 80psi	[2903.3.1]	{608.2}[27]
☐ Strainer req'd ahead of regulator F44	[n/a]	{608.2}
☐ Regulator & strainer accessible without removing piping	[manu]	{608.2}
☐ Regulated pressure computed at 80% of setting	[2903.7]	{608.2}
☐ Expansion tank req'd for regulators w/o integral bypass F45	[2903.4.1][28]	{608.3}[29]
☐ Expansion tank req'd for systems w/ supply check valves	[2903.4.2][30]	{608.3}

FIG. 44

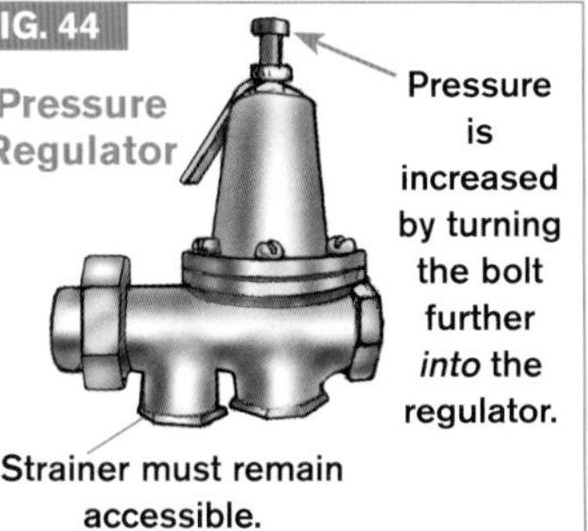
Pressure Regulator
Pressure is increased by turning the bolt further *into* the regulator.
Strainer must remain accessible.

FIG. 45

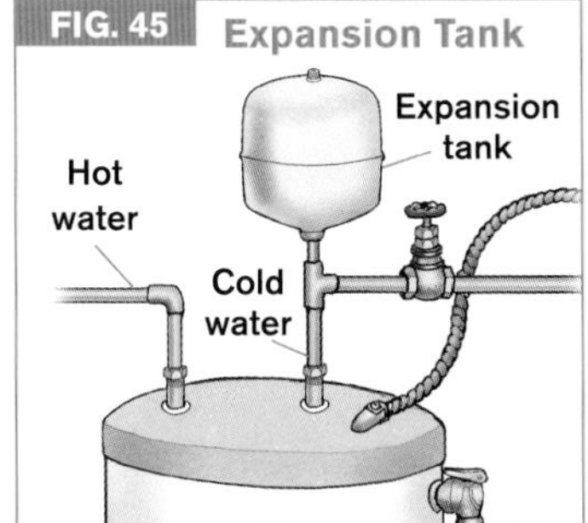
Expansion Tank
Expansion tank
Hot water
Cold water

TABLE 15 MINIMUM REQUIRED AIR GAPS [2902.2.1] & {T6-3}

Opening Diameter & Typical Fixtures	Not Affected by Side Walls (in.)		Affected by Side Walls[A] (in.)	
	IRC	UPC	IRC	UPC
≤1/2 in. (lav)	1	1	1 1/2	1 1/2
≤3/4 in. (LT)	1 1/2	1 1/2	2 1/2	2 1/4
≤1in. (BT)	2	2	3	3
>1 in. (pool)	2× dia.	3× dia.	2× dia.	3× dia.

A. Affected by side walls = any time the distance from the spout to the wall is <3× the dia. of the effective opening, or <4× the dia. for 2 intersecting walls.

FIG. 46

Air Gap

The air gap is the distance between the lowest edge of the faucet opening **D** & the top of the flood level.

2× **D** or 1 in. min.

CROSS-CONNECTION CONTROL

A fundamental principle of the plumbing codes is keeping potable water uncontaminated. Vacuum breakers prevent contaminants from entering through connected systems, such as lawn sprinklers. A physical separation in the form of an air gap prevents contamination at waste receptors, such as sinks. See pp. 108–109 for further examples of cross-connection controls at specific appliances and equipment.

General — IRC — UPC

- ☐ Prevent contamination of potable water supply ______ [2902.1] {602.1}
- ☐ Prevent cross-connections between individual potentially contaminated water supply & public water supply ____ [2902.1] {602.4}

Specific Types of Required Protection

Reduced-pressure principle backflow preventer (R-P PBP) OK for connection to:

- ☐ Boilers ______ [2902.5.1] {603.4.10}
- ☐ Fire-sprinkler systems w/ additives ______ [2902.5.4.1] {603.4.16.3}
- ☐ Lawn irrigation systems w/ chemical injectors ____ [2902.5.3] {603.4.6.1}
- ☐ Solar heating piping w/ additives ______ [2902.5.5] {603.4.4.1}

Atmospheric vacuum breakers OK for:

- ☐ Hose connections (n/a WH, boiler, & CW drains) __ [2902.4.3] {603.4.7}
- ☐ Reverse-osmosis drinking water treatment units______ [2907.2] {603.4.13}
- ☐ Swimming pool inlets w/o an air gap ______ [2902.1] {603.4.5}
- ☐ Irrigation system, 6in above highest head F74 ____ [2902.5.3] {T6-2}

Fixture outlet receptor air gaps must be:

- ☐ Min 2× dia of outlet & per table T15, F46 ______ [2902.3.1] {603.2.1}

WATER SUPPLY SIZING

Traditional water supply systems are designed with larger pipes at the inlet and progressively smaller ones toward the most remote outlets. Each portion of these systems can be sized using a system of fixture units and tables T18–22. In modern plastic piping systems, an alternative method of parallel distribution is used, with a central manifold and individual tubes to each outlet. The UPC requires that parallel distribution systems be installed according to their listing {610.4}, and the IRC provides a table [T16] for determining the demands on the parallel system based on the required flow rate and pressure that are needed to maintain the sanitary needs of the fixture. Water flow to a manifold must be sufficient to meet peak demand.

TABLE 16 — MANIFOLD CAPACITY (Parallel Distribution Systems) [T2903.6(1) & T2903.8.1]

Inlet Pipe Size (in.)	Max. GPM	
	Velocity (& Fixt. Units) at 12 ft./sec.–Plastic	Velocity (& Fixt. Units) at 8 ft./sec.–Metal
3/4	17 (14)	11 (7)
1	29 (45)	20 (20)
1 1/4	46 (>50)	31 (>50)
1 1/2	66 (>50)	44 (>50)

TABLE 17 — RECOMMENDED MIN. BEND RADIUS FOR PEX

Tubing Size (in. nominal)	Tubing: Outside Diameter (in.)	Bend Radius (in.)
3/8	1/2	4
1/2	5/8	5
3/4	7/8	7
1	1 1/8	9

FIG. 47 PEX Manifold

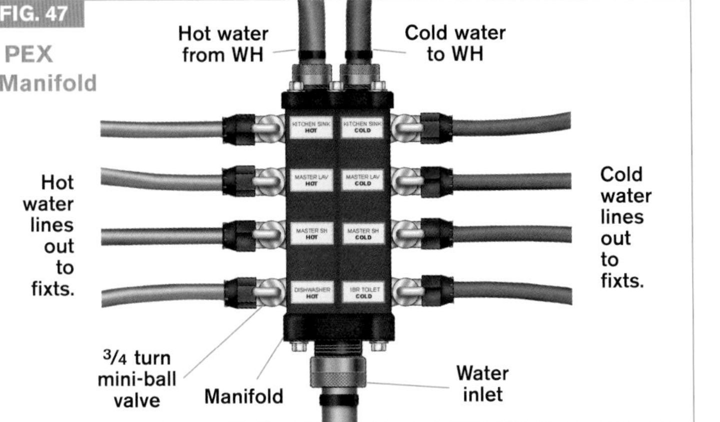

FIG. 48 PEX Support

PEX must be supported to prevent stress on connectors & per T1.

FIG. 49 PEX Distance to Water Heater

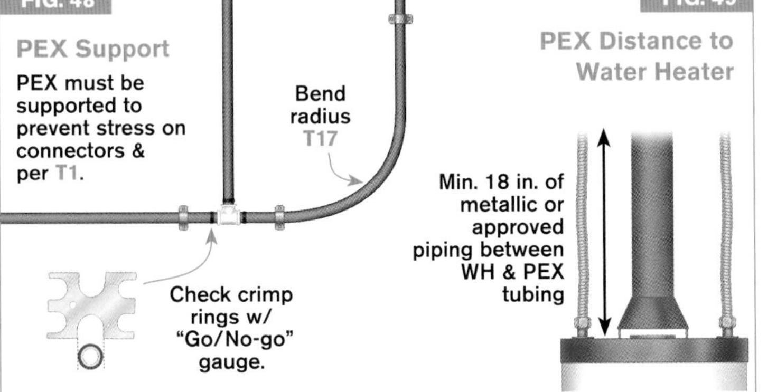

TABLE 18 WATER SUPPLY FIXTURE UNITS FILL-IN TABLE
[T2903.6] & {T6-4}

Fixture	IRC			UPC	#	Extension
	Hot	Cold	Comb.	Comb.		
BT	1	1	1.4	4		
CW	1	1	1.4	4		
DW	1.4	–	1.4	1.5		
Hose bibb	–	2.5	2.5	2.5		
KS	1	1	1.4	1.5		
Lav	0.5	0.5	0.7	1		
LT	1	1	1.4	1.5		
Stall shower	1	1	1.4	2.0		
WC	–	2.2	2.2	2.5		
Laundry group[A]	1.8	1.8	2.5	–		
Kitchen group[B]	1.9	1	2.5	–		
Half-bath group[C]	0.5	2.5	2.6	–		
Full-bath group[D]	1.5	2.7	3.6	–		
Total Demand						

A. Laundry group = CW, standpipe, & LT.
B. Kitchen group = DW & sink w/ or w/o garbage disposer.
C. Half-bath group = WC & lav.
D. Full-bath group = WC, lav, & tub (w/ or w/o shower) or shower stall.

TABLE 19 MIN. CAPACITIES AT FIXTURE SUPPLY OUTLETS
[T2903.1]

Fixture Outlet	GPM	Flow psi
BT	4	8
Bidet	2	4
DW	2.75	8
LT	4	8
Lav	2	8
Shower	3	8
Shower (temp. controlled)	3	20
Hose bibb	5	8
KS	2.5	8
WC (tank type)	3	8
WC (flushometer)	1.6	15
WC (one piece)	6	20

TABLE 20 — SIZING A WATER SERVICE FILL-IN TABLE [2903.7] & {610.8}

1. Determine fixt. unit demand (total from T18).	
2. Min. daily static pressure at meter or source.	
3. Subtract (add) 1/2 lb. pressure per ft. of rise (fall).	
4. Deduct pressure losses for filters, regulators, etc.	
5. Find pressure range group in T22 or T23.	
6. Find column for developed length to most remote fixt.	
7. Find row meeting fixt. unit demand req (total from T21).	
8. Find req'd meter & pipe size in left columns of T22 or T23.	

Note: The same procedure can be used for branches.

TABLE 21 — WATER SIZING FILL-IN TABLE

Section	Length	Fixture Units	Pipe Size	Section	Length	Fixture Units	Pipe Size

TABLE 22 — IRC WATER SIZING TABLE [T2903.7]

Meter	Supply	Units Allowed per Lengths[A] of Pipe					
40–49 psi		40 ft.	60 ft.	80 ft.	100 ft.	150 ft.	200 ft.
3/4 in.	1/2 in.[B]	3	2.5	2	1.5	1.5	1
3/4 in.	3/4 in.	9.5	9.5	8.5	7	5.5	4.5
3/4 in.	1 in.	32	32	32	26	18	13.5
1 in.	1 in.	32	32	32	32	21	15
1 in.	1 1/4 in.	80	80	80	80	65	52
50–60 psi		40 ft.	60 ft.	80 ft.	100 ft.	150 ft.	200 ft.
3/4 in.	1/2 in.[B]	3	3	2.5	2	1.5	1
3/4 in.	3/4 in.	9.5	9.5	9.5	8.5	6.5	5
3/4 in.	1 in.	32	32	32	32	25	18.5
1 in.	1 in.	32	32	32	32	30	22
1 in.	1 1/4 in.	80	80	80	80	68	57
over 60 psi		40 ft.	60 ft.	80 ft.	100 ft.	150 ft.	200 ft.
3/4 in.	1/2 in.[B]	3	3	3	2.5	2	1.5
3/4 in.	3/4 in.	9.5	9.5	9.5	9.5	7.5	6
3/4 in.	1 in.	32	32	32	32	32	24
1 in.	1 in.	32	32	32	32	32	28
1 in.	1 1/4 in.	80	80	80	80	80	80

A. First multiply actual length to most remote fixt. by 1.2 to compensate for loss at fittings.
B. Min. bldg. supply is 3/4 in.

Worksheet Area

Elevation = ____________

Distance = ____________

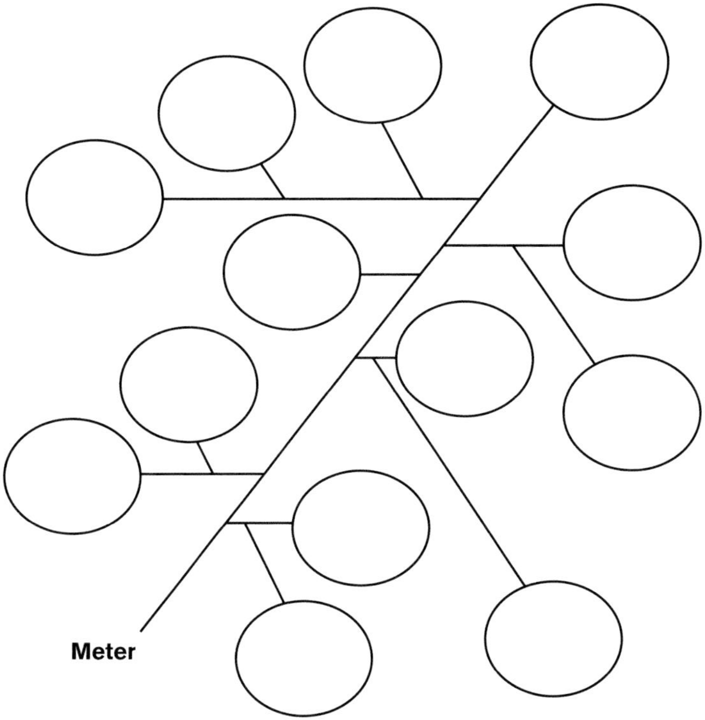

Meter

TABLE 23 UPC WATER SIZING TABLE {T6-5}

Meter	Supply	Units Allowed per Lengths of Pipe					
30–45 psi		40 ft.	60 ft.	80 ft.	100 ft.	150 ft.	200 ft.
3/4 in.	1/2 in.[A]	6	5	4	3	2	1
3/4 in.	3/4 in.	16	16	14	12	9	6
3/4 in.	1 in.	29	25	23	21	17	15
1 in.	1 in.	36	31	27	25	20	17
1 in.	1 1/4 in.	54	47	42	38	32	28
46–60 psi		40 ft.	60 ft.	80 ft.	100 ft.	150 ft.	200 ft.
3/4 in.	1/2 in.[A]	7	7	6	5	4	3
3/4 in.	3/4 in.	20	20	19	17	14	11
3/4 in.	1 in.	39	39	36	33	28	23
1 in.	1 in.	39	39	39	36	30	25
1 in.	1 1/4 in.	78	78	76	67	52	44
over 60 psi		40 ft.	60 ft.	80 ft.	100 ft.	150 ft.	200 ft.
3/4 in.	1/2 in.[A]	7	7	7	6	5	4
3/4 in.	3/4 in.	20	20	20	20	17	13
3/4 in.	1 in.	39	39	39	39	35	30
1 in.	1 in.	39	39	39	39	38	32
1 in.	1 1/4 in.	78	78	78	78	74	62

A. Min. bldg. supply is 3/4 in.

GAS PIPING

The required size of gas piping depends on the appliance demand, gas pressure, Btus per cubic foot of gas, and the length of the run. In modern CSST systems, high pressure (approximately 2 psig) can sometimes be supplied to a central manifold and low pressure lines will run from the manifold to each appliance. Fuel gas piping must be protected from damage and leaks.

Pipe Installation General

Item	IRC	UPC
☐ Size per manu or T24–27	[2413.3]	{1209.4.3}
☐ No gas pipe in circulating air ducts, vents, or clothes chutes	[2415.1]	{1211.2.5}
☐ Pipe may not pass through 1 townhouse to another	[2415.1][31]	{n/a}
☐ Pipe must be new or previously used for gas only	[2414.2]	{1209.5.1.2}
☐ PE pipe: min bend radius 25× pipe dia	[2416.3]	{1211.6.2}
☐ Piping prohibited underground beneath bldg unless in conduit sealed in bldg & vented on ext F50	[2415.11]	{1211.1.6}
☐ Ext pipes aboveground securely supported & protected	[2415.7]	{1211.2.1}
☐ Ext pipes aboveground min 3 1/2in elevation	[2415.7]	{n/a}
☐ Ext pipes above roof min 3 1/2in elevation	[2415.7]	{n/a}
☐ Concealed piping not allowed in solid walls [unless in chase or casing]	[2415.2]	{1211.3.3}
☐ Tubing <1 1/2in {0 in} from framing edge req's shield plates to extend min 4in past studs & plates	[2415.5]	{1211.3.4}
☐ Cap unused gas outlets	[2415.12]	{1211.8.2A}
☐ Slope piping min 1/4in/15ft for other than dry gas	[2419.1]	{1211.2.4}
☐ Accessible drip leg req'd if water vapor in gas F50	[2419.2,3]	{1211.7.1}
☐ Sediment trap req'd near appl inlet EXC ranges, clothes dryers, gas lights, & grills F59	[2419.4]	{1212.7}
☐ Pipes not in same room as equipment req'd to be labeled "Gas" every 5ft for other than steel pipe	[2412.5][32]	{n/a}

Pipe & Tubing Materials

Item	IRC	UPC
☐ Steel (galvanized or black pipe) OK	[2414.4.2]	{1209.5.2}
☐ Pipe must be brushed & de-burred	[2414.7]	{1209.5.5}
☐ Replace (not repair) defective pipes/fittings	[2414.7]	{1209.5.5}
☐ No CI	[2414.4.1]	{1209.5.2.1}
☐ Plastic only OK underground & outside	[2414.6]	{1209.5.4}
☐ Cu & brass tubing OK if gas <0.3 grains hydrogen sulfide per 100 cu.ft (check with supplier)	[2414.5.2]	{1209.5.2.3}
☐ CSST PMI F51	[2414.5.3]	{1209.5.3.4}

Joints in Pipe or Tubing

Item	IRC	UPC
☐ Fe req's threaded, flanged, welded, or brazed	[2414.10.1]	{1209.5.8.1}
☐ Tubing req's approved fittings or brazing	[2414.10.2]	{1209.5.8.2}
☐ Plastic heat-fused or mechanical connectors	[2414.11]	{1209.5.9}
☐ No unions or bushings in concealed locations	[2415.3]	{1211.3.2}

Corrugated Stainless Steel Tubing*

- ☐ Support per manu tables (manu)
- ☐ Bending radius per manu tables (manu)
- ☐ No direct burial–routing through conduit OK (manu)
- ☐ Striker plates: per manu F52 (manu)
- ☐ Avoid kinking, twisting, or contact w/ sharp objects (manu)
- ☐ Protect where passing through sheet metal (manu)
- ☐ Regulators in vented area or w/ vent limiters F51 (manu)

**CSST should be installed only by workers who have successfully completed a training program offered by the CSST manu.*

TABLE 24 PROCEDURES FOR SIZING GAS PIPE [2413.4.2] & {1217.2.}

1. Determine Btu/cu.ft. gas from local supplier (usually 950–1100).
2. Divide appl. Btu by Btu/cu. ft. to obtain appl. demand.
3. Measure developed length to most remote fixt.
4. Use column from T25 for most remote fixt. for all fixts.
5. Select row for pipe size equaling or exceeding demand for each section.

TABLE 25 SIZE OF SCHEDULE 40 GAS PIPING [T2413.4(1)]

Pipe (in.)	Length (ft.)										
	10	20	30	40	50	60	70	80	90	100	150
	Demand Capacity (cu. ft./hr.)										
1/2	172	118	95	81	72	65	60	56	52	50	40
3/4	360	247	199	170	151	137	126	117	110	104	83
1	678	466	374	320	284	257	237	220	207	195	157
1 1/4	1390	957	768	657	583	528	486	452	424	400	322
1 1/2	2090	1430	1150	985	873	791	728	677	635	600	482

{UPC T12-3} is very similar & in almost all cases will yield the same results.

TABLE 26 GAS-APPLIANCE DEMAND FILL-IN TABLE [T2413.2][A]

Appliance[B]	Typical kBtu/hr.	Actual Btu/hr.	Typical cu. ft./hr.[C]	Actual cu. ft./hr.	#	Extension
Clothes dryer	35		32			
Gas range	65		59			
Recessed oven	25		22			
Recessed top burner	40		36			
Log lighter	80		72			
WH (30–40 gal.)	35		32			
WH (50 gal.)	50		45			
WH, tankless 2 gpm	143		129			
Central furnace/ boiler	100		90			
Barbecue	40		36			
Other						
Total cu.ft./hr. gas demand:						

A. Appendix A of the IRC includes examples & procedures.
B. Typical appl. demands–use actual nameplate ratings.
C. Based on 1,100 Btu/cu.ft.–refer to gas provider for actual values.

Underground

	IRC	UPC
☐ Piping through basement wall must be sleeved	[2415.4]	{1211.1.5}
☐ Fe pipe req's protection–not just zinc coating	[2415.8]	{1211.1.3}
☐ Fe wrapping must be factory applied EXC	[2415.8.2]	{n/a}
Field wrapping OK where stripped for threading	[2415.8.2X]	{1211.1.3}
☐ Dielectric fitting req'd between underground & aboveground metal gas pipe	[2410.1]	{1211.14B}
☐ Min cover depth 12in {18in} EXC	[2415.9]	{1211.1.2A}
Individual gas lines to lights, grills, etc; 8in OK	[2415.9.1]	{Ø}
☐ 12in cover only OK if damage unlikely & pipe shielded	[n/a]	{1211.1.2A}
☐ Insulated Cu tracer min 18AWG {14AWG} wire buried w/ plastic pipe	[2415.14.3]	{1211.1.7C}
☐ Fe req'd for vert riser pipe from plastic EXC	[2415.14.1]	{1211.1.7A}
Listed risers PMI or sleeved wall head adapter	[2415.14.1X]	{1211.1.7X}

FIG. 50

Gas Pipe under Slab

Gas Piping in or under Slab

	IRC	UPC
☐ Piping prohibited underground beneath bldg EXC	[2415.11]	{1211.1.6}
OK in conduit sealed in bldg & vented on ext F50	[2415.11]	{1211.1.6}
☐ Interior end of conduit sealed F50	[2415.11]	{1211.1.6}
☐ Ext end of conduit vented above grade outside F50	[2415.11]	{1211.1.6}
☐ Ext end of conduit min 4in outside bldg & sealed F50	[2415.11]	{1211.1.6}
☐ Piping in solid floor OK in accessible channel OR	[2415.6]	{n/a}
Piping in casing 2in beyond point floor emergence	[2415.6]	{n/a}

Valves & Shutoffs

	IRC	UPC
☐ Valve req'd ahead of meter {regulator}	[2420.2]	{1211.10.1}
☐ Meter valve must be accessible {on bldg ext}	[2420.1.3]	{1209.6.2}
☐ Main valve must be plainly marked as such	[n/a]	{1211.10.3
☐ All valves req ready access	[2420.1.3]	{1212.4}
☐ Outdoor shutoff req'd each separate bldg	[2420.3]	{1211.10.2B}
☐ Each appl req's shutoff valve	[2420.5]	{1212.4}
☐ Valve within 6ft of appl [& in same room]	[2420.5]	{1212.4}
☐ Shutoff valves inside decorative fireplaces PMI	[2420.5X]	{1212.4}
☐ Shutoff valves inside solid-fuel burning fireplaces PMI {prohibited}	[2420.5.1]	{1211.8.2B}

TABLE 27 GAS SIZING FILL-IN TABLE

Section	Length	Cu. ft./hr.	Pipe Size

Appliance Connectors

	IRC	UPC
☐ Shutoff req'd ahead of connector **F59**	[2422.1.2.4]	{1212.4}
☐ Rigid metal OK	[2422.1]	{1212.1}
☐ Semi-rigid metal tubing {UPC only}	[Ø]	{1212.1}
☐ L&L flex connectors OK PMI	[2422.1]	{1212.1}
☐ CSST OK PMI	[2422.1]	{1212.1}
☐ Flex connectors may not be ganged together	[2422.1.2.1]	{1212.1}
☐ Connectors entirely in same room as appl	[2422.1]	{1212.1}
☐ No connectors through wall, floor, ceiling, or appl housing EXC	[2422.1.2.3]	{1212.1}
OK through fireplaces w/ factory-installed protective grommet PMI	[2422.1.2.3X]	{n/a}
☐ Aluminum alloys OK only indoors	[n/a]	{1212.1}
☐ Max length: 3ft EXC	[2422.1.2]	{n/a}
6ft OK for ranges & dryers	[2422.1.2.1	{1312.1}
☐ Min dia = inlet of appl	[2422.1.2.2]	{n/a}
☐ Gas hose OK only for outdoor portable appl	[n/a]	{1212B}

FIG. 51

CSST Manifold

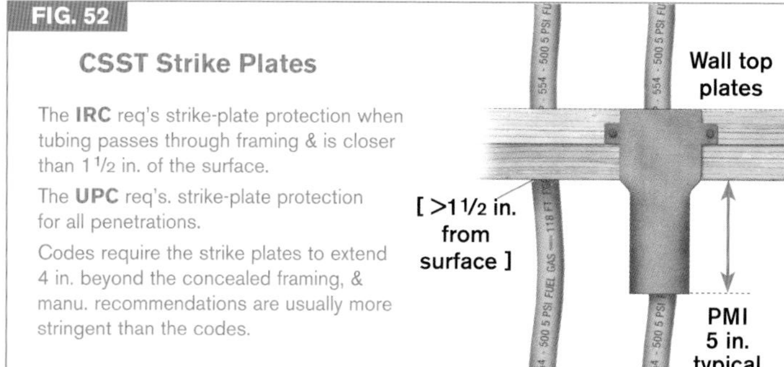

FIG. 52

CSST Strike Plates

The **IRC** req's strike-plate protection when tubing passes through framing & is closer than 1 1/2 in. of the surface.

The **UPC** req's. strike-plate protection for all penetrations.

Codes require the strike plates to extend 4 in. beyond the concealed framing, & manu. recommendations are usually more stringent than the codes.

Wall top plates

[>1 1/2 in. from surface]

PMI 5 in. typical

WATER HEATERS

Water heaters must supply hot water while safely controlling the energy that it takes to heat the water. Most water heaters have storage tanks, though instantaneous water heaters–known as "on-demand" heaters–are becoming more popular. Water heaters that are part of a boiler system providing space heating are discussed in *Code Check HVAC*.

Water Heaters General	**IRC**	**UPC**
☐ Valve req'd on cold-water supply at or near WH____	[2903.9.2]	{605.2}
☐ Valve must be full-bore type F41,42______________	[2903.9.2]	{605.2}
☐ WH also used for space heating must be L&L for both	[2448.2]	{n/a}
☐ Systems also used for space heating req master mixing valve to temper domestic water to 140°F {120°F} or less_____	[2802.2]	{n/a}
☐ Unions req'd {≤12in from heater} to allow removal____	{2801.3]	{609.5}
☐ Electric req's in-sight or lockable disconnect_	[1307.5,T4001.5]	{506.1}

Special Locations		
☐ Fuel-fired WH prohibited in storage closets _________	[2005.2]	{n/a}
☐ Not in BRs, BAs, or their closets EXC_______________	[2005.2]	{505.1}
Direct-vent WH OK w/o enclosure, OR __________	[2801.4X1]	{505.1}
WH OK in dedicated enclosure w/ solid, weatherstripped, self-close door & all combustion air from ext_______	[2801.4X2]	{505.1}[33]
☐ Ignition source min 18in above garage floor F53 _____	[2801.6]	{508.14}
☐ Ignition source elevation req'd if in room open to garage	[1307.3]	{508.14}
☐ Elevation not req'd for FVIR WHs F59 _____________	[2408.2X]	{508.14}
☐ Min 18in above floor in area where flammables stored (basements) ______________________	[2801.6, manu]	{508.13}
☐ On the ground 3in pedestal (concrete, etc) req'd ____	[1308.3]	{508.3}
☐ SDC D (& SDC C for townhouses): 1 strap req'd in upper 1/3 of tank & 1in lower 1/3 of tank (4in above controls) F53____	[2801.7][34]	{508.2}
☐ Barrier or elevation req'd in garage or carport F53 _	[1307.3.1]	{508.14}

FIG. 53 Water Heater in Garage

TABLE 28	MINIMUM HOT-WATER CAPACITY {5-1}	
# of BAs	# of BRs	1st-Hr. Rating[A] (gals.)
1–1½	1	42
	2	54
	3	54
2–2½	2	54
	3	67
	4	67
	5	80
3–3½	3	67
	4	80
	5	80
	6	80

A Non-storage & solar water heaters shall be sized to meet the appropriate 1st-hr. rating as shown in the table.

Water Heater Size — IRC — UPC

☐ Must be sufficient for cleaning & cooking T28 ______ [2801.1] {501.0}

Access & Working Space

☐ Remain accessible for service, inspection, & removal _[2801.3] {505.3.1}
☐ Appl must fit through attic door __________ [2005.1,1305.1.3] {509.4.1}[35]
☐ Attic hatch or door min 22in by 30in ______ [2005.1,1305.1.3] {509.4.1}[35]
☐ Attic min 24in passageway, solid floor to WH [2005.1,1305.1.3] {509.4.3}[35]
☐ Attic WH max 20ft from access
{if ceiling <6ft} ________________________ [2005.1,1305.1.3] {509.4.2}[35]
☐ Attic req's light & receptacle near WH ___[2005.1,1305.1.3.1] {509.4.5}[35]

Access & Working Space (cont.) — IRC — UPC

☐ Light switch req'd at entrance to attic eqpmt space________________ [2005.1,1305.1.3.1,3803.4] {509.4.5}[35]

Tankless Water Heater

☐ Type III vent typically req'd F54 ____________ [manu] [manu]
☐ TPRV/PRV if req'd by manu ____________ [manu] [manu]
☐ Size gas line to max Btu rating F54 ____________ [manu] [manu]

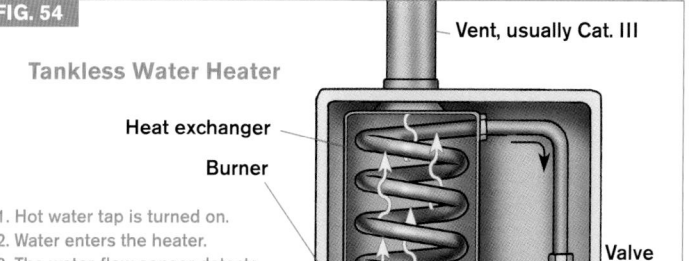

FIG. 54

Tankless Water Heater

Vent, usually Cat. III
Heat exchanger
Burner
Valve
Hot water out
Gas
Fan
Flow sensor
Cold water in

1. Hot water tap is turned on.
2. Water enters the heater.
3. The water-flow sensor detects the entry of water into the unit, switching on computer.
4. The computer ignites the burner.
5. Water circulates through the heat exchanger.
6. The heat exchanger heats water.
7. When the tap is shut off, the unit shuts down.

Gas line must be sized to max. Btu rating to deliver max. hot water.

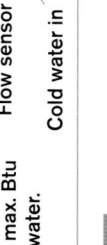

TEMPERATURE & PRESSURE-RELIEF VALVES

FIG. 55

Why TPRV?

Exploding heaters have been known to reach heights of 500 ft. and lift houses off their foundations.

Temperature & Pressure Relief Valves General	IRC	UPC
☐ All WHs req pressure-relief device **F55–56,58**	[2803.1]	{505.4}
☐ All WHs req temp-limiting device **F55–57**	[2803.1]	{505.5}
☐ Devices may be {must include} combination TPRV **F56**	[2803.5]	{506.2,608.3}
☐ Temp probe [in top 6in of tank] {per manu}	[2803.4]	{505.6,608.3}
☐ Setting shall not be >150psi [or 210°F]	[2803.4]	{608.4}
☐ Watts 210 also req's PRV **F57,58**	[2803.1]	{505.6}

FIG. 56

Temperature & Pressure Relief Valve

When the water heater is in a basement or below grade, it may not be possible to arrange for a gravity drain of the TPRV. A Watts 210 valve (**F44**) can be installed. The temp-sensing bulb of the valve goes in the upper portion of the tank, & the gas piping runs through the valve. The Watts 210 shuts off the gas if the temp. is excessive. In addition, a separate water-pressure-relief valve (**F58**) must be installed in the piping & must drain as in **F53**. Check w/ the local Administrative Authority to see if this method is accepted in your area.

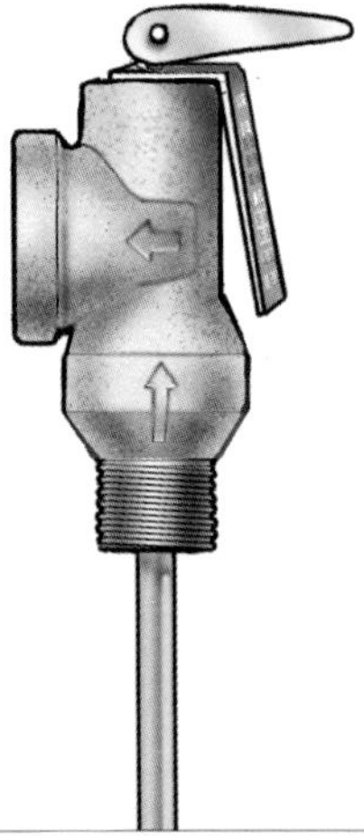

FIG. 57

Watts 210 Gas Shutoff

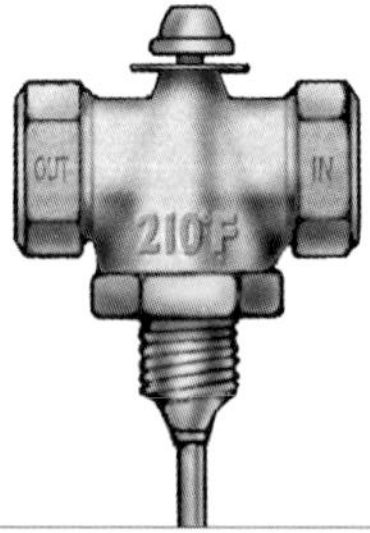

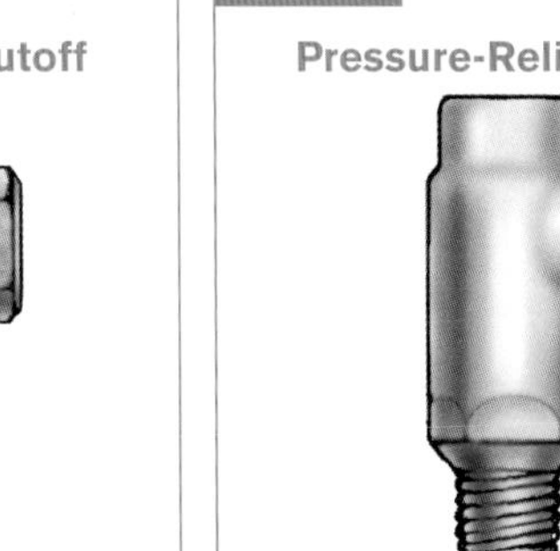

FIG. 58

Pressure-Relief Valve

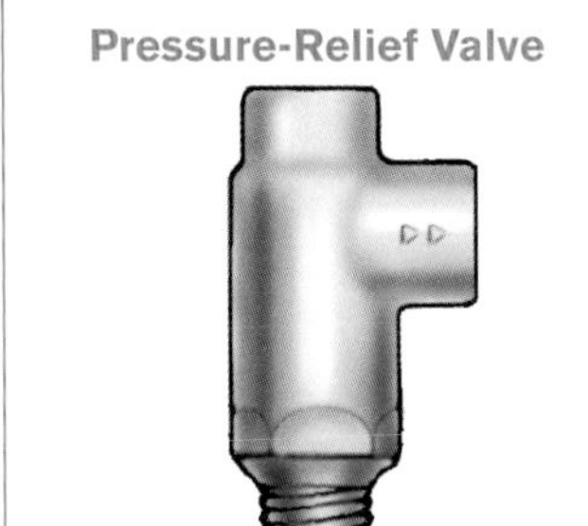

TPRV Drain Piping

	IRC	UPC
☐ Drain must end at floor, ext, or indirect waste	[2803.6.1]	{608.5}
☐ Piping ends outside or [floor or indirect waste] {approved located} F53	[2803.6.1]	{608.5}
☐ Terminate within 6in {6–24in} from ground, [floor] or receptor	[2803.6.1]	{608.5}
☐ Drain size at least same as outlet of valve	[2803.6.1]	{608.5}
☐ Must drain by gravity–not uphill or trapped	[2803.6.1]	{608.5}
☐ No kinks or restrictions in pipe	[2803.6.1]	{608.5}
☐ End of pipe visible & no threads on end F53	[2803.6.1]	{608.5}
☐ No shutoff valves before or downstream of TPRV	[2803.6]	{505.6}
☐ Same materials as allowed for water distribution {only galvanized steel, hard-drawn Cu, CPVC} T14	[2803.6.2]	{608.5}
☐ Protect from freezing (terminate through air gap to indirect receptor located in a heated space)	[2803.6.1]	{608.5}
☐ TPRV may {not} discharge into pan	[2803.6.1]	{508.5}
☐ May not drain to crawl space	[2803.6.1]	{608.5}[36]
☐ Max of 4 90° elbows & max length 30ft	[manu]	{manu}

Required Pans & Drain

	IRC	UPC
☐ Watertight, corrosion-resistant pan req'd for WHs in attics or where leakage could cause damage F53	[2801.5]	{508.4}
☐ Pan drain size min 3/4in F53	[2801.5.1]	{508.4}
☐ Drain req'd to terminate [in indirect waste or outdoors 6–24in above grade] {to approved location} F56	[2801.5.2]	{508.4}
☐ Pan min 1 1/2in deep F53	[2801.5.1]	{n/a}
☐ TPRV may {not} discharge into pan F56	[2803.6.1]	{508.5}

FIG. 59

FVIR Water Heater

Air enters through the vents & passes through the flame-arrestor plate into the sealed combustion chamber.

Since July 2003, all 30–50 gal. gas-fired residential WHs must conform to a new standard that mandates FVIR. This addresses the problem caused by improper storage of gasoline or other flammable liquids near gas-fired WHs. The new standard includes testing for lint, dust, & oil (LDO) on arrestor plate designs. Since the chamber is sealed, the arrestor plate must be self-cleaning.

COMBUSTION AIR

An adequate supply of combustion air is necessary for all fuel-burning water heaters. Most water heaters vent by gravity–their flue gases are lighter than the air in the environment in which the combustion occurs–so they naturally rise up in a vent that is open to atmosphere at the top. The open draft hood on the top of the water heater allows additional air to dilute the flue gases. Insufficient combustion air is hazardous. If there is not sufficient oxygen to fully burn the fuel at the correct temperature, deadly carbon monoxide will also be a product of combustion. If the air pressure in the water heater space is lower than that in the vent, products of combustion might "spill" out of the draft hood & enter the interior environment. As houses are built tighter and to higher standards of energy efficiency, infiltration alone will not supply combustion air to an appliance enclosure. When the appliance space has less than 0.40 ACH (40% air changes per hour) from infiltration, an outside source must be provided to that space. An alternative is to use a direct vent water heater, which has a pipe that brings combustion air from outside directly to the combustion chamber.

General

Note: IRC Chapter 17 is for oil-burning or solid-fuel appls & Chapter 24 is for gas-burning appls.

General	IRC	UPC
☐ Req'd for natural-draft WHs EXC	[1701.1, 2407.1]	{507.1.1}
Direct vent WH PMI	[1701.1, 2407.1]	{507.1.2}
☐ Draft hood in same space as appl	[2407.3]	{507.1.3}
☐ Consider effect of exhaust fans (kitchen, bath)	[2407.4]	{507.1.4}

Openings & Ducts

Openings & Ducts	IRC	UPC
☐ Oil WH openings req screens w/ 1/4–1/2in mesh	[1703.5]	{n/a}
☐ Gas WH openings screen mesh min 1/4in	[2407.10]	{507.8B}
☐ Net area of louvered openings 75% if metal, 25% if wood	[1701.5, 2407.10]	{507.8A}
☐ Ducts to outdoor air min cross-section dimension 3in	[1703.2.1, 2407.6]	{507.4}
☐ Oil WH no volume dampers in ducts	[1701.3]	{n/a}

Openings & Ducts (cont.)

Openings & Ducts (cont.)	IRC	UPC
☐ Motorized louvers req interlock to prevent burner ignition if louvers do not open	[2407.10]	{507.8C}
☐ Joist/stud space OK as combustion air duct if no more than 1 fireblock removed	[2407.11X]	{507.9X}
☐ Ext openings min 12in above grade	[2407.11]	{507.9}
☐ Ducts may serve only 1 enclosure	[2407.11]	{507.9}
☐ 2 direct ext openings min 1sq.in/4kBtu F61	[2407.6.1]	{507.4.1}
☐ 2 direct ext openings upper & lower 12in F61	[2407.6.1]	{507.4.1}
☐ Horiz ducts to upper part of enclosure may not slope down to source (upper duct not to originate from below) F62	[2407.11]	{507.9}
☐ Upper & lower ducts remain separate to source F63	[2407.11]	{507.9}
☐ Mechanical combustion air supply OK if min 35cu.ft/minute/kBtu	[2407.9]	{507.7}
☐ Appl interlock req'd if combustion air mechanically supplied	[2407.9.2]	{507.7.2}

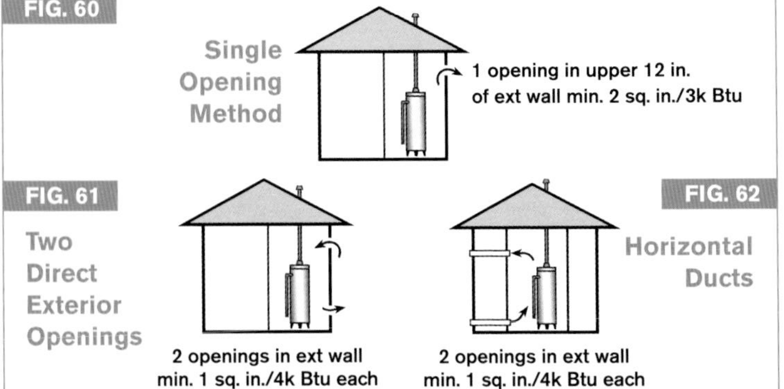

FIG. 63

Vertical Ducts to Outdoors

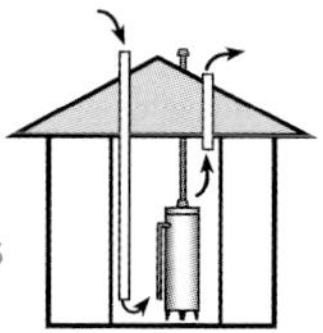

2 vert ducts
min. 1sq. in./4k Btu each

FIG. 64

Vertical Ducts to Attic

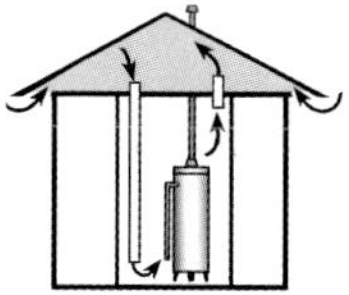

2 openings in ventilated attic
min. 1sq. in./4k Btu each &
sleeved min 6 in. above joist

Single-Opening Method

	IRC	UPC
☐ Single opening OK at upper 12in of enclosure direct to ext, min 1sq.in/3kBtu F60	[2407.6.2]	{507.4.2}
☐ Single opening must be ≥ size of all vent connectors	[2407.6.2]	{507.4.2}

Attic Space Source

	IRC	UPC
☐ Attic freely communicating w/ outdoors considered equivalent to outdoors (ventilated attic)	[1703.2, F2407.6.1(1)]	{F5-9}
☐ Min 1sq.in/4kBtu F64	[1703.2.1, 2407.6.1]	{507.4.1}
☐ Oil WH galvanized metal sleeve req'd min 6in above joist	[1703.3]	{n/a}
☐ Screens prohibited in duct to attic	[1703.3, 2407.11]	{507.9}

FIG. 65

All Air from Indoors by Infiltration

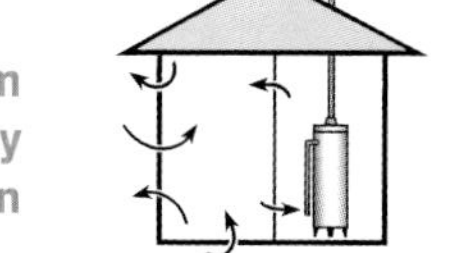

Openings from enclosed WH
space to bldg. interior:
min. 100 sq. in. each,
1 in upper 12 in. &
1 in lower 12 in. of enclosure

Indoor Air Source

	IRC	UPC
☐ Indoor air source alone only OK if ACH >0.40	[2407.5]	{507.2}
☐ Indoor air volume includes rooms directly communicating w/ appl space F65	[2407.5]	{507.2}
☐ Min volume of space 50cu ft/kBtu/hr F65	[2407.5.1]	{507.2.1}
☐ Openings connecting indoor spaces req'd to be located in upper & lower 12in of appl space F65	[2407.5.3.1]	{507.3}
☐ Openings connecting indoor spaces min 100sq.in each; min 1sq.in/kBtu if on same level or 2sq.in if on different levels	2407.5.3]	{507.3}
☐ Min volumes for known air-infiltration method:		
• Non-fan-assisted appl = (21cu. ft/ACH)kBtu	[2407.5.2]	{507.2.2}
• Fan-assisted appl = (15cu. ft/ACH)kBtu	[2407.5.2]	{507.2.2}

Crawl Space Source

	IRC	UPC
☐ Underfloor area freely communicating w/ outdoors considered equivalent to outdoors (ventilated crawl)	[1703.2, F2407.6.1(1)]	{F5-8}
☐ OK only for lower combustion air F66	[1703.2, 2407.11]	{507.9}
☐ Opening min 1sq.in/4kBtu F66	[1703.2.1, 2407.6.1]	{507.4.1}
☐ Oil WH crawl vents 2× req'd combustion air	[1703.4]	{n/a}

FIG. 66

Crawl Space & Attic Openings

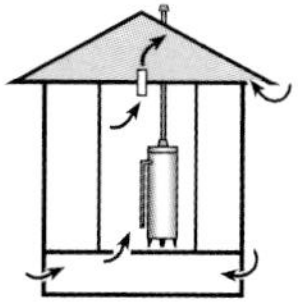

Attic & crawl space
min. 1sq. in./4k Btu each

FIG. 67

Crawl Space Cannot Be the Only Source

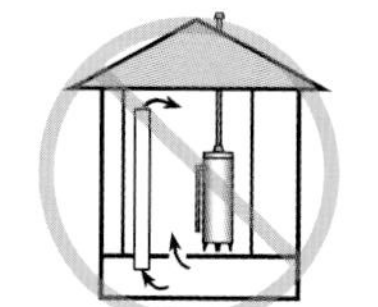

Crawl space may not provide
upper combustion air source

VENTS & FLUES

Vent pipes carry the combustion products outside the building and are designed for the fuel used (usually type L for oil and type B for gas). Proper size, clearances, and connections are necessary for a safe installation. Vents must terminate in an area where combustion products will not re-enter the building and proper flow of flue gases will not be impeded by winds acting on the building. Direct-vent (DV) appliances sometimes allow side wall venting and lesser clearances to building openings than Type B or L vents. The IRC contains rules for oil-burning appliances in Chapter 17 and gas-burning appliances in Chapter 24.

General

Item	IRC	UPC
☐ Draft hood or barometric damper in same WH space	[1802.1, 2407.3]	{507.1.3}
☐ Only 1 draft hood	[2427.12.2]	{501.0}
☐ Provide proper support for venting system	[2426.6]	{510.6.5}
☐ Properly support offsets	[2426.6]	{5106.5}
☐ Bends up to 45° OK, 1 60° OK if not using tables F68	[2427.6.9.2]	{510.6.1.1}
☐ No passing through circulating air duct or furnace plenum	[2427.3.4]	{510.3.6}
☐ Protect w/ steel strike plate extending 4in beyond framing plate when inside wall	[2426.7][37]	{n/a}

Size

Item	IRC	UPC
☐ Min size 3in dia & size of draft hood	[2428.2]	{ 510.6.3}
☐ When using tables supplied w/ WH & table allows offsets: subtract 5% for each bend up to 45° & 10% for each elbow up to 90°	[2428.2.3]	{511.1.2}
☐ Insulation shield req'd to 2in. above insulation	[2426.4]	{n/a}

Vent Connectors General

Item	IRC	UPC
☐ Min slope 1/4in/ft, no dips or sags F68	[2427.10.8]	{510.10.8}
☐ Req'd to be as short & straight as possible	[2427.10.6]	{510.10.6}

Vent Connectors General (cont.)

Item	IRC	UPC
☐ 2 WHs common connector size ≥ larger WH + 50% req size of smaller WH	[2427.10.3.4]	{510.10.3.4}
☐ Multiple appl should join vent as high as possible	[2427.10.3.4]	{510.10.3.4}

Connections to Masonry Chimneys

Item	IRC	UPC
☐ Must enter above bottom [min 12in] of chimney	[2425.9]	{510.10.11}
☐ CO or access must be provided	[2425.7.1]	{510.10.12}
☐ Chimney cross-sectional area max 7× connector area	[2428.2.8]	{511.1.9}[38]
☐ No connection to solid-fuel fireplace flue	[2427.5.6.1]	{510.10.13}
☐ Masonry chimneys: must be lined	[2427.5.2]	{510.5.1.3}

Single Wall Connectors

Item	IRC	UPC
☐ Clearance from combustibles 6in F68	[2427.7.7]	{T5-3}
☐ All joints & connections fastened by screws or other approved means	[2427.10.7]	{510.10.7}
☐ Allowed only in same room as WH–may not pass through any wall, floor, or ceiling	[2427.10.14]	{510.10.14.1}
☐ Not in unoccupied attic or concealed space	[2427.7.6]	{510.7.4.2}
☐ Not in cold unconditioned space	[2427.10.2.2]	{510.10.2.2}
☐ Max horiz length 75% of total vert height F68	[2427.10.9]	{T5-3}

Double Wall (B or BW) Connectors

Item	IRC	UPC
☐ Clearance from combustibles per listing (1in typical)	[T2427.7.7]	{510.10.5}
☐ Max horiz length 100% of total vert height	[2427.10.9]	{510.10.9.3}

Vent Terminations

Item	IRC	UPC
☐ Direct vent ≤50kBtu min 9in to bldg openings	[2427.8]	{510.8.3}
☐ Direct vent ≥50kBtu min 12in to bldg openings	[2427.8]	{510.8.3}
☐ Listed cap req'd	[2427.6.3]	{510.6.2.5}

B Vent Terminations

Item	IRC	UPC
☐ Min 5ft above draft hood F68	[1804.2.3,2427.6.5]	{510.6.2.1}
☐ Min 1ft above sloped roof & per T29	[2427.6.4]	{510.6.2}
☐ Min 2 ft above vert walls within 8ft horiz	[2427.6.4]	{510.6.2

FIG. 68

Common Venting Example

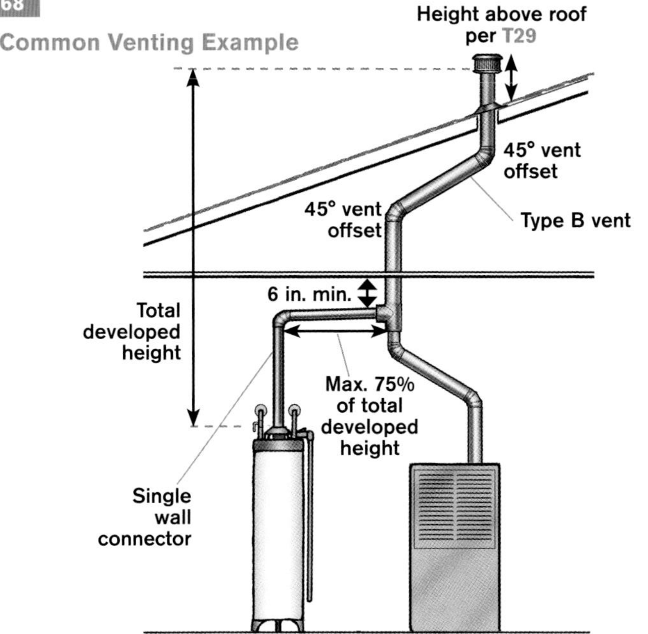

Appl. vent connectors should be as short & straight as possible; & when practical, it is best to have an individual vent for each appliance. Structural conditions often dictate that they be joined & that offsets are needed. When joining two connectors, the appl. w/ the lower rating should join above the one w/ the higher rating.

For further details on venting req's., refer to *Code Check HVAC.*

TABLE 29 — TERMINATION OF B & BW VENTS
[F2227.6.4] & {F 5-2}

Roof Slope	Height above Roof
up to 6/12	1 ft.
>6/12 to 7/12	1 ft. 3 in,
>7/12 to 8/12	1 ft. 6 in,
>8/12 to 9/12	2 ft.
>9/12 to 10/12	2 ft. 6 in,
>10/12 to 11/12	3 ft. 3 in,
>11/12 to 12/12	4 ft.
>12/12 to 14/12	5 ft.
>14/12 to 16/12	6 ft.
>16/12 to 18/12	7 ft.
>18/12 to 20/12	7 ft. 6 in,
>20/12 to 21/12	8 ft.

FIXTURES

Fixtures–showers, sinks, toilets, hose bibbs, and other equipment must be designed and maintained to prevent cross-connections between the supply and the waste. They must be capable of performing their intended function without accumulating contaminants, and they must not create points of leakage into the rest of the building. Fixtures are among the last components installed in the plumbing system and are part of the final inspection.

Size	IRC	UPC
☐ Fixts req'd to be smooth, impervious, & free from concealed fouling areas F69	[2701.1]	{401.1}
☐ Watertight seal req'd at contact area between fixts & walls or floors (caulk base of toilet)	[2705.1]	{407.2}
☐ Separate accessible shutoff req'd for each fixt exc BTs & showers	[2903.9.3]	{605.5}
☐ Shutoffs at manifolds must be labeled F47	[2903.8.5]	{605.5}
☐ Hot on left; cold on right	[2722.2]	{415.0}
☐ All fixts exc toilets {& urinals} req drain strainers	[2702.1]	{404.1}
☐ Tailpiece for lavs & bidets min 1 1/4in	[2703.1]	{404.3}
☐ Tailpiece for other fixts min 1 1/2in	[2703.1]	{404.3}
Laundry		
☐ Standpipe receptor ≥18in & ≤42in {30in} above trap F70	[2706.2]	{804.1}
☐ Must drain through air break (no pressurized waste)	[2718.1]	{805.0}
☐ No trap below floor	[n/a]	{804.1}
☐ Trap ≥ 6in & ≤ 18in above floor F70	[n/a]	{804.1}
☐ CW may drain directly into LT	[2706.3X2)]	{T7-3}
☐ LT may drain into washer standpipe if within 30in horiz	[2706.2.1]	{Ø}

FIG. 69 Concealed Fouling Areas

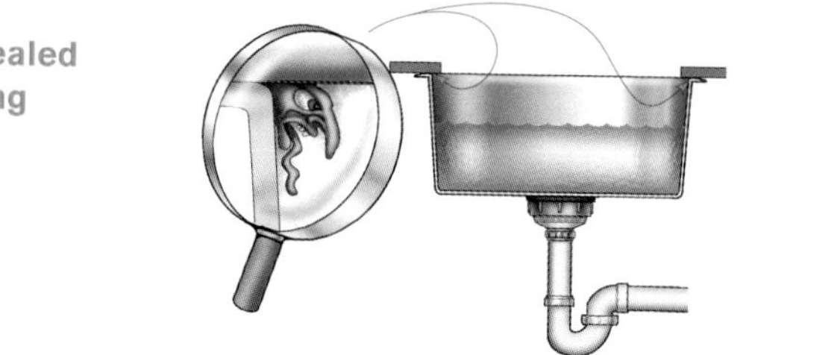

FIG. 70 Standpipe

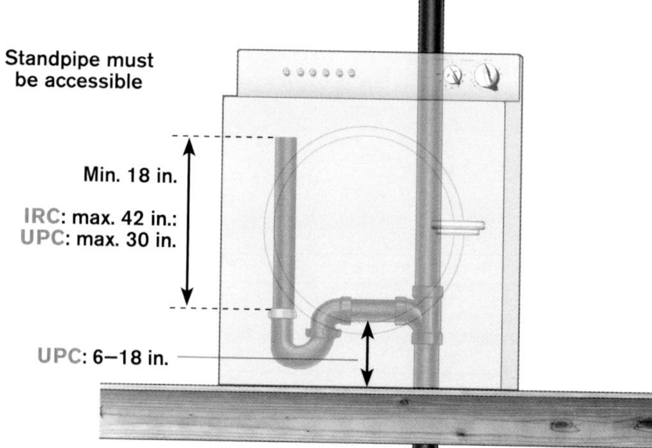

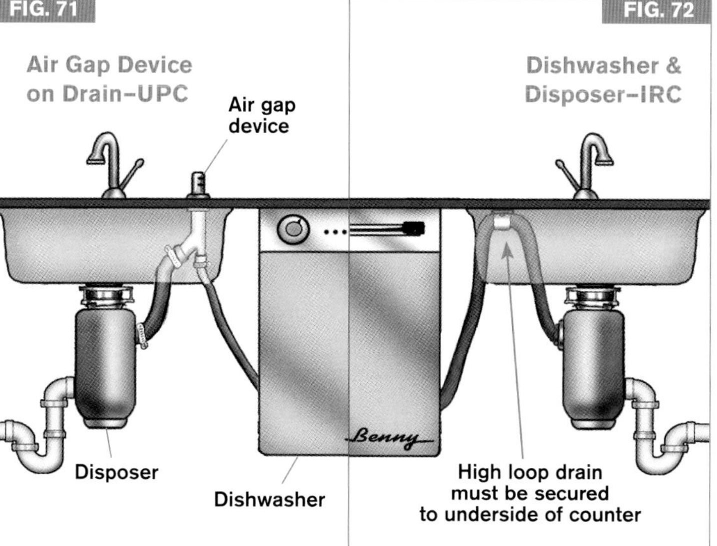

FIG. 71 Air Gap Device on Drain–UPC

FIG. 72 Dishwasher & Disposer–IRC

Kitchens

	IRC	UPC
☐ DW supply req's air gap or integral backflow device	[2717.1]	{301.1.1}
☐ May discharge directly to a trap, trapped fixt, or branch wye tailpiece on KS or directly to a garbage disposer	[2717.2,3]	{Ø}
☐ Secure drain hose to underside of counter F72	[2717.2,3]	{n/a}
☐ Air gap fitting above sink flood level req'd for DW drain F71	[n/a]	{807.4}
☐ No connection to discharge side of disposer F71,72	[n/a]	{404.4}
☐ Reverse-osmosis systems to recognized standards	[2907.1]	{603.4.13}
☐ Reverse-osmosis systems req air gap F73	[2907.2]	{603.4.13}
☐ No saddle fittings or tapping/drilling of drain line	[3003.2]	{311.2}

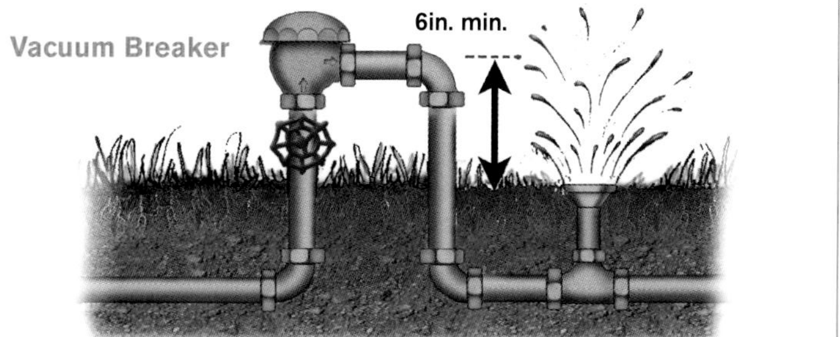

FIG. 73 Reverse-Osmosis Water Treatment

Outdoors & Irrigation Systems

	IRC	UPC
☐ Hose bibbs req backflow or vacuum breaker	[2902.4.3]	{603.4.7}
☐ Irrigation vacuum breakers {6in} above highest head F74	[2902.5.3]	{T6-2}

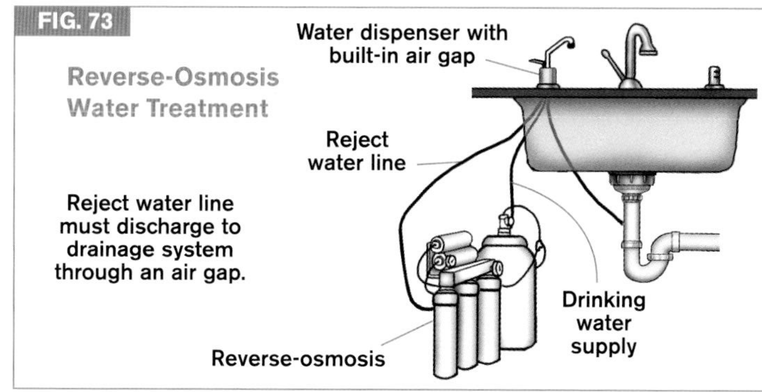

FIG. 74 Vacuum Breaker

FIG. 75 Toilet Flanges

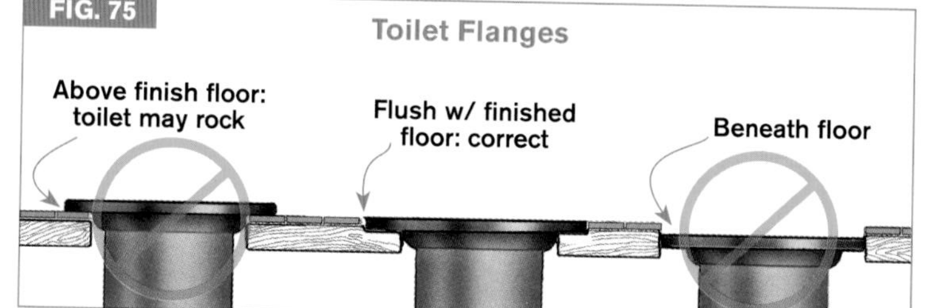

Toilets & Bidets

	IRC	UPC
☐ Floor flanges req'd for floor outlets F75	[2705.1]	{704.4.1}
☐ Secure floor flange w/ corrosion-resistant fasteners	[2705.1]	{704.4.5}
☐ Toilet or bidet req's 15in clearance from center to side walls or outer rim of adjacent toilets (toilet & bidet min 30in O.C.) F76	[2705.1]	{407.6}
☐ Min clear 21in {24in} in front of [lav] toilet or bidet F77	[2705.1]	{407.6}
☐ No offset or reducing floor flanges	[3002.3.1]	{701.2}
☐ Closet ring to vent max length [not limited] {6ft}	[3105.1]	{T10-1}
☐ Ball cock critical level ≥ 1in above overflow pipe F78	[2712.4]	{603.4.2}[39]
☐ Bidet water temp max 110°F	[2721.2][40]	{n/a}

FIG. 76 Toilet Clearances

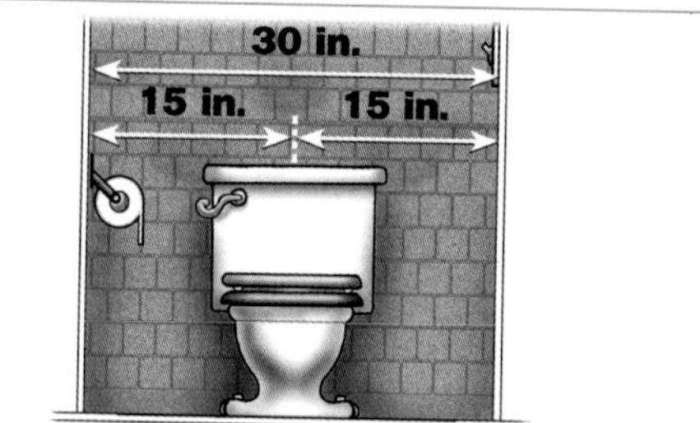

FIG. 77 Toilet Rough Layout

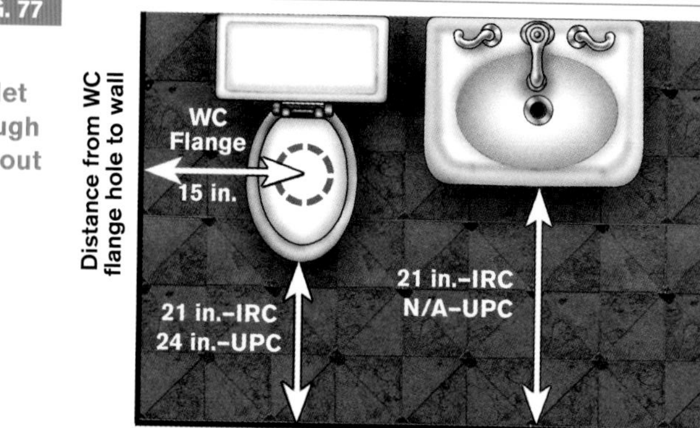

FIG. 78 Toilet Fill Valve

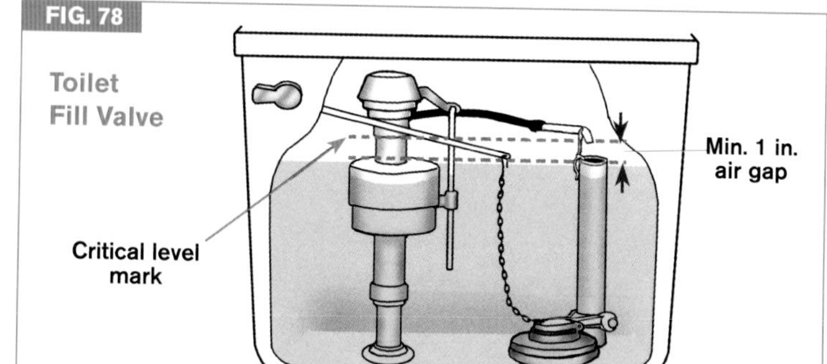

Tubs

	IRC	UPC
☐ Slip joints accessible w/ min 12×12in door **F25**	[2704.1]	{404.2}
☐ Over-rim bath spout min air gap 2in from FLR	[T2902.3.1]	{T6-3}
☐ Overflow [req'd] min 1 1/2in dia **F25**	[2713.1]	{404.3}
☐ Tub or whirlpool max water temp 120°F	[2713.3][41]	{414.5}[41]
☐ Whirlpool tub access must allow pump removal [min 12×12in], 18in by18in if pump >2ft from access opening	[2721.2][42]	{n/a}

Showers

	IRC	UPC
☐ Min area 900sq.in {1024sq.in} min dia 30in measured from finished wall to center of threshold **F80**	[2708.1]	{411.7}
☐ Min shower area to be maintained to 70in above drain	[2708.1]	{411.7}
☐ Showerheads, valves, grab bars, & soap dishes: allowed to protrude into req'd min space	[2708.1]	{411.7}
☐ Shower walls watertight to min 72in above drain	[307.2]	{IS-4
☐ Finished threshold height min 1in below receptor & 2–9in above top of drain **F80**	[2709.1]	{411.6}
☐ Door must open outward **F79**	[2708.1][43]	{411.6}
☐ Door min 22in wide **F79**	[2708.1.1][44]	{411.6}
☐ Finished floor slope 1/4–1/2in/ft	[2709.1]	{411.6}
☐ Secure shower valve, head/riser to permanent structure	[2708.2]	{411.11}
☐ Showerhead not discharging directly at door	[2705.1]	{411.10}
☐ Listed antiscald/pressure-balance valve req'd 120°F max	[2708.3]	{418.0}

FIG. 79

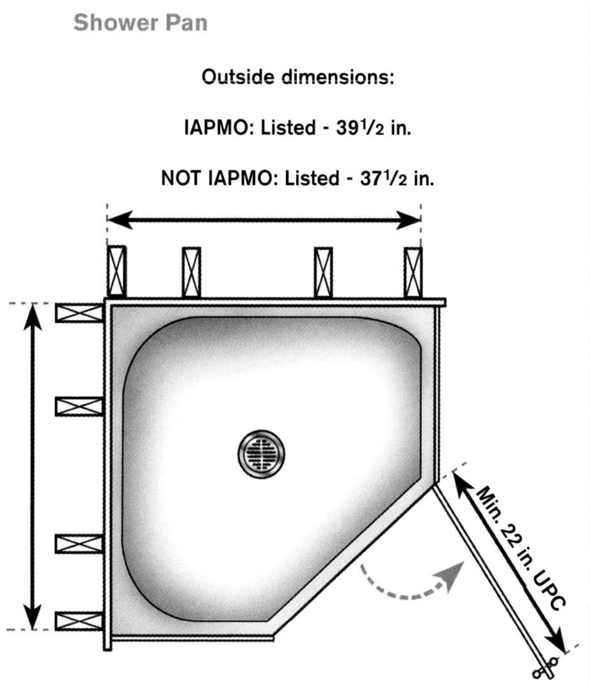

Shower Pan

Shower Pan & Liner

	IRC	UPC
☐ Rough pan min 32×32in	[2708.1]	{411.7}
☐ Must conform to approved standards	[2709.2]	{411.8}
☐ Slope underlayment $1/4$in/ft **F80**	[2709.3]	{411.8}
☐ Liner min 3in above dam **F80**	[2709.2]	{411.8}
☐ Pan liner plastic or 3 layers hot mop type 15 felt	[2709.2]	{411.8}
☐ Special attention to hot mop corner installation: extend 4in all directions from corner	[2709.2.3]	{411.8}
☐ PVC & CPE (chlorinated polyethelene) sheet lining: PMI	[2709.2.1]	{411.8}
☐ Weep holes at drain & must remain clear **F80**	[2709.4]	{411.8}
☐ No fasteners <1in above finished threshold	[2709. 3]	{411.8}
☐ Roll over top of rough threshold (no penetrations through top) & fastened to outside edge **F80**	[2709.3]	{411.8}

FIG. 80

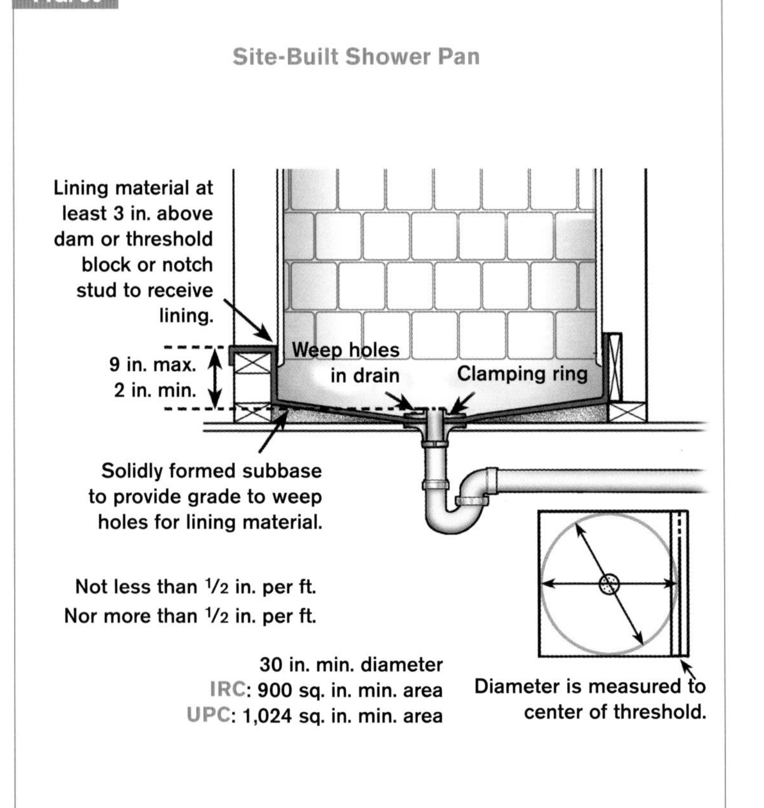

INSPECTIONS

General

	IRC	UPC
☐ Nothing concealed until inspected & approved	[2503.2]	{103.5.1.1}

Water Supply

	IRC	UPC
☐ Test all piping before cover or concealment	[2503.2]	{103.5.1.1}
☐ Water pipe test at working pressure 15minutes OR	[2503.6][46]	{609.4}
50psi air for other than plastic pipe 15 minutes	[2503.6][46]	{609.4}
☐ Water for testing must be from potable water source	[2503.6]	{609.4}
☐ R-P PBP devices tested at installation & annually	[2503.7.2]	{603.3.3}
☐ Test gauges must have increments of 1psi ≤100pounds		
☐ Test pressure or 2psi if >100pounds	[2503.8]	{319.2,3}

DWV Systems Rough Inspections

	IRC	UPC
☐ Water test 10ft head for 15minutes OR	[2503.5.1]	{712.2}
Air test 5psi (10in mercury column) 15minutes	[2503.5.1]	{712.3}

DWV Systems Finished Plumbing Inspections

	IRC	UPC
☐ Fill each drain, visually inspect each trap	[2503.5.2]	{712.1}
☐ [per BO] Gas test smoke 15minutes at 1in water column OR		
Peppermint test 2ounces in 10quarts water	[2503.5.2]	{n/a}

Gas

	IRC	UPC
☐ Leave all joints exposed until tested	[2417.3]	{1214.2.1}
☐ Test pressure min 1 1/2× working pressure	[2417.4]	{1214.3.2}
☐ Test pressure min 3psig	[2417.4]	{1214.3.2}
☐ Test min time 10 minutes	[2417.4]	{1214.3.3}
☐ Test medium air, nitrogen, or CO_2 (not oxygen)	[2417.2]	{1214.1.7}
☐ Test gauge scale not >5× test pressure	[2417.4.1}	{1214.3.1}
☐ Turn off {or cap} all outlets before pressure test	[2417.3.4]	{1214.2.5}
☐ Inspect for open fittings or valves before turning on gas	[2417.6.2]	{1214.5.2}
☐ Check for leakage after turning on gas	[2417.6.3]	{1214.5.3}

Gas (cont.)

	IRC	UPC
☐ Soapy water or gas detector OK for leak detection	[2417.5.1]	{1214.4.2}
☐ No matches for leak detection	[2417.5.1]	{1214.4.2}
☐ Purge appls before placing in operation	[2417.7.4]	{1214.6.4}

STRUCTURAL

Pipe Protection

	IRC	UPC
☐ Plastic or Cu pipe through framing: req's steel nail plate protection if pipe closer than 1 1/2in {1in} of edge of framing	[2603.2.1]	{313.9}
☐ Nail plate: must extend 2in {1 1/2in} beyond framing	[2063.2.1]	{313.9}[47]
☐ Gas pipe other than black or steel: steel nail plate protection if pipe closer than 1 1/2in of face of framing	[2415.5][48]	{313.9}
☐ Gas tubing strike plate protection through framing **F52**	[manu]	{1211.3.4}[49]

Structural Modifications–General

	IRC	UPC
☐ Structural members may not be damaged by repairs or additions to plumbing	[2603.1]	{313.2}
☐ Drilling & notching per **T30,31, F81**	[2603.2]	{313.2}

Notches in Joists & Rafters F81

	IRC
☐ Notches in sawn lumber max 1/6 depth of member **T30**	[502.8.1]
☐ Max length of notches in sawn lumber 1/3 depth of member	[502.8.1]
☐ Notches in sawn lumber not in middle 1/3	[502.8.1]
☐ End notches max 1/4 depth of member	[502.8.1]
☐ No notches in tension side of members ≥4in, exc at ends	[502.8.1]
☐ Holes 2in min to top or bottom or notch or other hole	[502.8.1]
☐ Engineered product (TJI®s) notches, holes, or cuts PMI only	[502.8.2]
☐ No field modification of trusses, ex: notching, cutting	[802.10.4]
☐ No alteration or loading of trusses w/o written concurrence of design professional	[802.10.4, 2405.2]

TABLE 30 JOIST SPANS, NOTCHING, & BORING [502.8.1]

Floor Joist Spans—40lb. Live Load				Notching		Boring
DF#2	12 in. O.C.	16 in. O.C.	24 in. O.C.	End	Outer 1/3	2 in. to edge
[2×6]	10 ft.9 in.	9 ft. 9 in.	8 ft. 1 in.	1 3/8 in.	7/8 in.	1 1/2 in.
[2×8]	14 ft. 2 in.	12 ft. 7 in.	10 ft. 3 in.	1 7/8 in.	1 1/2 in.	2 3/8 in.
[2×10]	17 ft. 9 in.	15 ft. 5 in.	12 ft. 7 in.	2 3/8 in.	1 1/2 in.	3 1/8 in.
[2×12]	20 ft. 7 in.	17 ft. 10 in.	14 ft. 7 in.	2 7/8 in.	1 7/8 in.	3 1/2 in.

FIG. 81

Notching & Boring Joists

Stud Notching & Boring

IRC

- ☐ Bearing or ext wall max notch 25% boring 40% T31 ____ [602.6]
- ☐ Boring to 60% OK if stud doubled & not >2 successive studs bored T31 ____ [602.6]
- ☐ Nonbearing studs notches OK to 40% T31 ____ [602.6]
- ☐ Nonbearing studs boring OK to 60% T31 ____ [602.6]
- ☐ Holes no closer than 5/8in to face of stud ____ [602.6]
- ☐ Holes not in same stud section as cuts or notches ____ [602.6]
- ☐ If top plate >50% removed for pipes, attach galvanized metal tie 1 1/2in wide 16gauge strap w/ 8 16d nails nails each side of notch EXC ____ [602.6.1][50]
  If entire side of wall w/ notch covered w/ wood sheathing ____ [602.6.1X]

TABLE 31 STUD NOTCHING & BORING [602.6]

Stud Size	2×4	3×4	2×6
Bearing Walls (to 10 ft. high)			
Notching	7/8 in.	7/8 in.	1 3/8 in.
Boring	1 3/8 in.	1 3/8 in.	2 3/16 in.
Boring 2 doubled consecutive	2 in.	2 in.	3 1/4 in.
Nonbearing Walls			
Notching	1 3/8 in.	1 3/8 in.	2 3/16 in.
Boring	2 in.	2 in.	3 1/4 in.

A

AAV (**air admittance valve**) One-way valve designed to admit air into the plumbing system during periods of relative negative pressure in the vent system to protect traps from siphonage and to remain sealed under zero or positive pressure. **F31,32**

ABS (**acrylonitrile-butadiene-styrene**) A plastic pipe that is usually black and used for water distribution, drain, waste, vent, and sewage.

Accessible Capable of being reached, but may first require removal of a door, access panel, or cover.

AHJ (**authority having jurisdiction**) An organization responsible for enforcing the code, typically the building department and its authorized representatives.

Airbreak A physical separation in which a discharge pipe from a fixture, appliance, or device drains indirectly into a receptor and enters below the flood level rim of a receptor, such as a clothes washer standpipe.

Air chamber A pressure surge-absorbing device operating through the compressibility of air.

Air gap (for drainage systems) Unobstructed vertical distance through free atmosphere between the outlet of a waste pipe and the flood level rim of the receptacle into which the waste pipe is discharging. **F46,71,73**

Air gap (for water distribution systems) An unobstructed vertical distance through the free atmosphere between the lowest opening from any pipe or faucet supplying water to a tank, plumbing fixture, or other device and the flood level rim of the receptacle.

Antisiphon valves Valves or devices that prevent siphoning, typically by an opening to the atmosphere. **F74**

Approved Accepted by the authority having jurisdiction.

B

Backflow A flow of water or other liquids, mixtures, or substances into the distributing pipes of a potable supply of water from any source other than its intended source.

Backflow connection Any arrangement whereby backflow can occur.

Backflow preventer A device or means to prevent backflow into the potable water system.

Backpressure A potential backflow condition, created by pressure in the potential backflow source higher than the pressure in the water main.

Backsiphonage Backflow caused by a loss of supply pressure.

Backwater valve A device that prevents the backflow of sewage. **F13,15,16**

Ball cock A valve in a toilet tank to control the supply of water into the tank. **F78**

Bathroom group A group of fixtures, located on the same floor and including a water closet, a lavatory, and a bathtub or shower.

Branch Any part of the drains except a main, riser, or stack.

Branch interval A vertical measurement of distance, at least 8ft., between connections of horizontal branches to a drainage stack, measured down from the highest horizontal branch connection.[45] **F14**

Branch vent A vent connecting two or more individual vents with a vent stack or stack vent. **F29**

Building drain Part of the lower piping of a drainage system that receives the discharge from soil, waste, and other drainage pipes inside the walls of the building and conveys it to the building sewer and ends 30in. {24in.} outside the building wall. **F1**

Building sewer Part of the drainage system that extends from the end of the building drain to a public or private sewer or other point of disposal. **F1**

Chimney A structure containing one or more flues to convey gaseous products of combustion to the atmosphere

Chimney connector A pipe connecting a fuel-burning appliance to a chimney flue.

CO (cleanout) An opening for cleaning the drainage system. See p.73

Common vent A pipe venting two trap arms on the same branch, either back-to-back or one above another. F35

Concealed Not exposed to view without removal of building surfaces or finishes.

Contamination Impairment of potable water quality that creates a health hazard.

Continuous waste A drain from two or more adjacent fixtures connecting the compartments of a set of fixtures to a trap or connecting other permitted fixtures to a common trap. F18

CPVC (chlorinated polyvinyl chloride) Plastic pipe designed for hot and cold water.

Cross-connection Any piping connection or arrangement whereby a potable water system may be contaminated.

CWV (combination waste drain & vent system) A system employing horizontal and/or vertical pipes functioning as vents and drains for sinks and floors by providing for free movement of air above the water line in the pipe. F38

D

Developed length The distance measured along the centerline of a pipe.

DFU (drainage fixture unit) A value used to calculate a fixture's load on the drainage system. T3

Directional fittings Wye, combos, or tees with baffles. F9,19

Discharge pipe A pipe that conveys the discharge from a fixture or plumbing appliance.

Drain A pipe that conveys wastewater or water-borne waste.

Drainage system Includes all the piping within public or private premises, which conveys sewage to a legal point of disposal. F1

DWV (drain, waste, and vent) The system of piping and fittings that carries drainage, waste and sewer gases.

Effective opening The cross-sectional area of a water outlet, expressed in terms of the diameter of an equivalent circle. In the case of a faucet, measured at the smallest orifice or in the supply piping.

Fixture branch drain A drain serving two or more fixtures that discharges to another drain or to a stack.

Fixture branch supply A water supply pipe between the fixture supply pipe and the water distributing pipe.

Fixture drain A drain from a fixture trap to a junction with any other drain–also called a trap arm. F21

Fixture unit A unit of measure for the drain (DFU) or water supply (WSFU see WSFU) load from different fixtures.

FLR (flood level rim) A level on a fixture or the surface to which it is fastened at which water overflows. F46

Flue See Vent.

G

Grade A slope or fall of a line of pipe in reference to a horizontal plane. In drainage, it is usually expressed as the fall in vertical units compared to horizontal units or a fraction of an inch per foot length of pipe. H.

H

Hangers See Supports.

Horizontal branch drain A drainage pipe that extends from a stack or drain and which serves two or more fixtures.

Horizontal pipe Any pipe installed >45° from horizontal.

Hot water Water at a temperature ≥110°F {120°F}.

I

Indirect waste pipe A waste discharge into the drainage system through an air gap into a trap, fixture, or receptor, such as a clothes washer standpipe. F70

Individual vent A pipe that vents a fixture trap. F21

J

Joint A connection between two pipes. Types of joints:

- **Brazed** Any joint obtained by joining metal parts with alloys that melt at temperatures >840°F (449°C), but lower than the melting temperature of the parts to be joined.
- **Expansion** A loop, return bend, or return offset that accommodates pipe expansion and contraction.
- **Flexible** A joint that allows movement of one pipe without deflecting the other pipe.
- **Mechanical** A joint that employs compression to seal.
- **Slip** A joint that incorporates a washer or special packing material to accomplish a seal. F25
- **Soldered joint** A joint obtained by joining of metal parts with metallic mixtures or alloys that melt at a temperature <800°F (427°C) & >300°F (149°C).
- **Welded joint or seam** Any joint or seam obtained by the joining of metal parts in the plastic molten state.

Labeled Components that bear the label of a listing agency affirming that the product meets recognized standards.

Liquid waste A discharge from any fixture or appliance that does not receive fecal matter.

Listed Equipment included in a list published by an approved organization that tests and inspects sample products to determine their compliance with nationally recognized standards.

Listing agency An agency accepted by the authority having jurisdiction that maintains a periodic inspection program on current production of listed models and that makes available a published report of such listings.

Main A principal artery of any system of continuous piping to which branches may be connected.

Main vent A principal artery of the venting system to which vent branches may be connected. F30

Offset A combination of elbows or bends in a line of piping that brings a section of the pipe out of line, but into a line parallel with the other section. F26

PEX tubing Water-supply tubing made of cross-linked polyethylene, typically found in parallel distribution systems. See p.92

Plumbing system The water-supply system, drainage system, storm drains, sewers, connected fixtures, supports, appurtenances, and appliances. F1

Potable water Water fit for human consumption.

Pressure balancing valve A mixing valve that senses incoming hot and cold water pressures and compensates for fluctuations in either to stabilize outlet temperature.

Public sewer Sewer controlled by public authority.

Public water main Water pipe controlled by public authority.

Readily accessible Access that does not require removing a panel or door.

Relief valve:

- **PRV** (**pressure relief valve)** A mechanical device that relieves excess pressure in the system served. F56,58
- **TPRV (temperature and pressure relief valve)** A relief device that is actuated by either excess pressure or temperature. F56
- **Temperature relief valve** A mechanical device that relieves excessive temperature in the system served. F56,57

Relief vent A vent providing air circulation between vent and drainage systems.

Rim An unobstructed open edge of a fixture. F46

Riser A water or gas supply pipe that extends one or more stories.

Rough-in Part of the plumbing system that is installed in a structure before fixture installation.

Sewage Liquid waste that contains chemicals or animal or vegetable matter.

Sewage ejector A device for lifting sewage at high velocity with air or water. F15

Sewage pump A device, other than an ejector, for lifting sewage from a sump.

Shielded coupling An approved elastomeric sealing gasket with an approved outer shield and a tightening mechanism (i.e., no hub coupling).

Slope Fall or pitch along a line of pipe.

Soil pipe A pipe that conveys waste including fecal matter.

Stack A vertical drain line that extends one or more stories. F14

Stack vent A vent that extends from a stack.

Stack venting A method of venting fixtures through the soil or waste stack. F14

Static pressure The pressure existing without any flow.

Sump A tank or pit that receives waste and is discharged by mechanical means. F15

Supports Devices used to support or secure pipes, fixtures, or equipment.

Tailpiece A pipe or tubing connecting the outlet of a plumbing fixture to a trap. F20

Trap A fitting or device that employs a liquid seal to prevent the escape of sewer gas from the plumbing system. F17

Trap arm A horizontal pipe between a trap and the connection to the drain and vent system–also called a fixture drain. F21, 22

Trap seal Vertical distance between the crown weir and the top dip of the trap. F17

- **Crown weir** (**trap weir**) Lowest point in the cross section of the horizontal waterway at the exit of the trap. F17

• **Top dip** (**of trap**) Highest point in the internal cross section of the trap at the lowest part of the bend. By contrast, the bottom dip is the lowest point in the internal cross section. F17

Tubular brass Traps, waste bends, and tailpieces with slip-joint connections.

Vent A passageway for conveying flue gases from fuel-burning appliances to the atmosphere.

Vent stack A vertical vent pipe that provides air circulation for the drainage system.

Vent system Piping that prevents trap siphonage and back pressure, or equalizes the air pressure within the drainage system.

Vertical pipe Any pipe that makes an angle of 45° or less from the vertical.

Waste Drainage discharge that does not contain fecal matter.

Waste pipe A pipe that conveys only waste.

Water main A water supply controlled by a public entity or utility.

Water pipe A pipe that conveys water to fixtures and outlets.

Water supply system Pipes, valves, fittings, and supports that supply water to and throughout a residence and its accessories, such as sprinkler piping.

Weir See Trap seal.

Wet vent A vent that also serves as drain. F36,38

Whirlpool bathtub (**hydromassage tub**) A bathtub fixture equipped and fitted with a pump and circulating piping and that is drained after each use.

WSFU (Water supply fixture unit) A measure of the probably estimated normal demand on the water supply by various types of plumbing fixtures. T17

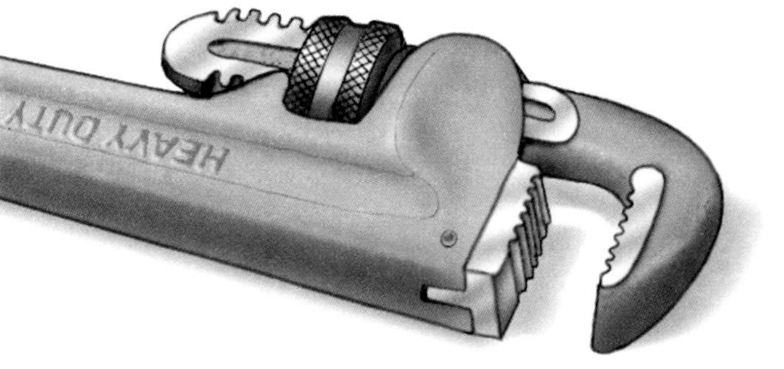

TOP 50 PLUMBING CODE CHANGES

1 The req to sleeve pipes through concrete floors is new in the 2006 UPC, & the UPC now exempts pipes through openings that are drilled or bored.

2 Listed transition solvent cement between ABS & PVC is a new method in the 2006 UPC.

3 The horiz branch distance from a stack is new in the 2006 IRC–a similar rule has long been in the *International Plumbing Code.*

4 The 2006 IRC separately lists above- & below-ground materials–in the 2003 edition they are in the same table.

5 The 2003 IRC allows Cu type DWV for bldg sewer.

6 The 2003 IRC allows Cu type M for bldg sewer.

7 PVC DR-PS has more restrictions in 2006–consult with local BO & sewer authority.

8 The 2003 IRC req's all shower drains to be 2in. min.

9 The 2003 IRC does not req COs to be same size as pipe–it has a table similar to the UPC rules. The new rule affects only req'd COs & does not alter the rule allowing rodding through a removable fixt drain.

10 Due to the above rule (#9), the stack CO must be no more than 1 pipe size smaller than the pipe.

11 The 2003 IRC does not define the req'd proximity of CO.

12 The 2003 IRC does not allow 2-way COs at bldg drain/bldg sewer.

13 The 2003 IRC does not allow any offsets in waste stack vents.

14 The 2003 IRC req's the ext vent for the bldg drain to be through the roof–in 2006 it can be a side wall vent.

15 Because the req'd vent must be a dry vent, the 2006 IRC rule excludes a waste stack vent from being the only ext vent.

16 The 2003 IRC allows a "vertical leg waste fixture drain" to have a vent opening below the trap weir. A vertical leg fixture drain is a specified form of CWV & was eliminated in the 2006 IRC.

17 The 2003 IRC req's the ext vent for the drain to be through the roof, not a side wall.

18 The 2003 IRC does not specify a min vent distance above the roof. Instead this is left up to the local agency adopting the code.

19 The 2003 UPC does not allow any horiz wet venting. The 2006 UPC allows 1 or 2 BA groups on the same floor to be horiz wet vented.

20 2003 UPC does not allow any horiz wet venting (#19).

21 New wording in the 2006 IRC clarifies that each wet vented fixture must connect independently to the wet vent

22 The 2003 UPC does not allow any horiz wet venting (#19).

23 The 2006 IRC clarified where fixt drain distance is measured.

24 By eliminating the list of specific fixts from the req for CWVs, a floor drain could drain through a CWV with a vert pipe. In the very limited CWV applications of the UPC, vert pipes are not allowed, which is another reason the UPC seldom allows these for residential use.

25 The protection distance past the opening is new in the 2006 UPC.

26 The length of 3/8in tubing was not limited in the 2003 code exc by the flow rate & fixt unit demand tables. It is important to consider the specific

demands of the fixts. A 1-piece toilet req's twice the flow rate as a 2-piece toilet.

27 The 2006 UPC clarified that the regulator must control all water outlets in the bldg.

28 The 2006 IRC reqs for thermal expansion protection are specific here for pressure reducing valves on systems up to 2in dia, & they are not specific about the form the protection must take. This section is performance-based more than prescriptive.

29 The 2003 UPC allows PRVs as an optional method of protection, & the 2006 UPC req's that it be a pressure-reducing regulator.

30 This specific section of the 2006 IRC addresses thermal expansion from storage water heating eqpmt.

31 A new rule in the 2006 IRC specifies that piping may not pass through 1 townhouse to serve another.

32 The 2006 IRC rule simplifies the req for labeling gas pipe to apply to all but steel–the 2003 rule is for all but black pipe.

33 The 2003 UPC did not allow WHs (other than direct vent) in a closet off BR. This code is helpful for applications such as basement conversions.

34 The rule for townhouses in SDC C is new in 2006 as is SDC D_0.

35 The 2003 UPC did not have a section addressing attic installations. The 2006 rules are very similar to those in effect in 2000.

36 The 2006 UPC explicitly prohibits draining the TPRV to a crawlspace.

37 Strike-plate protection for concealed vents is new in the 2006 IRC.

38 The 2003 UPC rules for chimney size differentiate between chimneys on the ext or partially on the interior, & refer to several tables. The 2006 UPC has a simple prescriptive rule that the chimney cross section may not be more than 7× the appl vent collar unless engineered.

39 This rule is found in a different UPC chapter (section 411.5) in 2003.

40 The water temp to a bidet is not limited in 2003.

41 The water temp to a tub or whirlpool is not limited in either code in 2003.

42 The 2003 code req's only 12in by 12in openings. The new req helps to assure that pumps & motors for whirlpool tubs are accessible.

43 The req for shower doors to swing outward appears to have been inadvertently omitted in 2003 & is back in the 2006 code.

44 The 2003 code does not specify minimum shower door width.

45 The definition of a branch interval has been clarified to show that it is the distance between horiz branches, not the distance into which horiz branches are discharged.

46 The 2003 IRC did not specify the duration of the test.

47 The 2003 UPC did not specify a distance that the strike plate must extend past the outer diameter of the pipe hole.

48 The distance from the framing surface to where strike-plate protection is req'd was increased in the 2006 IRC from 1 in. to 1 1/2 in.

49 The 2003 UPC does not specify strike plate reqs for gas tubing. CSST manus may have reqs that are more stringent than the code reqs.

50 The 2006 IRC clarifies that the plate across a notched top plate req's 8 nails at *each side*, not 8 nails total.

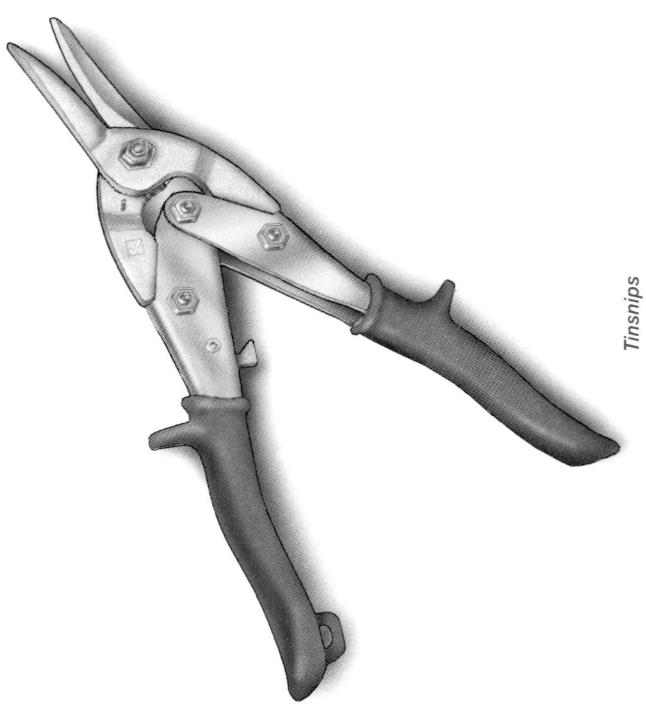

Tinsnips

Code ✓Check® HVAC Third Edition

BY DOUGLAS HANSEN AND REDWOOD KARDON

Code Check HVAC (heating, ventilation and air conditioning) is a condensed field guide to the most commonly used residential mechanical codes. There are essentially two groups of mechanical codes that apply to construction of one- and two-family dwellings. One is the International Residential Code (IRC) and the other is the alliance of the Uniform codes and the NFPA codes. We have used several of these codes to emphasize their similarities and their common safety principles and to clarify distinctions between them.

Check with your building department to determine the codes and editions applicable in your area. If your area is using an older version of the codes, the list of code changes on pp. 180–181 will help in finding the rules that apply today in your area.

MODEL CODES & ORGANIZATIONS

ICC = International Code Council www.iccsafe.org

IAPMO = International Association of Plumbing & Mechanical Officials www.iapmo.org

NFPA = National Fire Protection Association www.nfpa.org

TABLE 1 CODES REFERENCED IN CODE CHECK HVAC

Organization	Code	
ICC	2006 IRC	International Residential Code
IAPMO	2006 UMC	Uniform Mechanical Code
IAPMO	2006 UPC	Uniform Plumbing Code
NFPA	2006 NFPA 54	National Fuel Gas Code
NFPA	2004 NFPA 58	Liquified Petroleum Gas Code
NFPA	2006 NFPA 31	Standard for Installation of Oil-Burning Equipment
NFPA	2006 NFPA 211	Chimneys, Fireplaces & Solid Fuel-Burning Appliances
NFPA	2005 NFPA 70	National Electrical Code

The code changes referenced on pp. 180–181 compare the codes in this table to the 2003 IRC & the 2003 UMC.

KEY TO CODE CHECK HVAC & EXAMPLES

Code summaries are followed by two bracketed numbers, ex:

☐ Maintain system in proper working order_____________ [1202.3] {104.4}

The left column number, in square brackets, refers to the IRC section. The right column number, in braces, refers to the UMC section.

{UPC} at the beginning of a UMC code citation refers to the Uniform Plumbing Code. Detailed UPC regulations are found in *Code Check Plumbing.*

Some sections of the book use other codes in the right column. Look at the column heading at the top of the page to verify which code is referenced.

Code summaries without check boxes are alternates or exceptions to the preceding summary as shown in the two examples below:

Code summaries ending in the word OR mean that an alternative method of compliance is found in the following line, ex:

☐ Locate above design flood elevation OR____________ [1301.1] {308.2}[1]
Design equipment to prevent water entry & resist buoyancy [324.1.5X] {308.2}[1]

In the above lines, the "X" at the end of the IRC citation means that it refers to an exception in the code. The blue color for the UMC citation denotes a code change, and the superscript number refers to the list of code changes on ***pp. 180–181.*** *A line without a check box means it is contingent on the preceding line.*

Code summaries ending in the word EXC mean that an exception follows in the next line, ex:

☐ Min [level] 30in×30in work space on control side EXC_ [1305.1][2] {305.0}
18in-deep space OK for room heaters _____________ [1305.1] {305.0X}

These lines are saying that the basic rule is for a 30in × 30in work space, with the exception of room heaters, which require only an 18in work space.

Figure numbers and table numbers in code summary lines refer to the figures and tables in the book.

n/a as a code reference means that an item is not addressed by that particular code.

Ø as a code reference means that an item is prohibited by that particular code.

Brackets within a code summary mean that a particular item applies only to the code corresponding to the shape of the brackets, ex:

☐ Service receptacle [NOT] req'd within 25ft_____ [3801.11X][11] {309.0}

The word "NOT" in square brackets here refers only to the IRC. This line is saying that the IRC does not require an electrical receptacle within 25ft of an evaporative cooler, per a new exception to 3801.11, and the UMC does require a receptacle, per section 309.0.

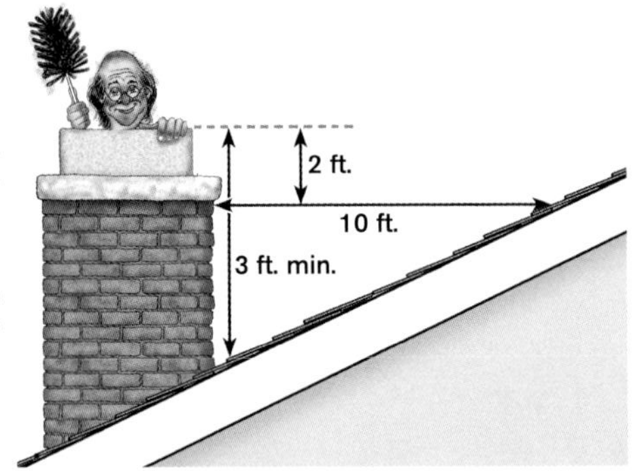

The chimney clearances shown here help ensure proper draft & provide separation from combustible materials.

Dirty chimneys are a fire hazard due to the accumulation of combustible creosote.

ABBREVIATIONS

ACH = air changes per hour
AHJ = Authority Having Jurisdiction
AMI = in accordance with manufacturer's instructions
AWG = American wire gauge
B (vent) = listed gravity gas double wall appliance flue
BO = building official
Btu = British thermal unit
BW (vent) = listed oval double wall furnace flue
cat = vented appliance category (see definition p. 179)
cfm = cubic feet per minute
CPVC = chlorinated polyvinyl chloride plastic pipe
CSST = corrugated stainless-steel tubing for fuel gas
Cu = copper
cu = cubic (ex: cu in = cubic inches)
eqpmt = equipment
ex = example, for example
exc = except
FAU = forced air unit
Fe = iron or steel pipe
ft = foot or feet (ex: 5ft)
FVIR = flame vapor ignition-resistant
gal = gallon
horiz = horizontal
hr = hour
in = inch or inches (ex: 14in)
k = 1,000
L&L = listed & labeled or listing & labeling
lb = pound, pounds
LP = liquefied petroleum (gas)
manu = manufacturer
min = minimum
max = maximum
PB = polybutylene
PEX = cross-linked polyethylene
PEX-AL-PEX = PEX aluminum PEX
PL = property line
PRV = pressure-relief valve
psi = pounds per square inch
psig = pounds per square inch gauge
req('s/'d) = require, (requires, required)
sq = square (ex: sq. in = square inches)
TPRV = combination temperature & pressure-relief valve
UL = Underwriters Labratories
vert = vertical
w/ = with
w/o = without
WH = water heater

In colonial America, most homes were warmed by building a fire in a fireplace. This method resulted in sending most of the heat up the chimney, using a lot of wood as well as causing many house fires. In 1742, Ben Franklin invented an iron furnace stove, equipped with loosely fitting iron plates through which air circulated and warmed before passing into the room. It warmed homes more efficiently, less dangerously, and with less wood–resulting in less air pollution. He named this furnace stove the "Pennsylvania Fireplace," although today it is known as the "Franklin Stove."

CODE CHECK HVAC CONTENTS

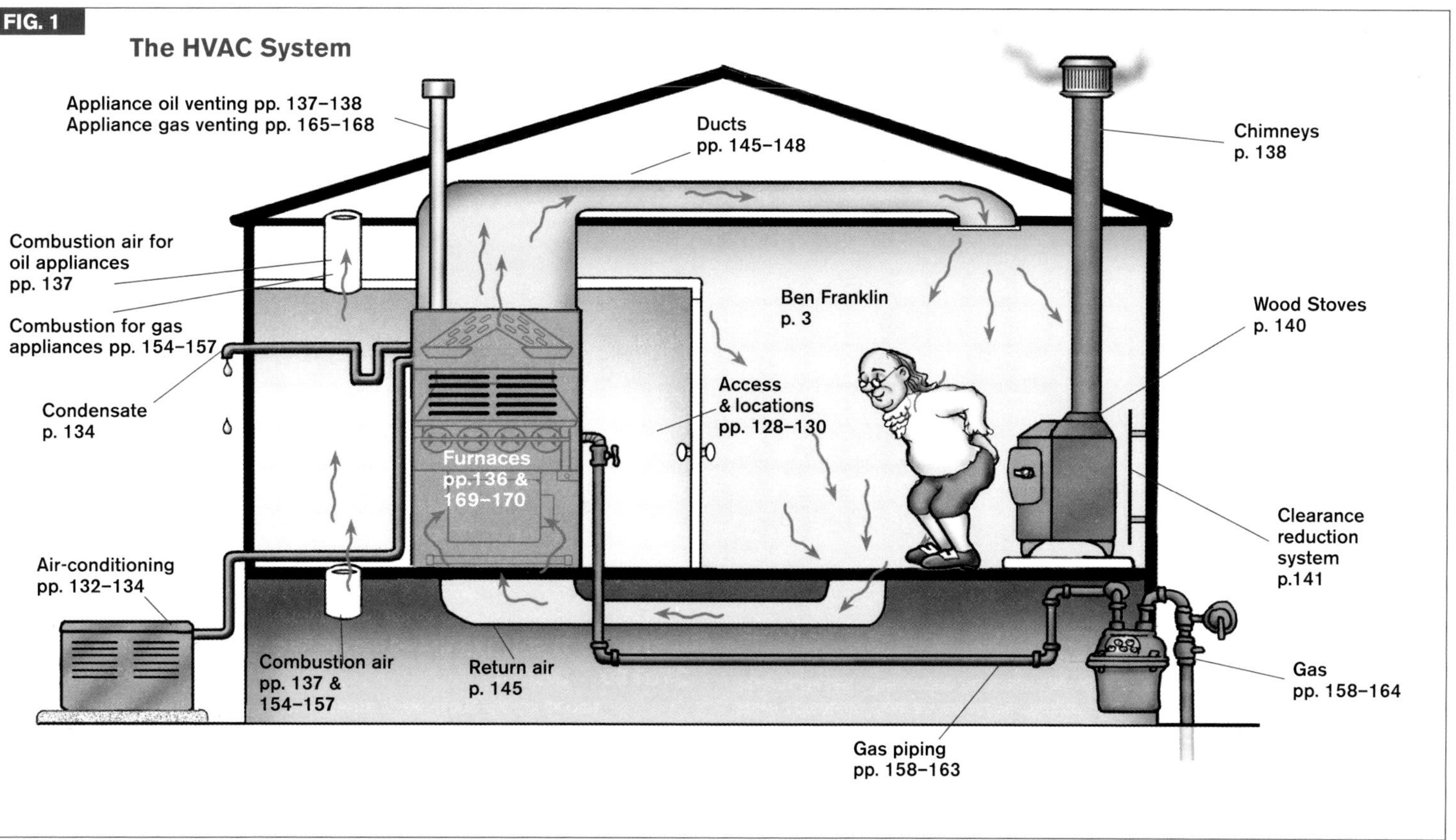
FIG. 1
The HVAC System
Appliance oil venting pp. 137–138
Appliance gas venting pp. 165–168
Ducts
pp. 145–148
Chimneys
p. 138
Combustion air for
oil appliances
pp. 137
Combustion for gas
appliances pp. 154–157
Ben Franklin
p. 3
Wood Stoves
p. 140
Access
& locations
pp. 128–130
Condensate
p. 134
Furnaces
pp.136 &
169–170
Clearance
reduction
system
p.141
Air-conditioning
pp. 132–134
Combustion air
pp. 137 &
154–157
Return air
p. 145
Gas
pp. 158–164
Gas piping
pp. 158–163

GENERAL REQUIREMENTS

The IRC and UMC have different approaches for acceptance of equipment and installations. The IRC requires all equipment to be listed and labeled, whereas the UMC accepts unlisted equipment. Check with your local building department to see if it has amendments to the administrative rules in the model codes

Administration	06 IRC	06 UMC
☐ New installations, alterations, & repairs req permits EXC	[105.1]	{112.1}
Portable equipment & minor part replacement exempt	[105.2]	{112.2}
☐ BO or AHJ may accept alternative materials & eqpmt	[104.11]	{105.0}
☐ Installations not covered by IRC must comply w/ International Mechanical Code or International Fuel Gas Code	[1301.1]	{n/a}
☐ Continued use of existing installations OK if safe & compliant w/ code at time of construction	[1202.2]	{104.2}

Listing & Labeling

☐ All appliances L&L {or approved by AHJ}	[1302.1]	{302.1}
☐ Install listed appliances AMI & per listing	[1307.1, 1401.1]	{304.1}
☐ Attach installation & operating instructions to appliance	[1307.1]	{304.1}
☐ Factory-applied nameplates must include:	[1303.1]	{307.1}

- Fuel-fired: hourly rating in Btu/hr, watts, type of fuel, req'd clearances, & maintenance actions [& listing agency]
- Electric appliances: volts, amps, phases, req'd clearances, & listing agency

Minimum Heating Requirements

☐ Installed heat source capable of min 68°F at 3ft above floor req'd for each habitable room (see glossary)	[303.8]	{n/a}
☐ Portable space heater not OK to meet above rule	[303.8]	{n/a}

Appliance Maintenance

☐ Maintain system in proper working order	[1202.3]	{104.4}
☐ BO may order reinspections to determine compliance	[1202.3]	{104.4}

APPLIANCE ACCESS & LOCATION

Appliances must remain accessible for inspection, service, repair, and replacement without the need to remove permanent construction. Appliances must be located or protected so they are not subject to flooding or damage. They must have adequate clearances for ventilation and for protection of adjacent combustible surfaces. The codes have special rules to help maintain access and working space for specific locations, such as underfloor areas and attics.

Flood Elevation	06 IRC	06 UMC
☐ Locate above design flood elevation OR	[1301.1.1]	[308.2][1]
Design to prevent water entry & resist buoyancy	[324.1.5X]	[308.2][1]
☐ No equipment mounted on breakaway flood walls	[324.1.5X]	[308.2.1][1]

General Accessibility

☐ Maintain accessibility for service {of gas appliances}	[1305.1]	{305.1}
☐ Min [level] 30in × 30in work space on control side EXC	[1305.1][2]	{305.0}
18in-deep space OK for room heaters	[1305.1]	{305.0X}

Alcoves or Closets

☐ Install w/ clearances per **F2** EXC	[1305.1.1]	{n/a}
Replacement appliances installed AMI OR	[1305.1.1X]	{n/a}
Gas appliances installed AMI	[2409.3,4]	{n/a}

See p. 169 for gas appliance clearances in rooms & alcoves

☐ Equipment room door min 24in wide **F2**	[1305.1.2]	{n/a}
☐ Equipment must fit through door **F2**	[1305.1.2]	{305.0}
☐ Min 30in deep work space w/ door open **F2**	[1305.1.2]	{305.0}

FIG. 2

IRC Central Furnace Alcove Clearances

Oil-burning furnace

6 in. front clearance of open firebox

3 in. min to sides, top, & back

Total interior space 12 in. wider than unit

24-in. min. doorway opening

Appliance must fit through door opening

30-in. min. working clearance w/ door open

Appliances in Attic

	06 IRC	06 UMC
☐ Access min rough-framed opening 22in × 30in **F3**	[807.1]	{904.11.1}[3]
☐ Appliance must fit through opening	[1305.1.3]	{904.11.1}[3]
☐ 20in × 30in clear opening OK if appliance fits	[1305.1.3]	{Ø}
☐ Solid floor min 24in wide from entrance to appliance **F3**	[1305.1.1]	{904.11.3}[3]

Appliances in Attic (cont.)

	06 IRC	06 UMC
☐ Max 20ft from access opening to appliance EXC **F3**	[1305.1.3]	{904.11.2}[3]
Passageway ≥6ft high OK for [50ft] {unlimited} length	[1305.1.3X2][4]	{904.11.2}[3]
☐ Min 30in × 30in level platform req'd for service access **F3** EXC	[1305.1.3]	{904.11.4}[3]
Platform not req if service possible from access opening	[1305.1.3X1]	{Ø}
☐ Luminaire & receptacle outlet near appliance **F3**	[1305.1.3.1]	{904.11.5}[3]
☐ Switch for luminaire at passageway entrance **F3**	[1305.1.3.1]	{904.11.5}[3]

FIG. 3 Attic Furnace

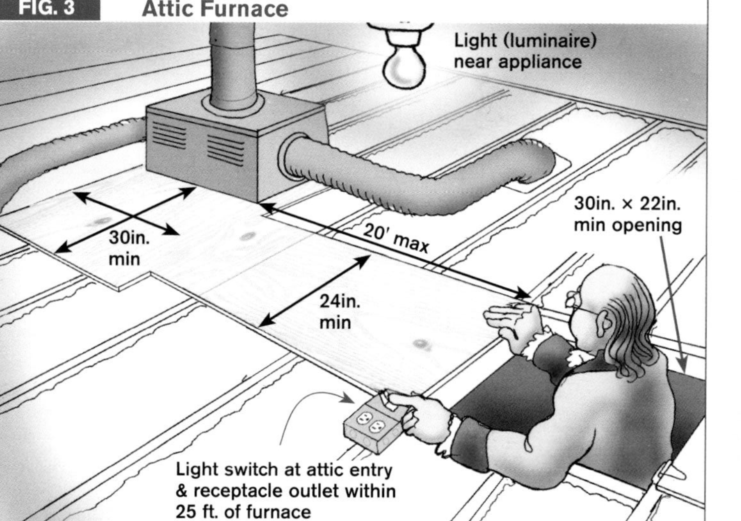

Appliances under Floors	06 IRC	06 UMC
☐ Access opening min rough-framed size 22in × 30in	[1305.1.4]	{n/a}
☐ Appliance must fit through opening	[1305.1.3]	{n/a}
☐ Passageway min 22in wide × 30in high	[1305.1.4]	{n/a}
☐ Passageway max 20ft long EXC	[1305.1.4]	{n/a}
Passageway ≥6ft high OK for unlimited length	[1305.1.4X2][5]	[n/a]
☐ Min 30in × 30in level space on service side **F4**	[1305.1.4]	{n/a}
☐ Support on concrete slab above ground OR	[1305.1.4.1]	{n/a}
Suspend from floor min 6in above ground **F4**	[1305.1.4.1]	{n/a}
☐ Excavations min 6in below appliance, 12in on sides, 30in on control side **F4**	[1305.1.4.2]	{n/a}
☐ If excavation for passageway or service space >12in below adjacent grade, line w/ concrete extending 4in above adjacent grade **F4**	[1305.1.4]	{n/a}
☐ Luminaire & receptacle outlet near appliance	[1305.1.4.3]	{n/a}
☐ Switch for luminaire at passageway entrance	[1305.1.4.3]	{n/a}

FIG. 4 Under-Floor Furnace

Heating & Cooling Appliances in Garages	06 IRC	06 UMC
☐ Ignition source min 18in above floor unless FVIR **F5**	[1307.3]	{308.1}
☐ Applies to spaces communicating w/ garage air	[1307.3]	{308.1}
☐ Protect appliances from impact **F5**	[1307.3.1]	{308.1}
☐ Ducts min 26gauge steel	[309.1.1]	{n/a}
☐ No duct openings into garage	[309.1.1]	{n/a}
☐ Openings around duct penetrations through common wall sealed w/ approved materials	[309.1.2]	{n/a}

FIG. 5 Garage Furnace

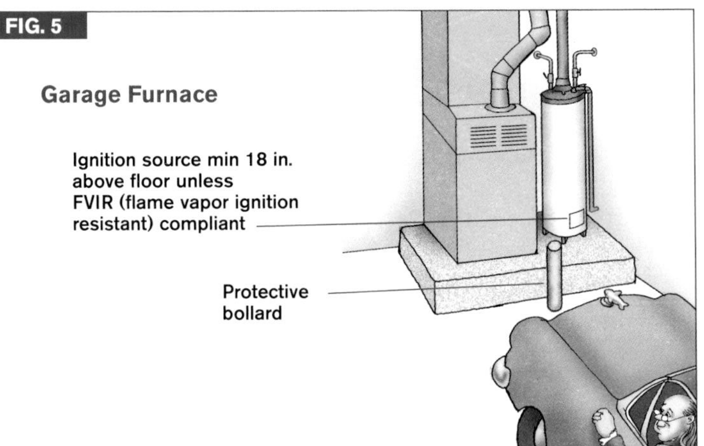

Outdoor Appliances		
☐ Outdoor equipment must be L&L for outdoors	[1401.4]	{n/a}
☐ Supports level & AMI	[1401.4]	{n/a}
☐ Equipment pad min 3in above finished grade **F9**	[1308.3][6]	{n/a}

ELECTRIC HEAT

Electric resistance heating can be in the form of central forced-air furnaces, baseboard heaters, radiant ceiling panels, duct heaters, and even exotic systems such as electric heat in ceramic tile bath floors. The circuits for electric heating must be sized to 125% of the load to ensure that they too do not become heaters.

General	06 IRC	05 NEC
☐ Circuits considered continuous load	[3602.10][7]	{424.3B}[7]
☐ Circuits for continuous loads must be sized to 125% of load	[3601.2]	{210.20A}
☐ All electric heating equipment must be L&L		

Central Electric Heat		
☐ Disconnect in sight of equipment OR	[4001.5]	{424.19A}
Breaker capable of being locked in *off* position	[4001.5]	{424.19A}

Baseboard Heaters		
☐ Must be L&L and installed AMI	[3303.3]	{424.6}
☐ Branch circuits for 2 or more units can be 15, 20, 25, or 30amps	[3602.10][7,8]	{424.3}
☐ No receptacles above heaters; integral receptacles w/ heaters can substitute for req'd receptacles in rooms **F6**	[1405.1]	{424.9}

FIG. 6

Electric Baseboard Heaters

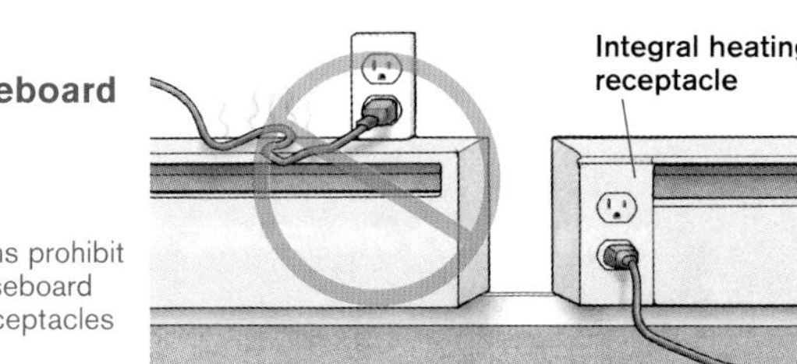

Listing instructions prohibit installation of baseboard heaters under receptacles

Electric Radiant Heat Systems	06 IRC	05 NEC
☐ Install AMI	[1406.1]	{424.93A1}
☐ Install panels parallel to framing	[1406.3]	{424.93B2}
☐ Fasteners >1/4in from heating element	[1406.3]	{424.93B3}
☐ Min 8in distance from surface-mounted fixture boxes	[n/a]	{424.93A3}
☐ Min 2in distance from recessed fixtures & trim	[n/a]	{424.93A3}
☐ No field modification of panels unless so listed	[1406.3]	{424.93B4}
☐ Wiring above heated ceiling min 2in clearance	[n/a]	{424.94}
☐ Wiring above heated ceiling considered as 50°C ambient, exc no temperature correction required if above min 2in thermal insulation	[n/a]	{424.94}

Electric Duct Heaters		
☐ Install AMI	[1407.1]	{424.66}
☐ If <4ft from heat pump/air-conditioning, both must be listed for such clearances	[1407.3]	{424.61}
☐ Must be accessible for servicing	[1407.4]	{424.66}
☐ Interlock req'd to prevent heat if fan not operating	[1407.5]	{424.63}

Heating Cables in Concrete or Masonry Floors		
☐ Min 1in spacing between cables	[n/a]	{424.44B}
☐ Leads protected where leaving floor	[n/a]	{424.44E&F}
☐ GFCI (ground-fault circuit interrupter) protection req'd for cables in bathroom floors	[n/a]	{424.44B}

FIG. 7

Heat Pump Operating in Cooling Mode

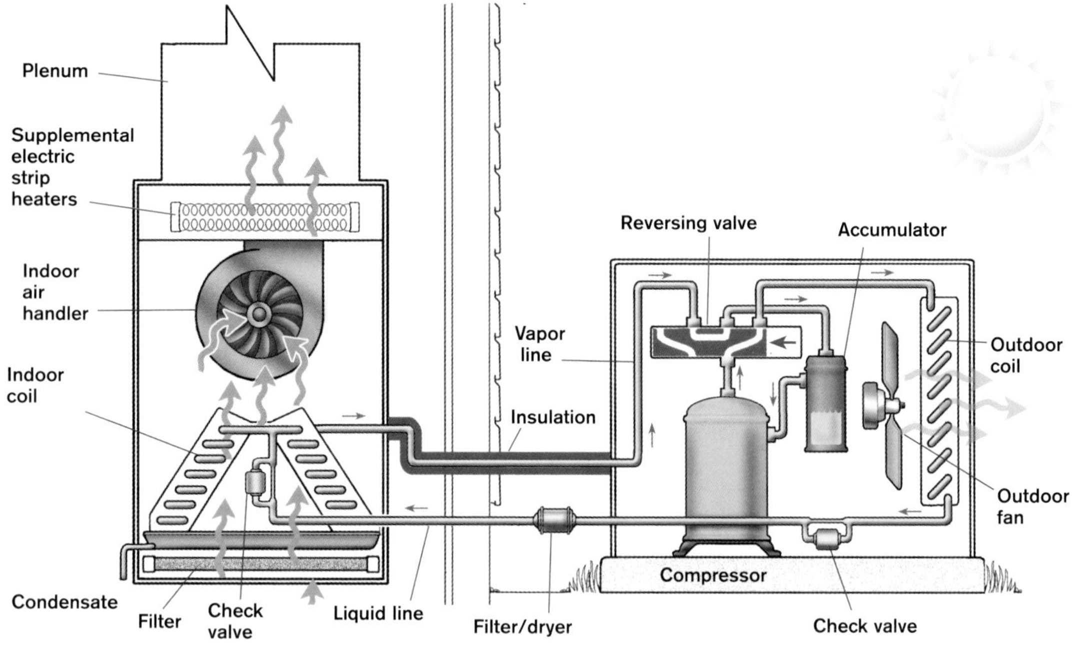

Plenum
Supplemental electric strip heaters
Indoor air handler
Indoor coil
Condensate
Filter
Check valve
Liquid line
Filter/dryer
Vapor line
Insulation
Reversing valve
Accumulator
Compressor
Check valve
Outdoor coil
Outdoor fan

FIG. 8

Reversing Valve in Heating Mode

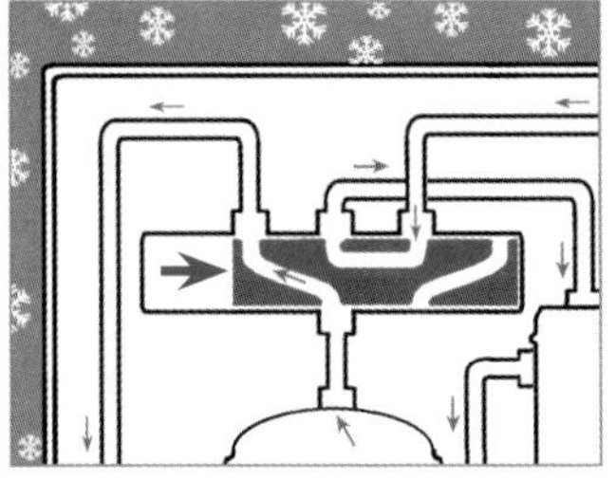

Heat pumps can be used for both heating and cooling.

A ***reversing valve*** determines the direction of refrigerant flow.

In ***heating mode,*** the outdoor coil extracts heat from the atmosphere and the indoor coil gives up that heat to the interior space.

When the outdoor temperature is below the balance point, supplementary electric strip heaters are activated in the indoor air handler.

In ***cooling mode,*** the system operates as in **F7**.

HEAT PUMPS & AIR-CONDITIONING

Cool air is heavier than warm air. Furnace blowers that also provide cooling must have 2-speed fans that overcome the resistance of the cooling coil and the weight of the cooled air. A retrofitted air conditioning coil should not be installed in an existing furnace unless it has the proper type of blower for the application.

Heat Pumps & Air-Conditioning — 06 IRC — 06 UMC

- ☐ Heat pump return air duct min 6sq.in per kBtu output **T2** ____ [1403.1] {manu}
- ☐ Outdoor unit on min 3in raised pad **F9** ____ [1308.3,1403.2][6] {1106.2}
- ☐ Furnace w/ cooling coil must have adequate pressure capacity [min 0.5in water column] or be L&L for cooling [1411.2] {904.8A&B}
- ☐ Cooling coil downstream from heat exchangers unless L&L for upstream (stainless-steel heat exchanger) ____ [1411.2] {904.8C}
- ☐ Central air-conditioner req's air filter **F7** ____ [1401.1] {312.1}
- ☐ Condenser not near clothes dryer vent ____ [manu] {manu}
- ☐ Refrigerant vapor (suction) lines insulated min R4 **F7** ____ [1411.5] {manu}
- ☐ Nail-plate protection for refrigerant piping close to [<1 1/2in] framing edge ____ [2603.2.1] {1111.3}
- ☐ Tubing secured ≤6ft of first 90° bend from compressor [manu] {1111.2}
- ☐ Tubing secured within 2ft of all other bends ____ [manu] {1111.2}
- ☐ Tubing supported at points ≤15ft apart ____ [manu] {1111.2}

TABLE 2 — HEAT PUMP RETURN DUCT MIN. SIZES [1403.1]

Btu rating	24k	30k	36k	42k	48k	60k
Tons	2	2.5	3	3.5	4	5
Blower (sq. in.)	144	180	216	252	288	360
Round duct diameter	14 in.	16 in.	18 in.	18 in.	20 in.	22 in.

Electrical Requirements — 06 IRC — 05 NEC

- ☐ Disconnect in sight of condenser **F9** ____ [4001.5] {440.14}
- ☐ Disconnect cannot be mounted on access panels ____ [T4001.5] {440.14}
- ☐ Working clearance req'd in front of disconnect **F9** ____ [3305.1] {110.26}
- ☐ Clear space min 30in wide × 36in deep **F9** ____ [3305.2] {110.26A1}
- ☐ Thermostat wire not inside power conduit **F9** ____ [4204.1] {725.55A}
- ☐ Size conductors & overcurrent per nameplate ____ [3602.11] {440.4B}

FIG. 9

Window & through-Wall Units	06 IRC	06 UMC
☐ Max cord length: 120V–10ft, 240V–6ft	[manu]	{440.64}
☐ Cord plug OK as disconnect if controls ≤6ft of floor	[manu]	{440.63}
☐ AFCI (arc fault circuit interrupter) or LCDI (leakage current detection interrupter) req'd in attachment plug	[manu]	{440.65}
☐ Max load rating 80% of individual circuit	[3602.12.1]	{440.62B}
☐ Max load rating 50% of shared circuit	[3602.12.2]	{440.62C}

Condensate

☐ Condensate may not drain to public way	[1411.3]	{310.1}
☐ Drainpipe min 3/4in w/ 1/8in/ft slope	[1411.3.2]	{310.1}
☐ Threaded female PVC fittings only on plastic male	[manu]	{310.5}
☐ May drain to indirect waste (lavatory tailpiece, tub overflow) **F10**	[1411.3]	{310.1}
☐ No direct connection to waste or vent pipe	[1411.3]	{310.1}
☐ Protection req'd if condensate stoppage would damage building components:	[1411.3.1]	{310.2}
• Secondary drain to conspicuous point of disposal **F10**	[1411.3.1]	{310.2}
• Auxiliary drain pan w/ drain to conspicuous point	[1411.3.1]	{310.2}
• Auxiliary drain pan w/ interlocked detector & no drain	[1411.3.1]	{n/a}
• Water level detection in primary w/ interlocked cutout	[1411.3.1][9]	{n/a}
☐ Down-flow units w/ no secondary & no means of installing auxiliary drain pan req internal primary drain blockage detector w/ interlocked cutout	[1411.3.1.1][10]	{n/a}
☐ No drilling (saddle fittings) of DWV (drain, waste, or vent) pipes to accept condensate drain	[3003.2]	{UPC}

FIG. 10

Condensate to Indirect Waste

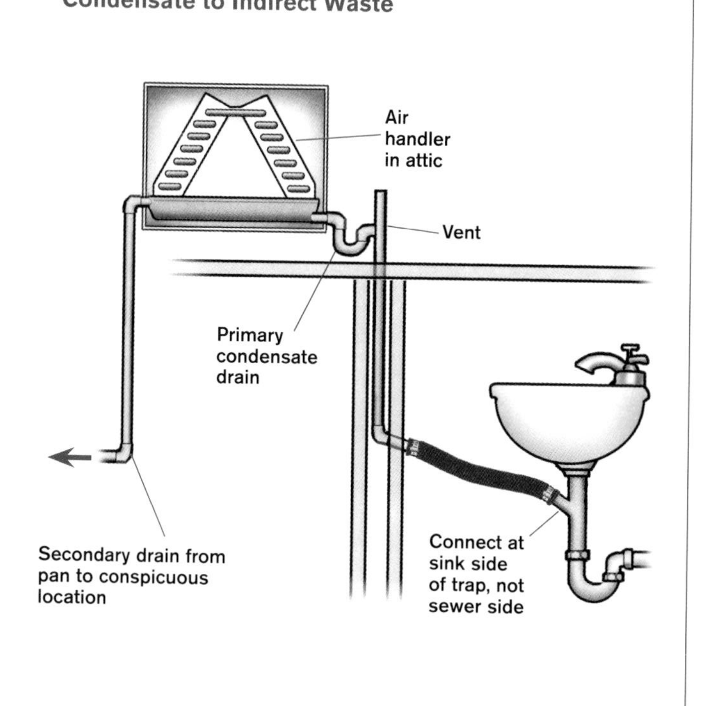

EVAPORATIVE (SWAMP) COOLERS

In dry climates, evaporative coolers can reduce the sensible temperature and provide fresh air to the building interior. Care must be taken in locating these to ensure that objectionable odors are not brought into the building.

General	06 IRC	06 UMC
☐ Install AMI	[1413.1]	{304.1}
☐ Ground-mounted units secured in place & level base min 3in above adjoining ground	[1413.1]	{405.3}
☐ Platform-mounted unit min 6in above adjoining ground	[n/a]	{405.3}
☐ Flash openings into building	[1413.1]	{405.3}
☐ *Off* switch or disconnect in sight if motor >1/8hp	[4001.5]	{309.0}
☐ Backflow protection on supply (internal air gap OK)	[1413.2]	{UPC}
☐ Min 10ft horiz clearance to plumbing or gas vents **F11**	[303.4.1]	{UPC}
☐ Service receptacle within 25ft [NOT] req'd	[3801.11X][11]	{309.0}

FIG. 11

Evaporative (Swamp) Cooler

Hot outside air is pulled through moist pads, where it is cooled by evaporation & circulated through the house or building by a large blower, leaving the air much cooler & slightly more humid than when it entered the cooler. Evaporative cooling is especially well suited for climates where the air is hot & humidity is low.

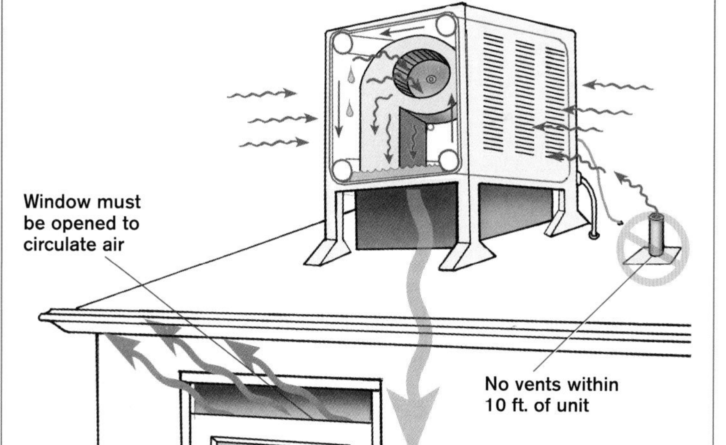

NON-CENTRAL OIL FURNACES & HEATERS

Oil-burning floor furnaces and wall furnaces are listed to UL standards. Vented room heaters can be oil fired, pellet stoves, or solid fuel. The appliances addressed in chapter 24 of the IRC are gas fired, as are those in the UMC. All appliances must be supplied with sufficient combustion air and compliant venting systems.

Floor Furnaces	06 IRC	NFPA 31
☐ Must be L&L & installed AMI	[1408.1]	{10.9.1}
☐ Floor register type min 6in from wall	[1408.3]	{10.9.4}
☐ Min. 12in from draperies or door in any position	[1408.3]	{10.9.4}
☐ Wall register type min 6in from inner corner	[1408.3]	{10.9.5}
☐ Not project into occupied {habitable} area under floor	[1408.3]	{10.9.9}
☐ [Foundation] opening to underfloor area min 18in × 24in	[1408.4]	{10.9.7.1}
☐ Trap door in floor–opening dimensions min 22in × 30in	[1408.4]	{n/a}
☐ Passageway to furnace min 22in × 30in {24in × 24in}	[1305.1.4]	{10.9.7.2}
☐ Furnace support req'd to be independent of grill	[1408.5]	{10.9.3}
☐ Min. 6in clearance from ground	[1408.5]	{10.9.6}
☐ Vent connector min 9in clear to combustibles or AMI	[1803.3.4]	{10.9.11}
Wall Furnaces		
☐ Must be L&L & installed AMI	[1409.1]	{10.13.3}
☐ Locate where fire hazard not created to walls, floors, furnishings, or doors	[1409.2]	{10.13.4}
☐ Doors cannot swing within 12in of inlet or outlet air	[1409.2]	{manu}
☐ Doorstops not OK to obtain req'd clearance	[1409.2]	{manu}
☐ Min 3ft from wall opposite the register	[manu]	{10.13.5}
☐ Panels, grills, & covers not attached to construction	[1409.4]	{10.13.6}
Vented Room Heaters		
☐ Must be L&L & installed AMI	[1410.1]	{n/a}

Vented Room Heaters (cont.)	06 IRC	NFPA 31
☐ Fasten or anchor in approved manner	[1307.2]	{n/a}
☐ Noncombustible floor min 18in beyond appliance EXC	[1410.2]	{n/a}
Listed floor protectors AMI or heater L&L for combustible floor	[1410.2X]	{n/a}

FIG. 12 Oil-Burning Appliance Vents

COMBUSTION AIR FOR OIL-BURNING APPLIANCES

When oil-burning appliances are installed in a building of unusually tight construction (see **p. 179**) an outside air source is required. In buildings of ordinary tightness, combustion air can be obtained from indoors or spaces communicating with indoors. When the appliances are in a confined space, the lower combustion air opening is considered the primary air source and the upper combustion air opening is considered the ventilation opening.

General	**06 IRC**	**NFPA 31**
☐ Provide air for space in which appliance is installed	[1701.1]	{5.2.1}
☐ Obtain air from outside building thermal envelope if building is unusually tight construction	[1701.1.1]	{5.2.3}
☐ Consider effect of exhaust fans (kitchen, bath)	[1701.2]	{5.7}
☐ Not from sleeping rooms or bathrooms EXC	[1701.4]	{n/a}
Solid fuel OK in unconfined space in ordinary construction	[1701.4X1]	{n/a}
☐ Closet w/ exterior openings & self-closing tight-fit door	[1701.4X2]	{n/a}
☐ Screen outside openings min 1/4in mesh max 1/2in	[1703.5]	{5.6.2}
☐ Consider effect of louvers:	[1701.5]	{5.6.3}
• Free area 75% for metal louvers {60–75%}		
• Free area 25% for wood louvers {20–25%}		

Indoor Air Source		
☐ Only OK for buildings of ordinary tightness	[1702.3]	{5.3.1}
☐ Infiltration sufficient for unconfined space	[1702.1]	{5.3.1}
☐ Unconfined space min 50cu. ft/kBtu/hr	[1702.1]	{3.3.59}
☐ Confined space req's openings to unconfined space	[1702.2]	{5.4.1.3}
☐ Openings min 100sq. in & min 1sq. in/kBtu/hr **F34**	[1702.2]	{5.4.1.2}
☐ Openings within 12in of top & bottom of confined space **F34**	[1702.2]	{5.4.1.2}

Outdoor Air Source	**06 IRC**	**NFPA 31**
☐ Openings to vented attic or crawl equivalent to outdoors	[1703.2]	{5.4.2.2}
☐ Ducts to upper opening level or extending upward **F36**	[1703.2]	{5.4.2.2}
☐ Direct openings or equivalent 1sq.in/4kBtu/hr	[1703.2.1]	{5.4.2.3}
☐ Vert ducts 1sq.in/4kBtu/hr **F37**	[1703.2.1]	{5.4.2.3}
☐ Horiz ducts 1sq.in/2kBtu/hr **F36**	[1703.2.1]	{5.4.2.4}
☐ Attic req's outdoor ventilation to be considered equivalent	[1703.3]	{5.4.2.2}
☐ Ducts to attic no screens, 6in above joists & insulation	[1703.3]	{n/a}
☐ Underfloor req's outdoor ventilation to be considered equivalent	[1703.4]	{5.4.2.2}

OIL-BURNING APPLIANCE VENTS

Section 801.2 of the UMC defers all requirements for oil appliance venting to NFPA 211. NFPA 31 is included by reference as part of the NFPA 211 Standard. Chapter 18 of the IRC also covers venting for pellet stoves.

General		
☐ Appliances vented to outdoors per L&L & AMI	[1801.1]	{6.2.1}
☐ Vent system shall satisfy draft requirements AMI	[1801.2]	{6.3.1}
☐ Draft regulator req'd if connected to chimney EXC	[1802.3]	{6.4.1}
Appliances listed for use W/O one	[1803.3]	{6.4.1}
☐ No manually operated dampers	[1802.2.1]	{6.4.2}
☐ Close or cap unused openings	[1801.10]	{n/a}
☐ Install L vents AMI	[1804.3]	{6.7.1.2}
☐ Draft regulator in same room as appliance **F12**	[1802.3.1]	{n/a}

Connectors		
☐ Connector rise min 1/4in per ft **F12**	[1803.3]	{6.5.10}
☐ Connectors short & straight as possible **F12**	[1803.3]	{6.5.1}
☐ Connector [min 26gauge] {steel or refractory}	[T1803.2]	{6.5.8}

Connectors (cont.)

	06 IRC	NFPA 31
☐ Provide adequate support & fastening	[1803.3]	{6.5.13}
☐ Connector diameter ≥ flue collar of appliance	[1803.3.3]	{6.5.7}
☐ Horiz length max 75% of vert vent above connector	[1803.3.2]	{6.5.1.2}
☐ Horiz type L vent max 100% of vert	[1803.3.2]	{Ø}
☐ Max horiz length 10ft unless draft fan used	[n/a]	{6.5.1.1}
☐ Connectors to common vent on same floor	[1801.11]	{6.5.24}
☐ Connector inlets offset (& smaller highest)	[1801.11]	{6.5.21}
☐ No connectors through floor or ceiling	[1803.3.1]	{6.5.2}
☐ Connectors through wall AMI or thimble	[1803.3.1]	{6.5.4.1}

Connector Clearances

☐ Single-wall connector 18in to combustibles EXC	[T1803.3.4]	{6.5.15}
Single wall 9in OK if appliance listed for L vent	[T1803.3.4]	{n/a}
☐ Type L connector 9in to combustibles EXC	[T1803.3.4]	{n/a}
3in OK {or AMI} if appliance listed for L vent	[T1803.3.4]	{n/a}
☐ Clearance reduction systems allowed **T3**	[1803.3.4]	{6.5.15

Type L Vent Terminations

☐ Min 5ft height from collar	[1804.2.3]	{n/a}
☐ Type L vent req's listed cap	[1804.2.4]	{n/a}
☐ 2ft min above roof or anything within 10ft	[1804.2.4]	{6.7.1.4}
☐ Direct vent [AMI], 9in to building openings if ≤50kBtu, 12in to openings if >50kBtu	[1804.2.5]	{6.7.3.4&5}

Chimneys

☐ No common connection w/ solid-fuel chimney	[1801.12]	{6.5.25}
☐ OK to vent gas & oil to same chimney	[2427.5.6.2]	{6.5.26}
☐ Masonry chimneys req cleanout	[1801.3.3]	{6.6.1}
☐ Masonry chimneys req liner	[1805.1]	{6.6.8}

Connecting to Existing Chimneys

	06 IRC	NFPA 31
☐ Resize as necessary to control condensation	[1801.3.1]	{6.3.1}
☐ Clean & inspect flue passageway	[1801.3.2]	{6.6.7}
☐ Recommend NFPA 211 Inspection if damage found	[1801.3.2]	{6.6.7.2}

OIL TANKS & PIPING

Oil Tanks

☐ Must be L&L	[2201.1]	{7.2.9.2}
☐ [Inside] tanks require oil-level gauge	[2201.5]	{8.8.1}
☐ Inside tanks req device to indicate fill point	[2201.5]	{8.8.3}
☐ Inside tanks installed at lowest floor of building	[n/a]	{7.5.5}
☐ Max fuel storage inside or above ground: 660gal	[2201.2]	{7.5.5}
☐ Support tanks to prevent shifting	[2201.2]	{7.3.2}
☐ Restrain against movement from earthquakes	[n/a]	{7.3.6}
☐ Inside tanks >10gal min 5ft from any fire or flame	[2201.2.1]	{7.5.8}
☐ Outside tanks min 5ft from PL {10ft if >275gal}	[2201.2.2]	{7.9.2}
☐ Outside tanks protected from weather & damage	[2201.2.2]	{7.9.6}
☐ Underground tanks & piping corrosion resistant	[2203.7]	{7.4.6}
☐ Buried tanks >1ft from house or PL & not undermining foundation **F13**	[2201.3]	{7.4.5}
☐ Buried tanks min 1ft earth cover **F13**	[2201.3]	{7.4.6}
☐ Cross-connected tanks limit 660gal total **F14**	[2203.6]	{7.7.1}
☐ Cross-connected tanks on same level **F14**	[2203.6]	{7.7.1}
☐ Swing joints or flex connectors to buried tanks **F14**	[2203.1]	{8.2.11}
☐ Install above flood line or anchor to prevent floating	[2201.6]	{7.2.10}
☐ Abandoned buried tanks to be removed or filled	[n/a]	{7.13.1}
☐ Remove exterior piping of abandoned tanks	[2201.7][12]	{7.13.1}

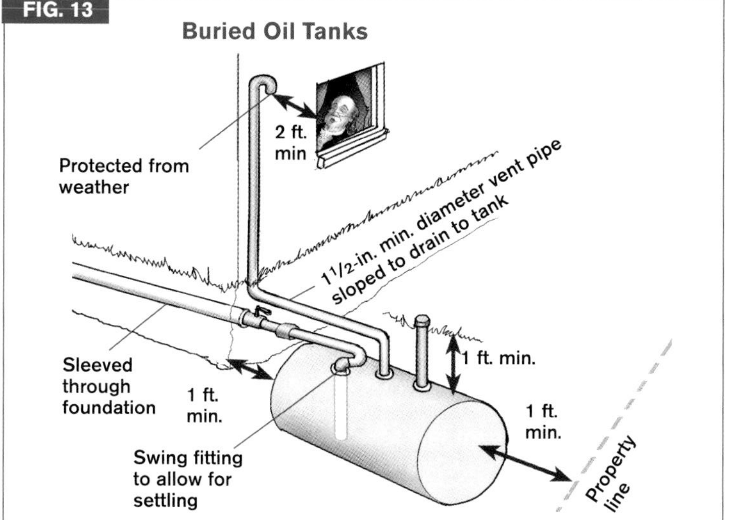

FIG. 13

FIG. 14

Indoor Oil Tanks

Vent pipe

Fill pipe

Cross-connector

Shutoff valves

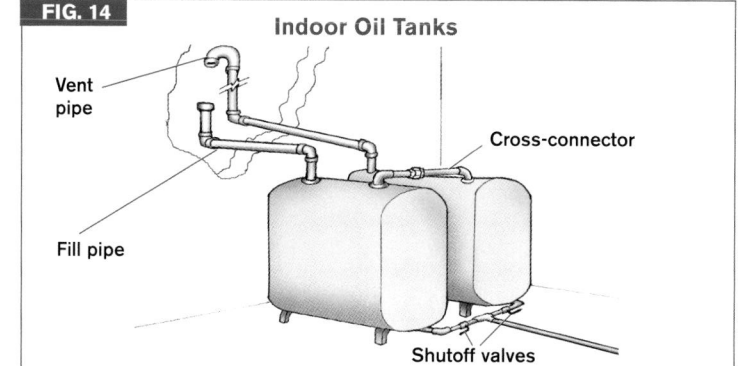

Fill & Vent Piping	06 IRC	NFPA 31
☐ Fill pipes outside & min 2ft from building openings & equipped w/ tight metal cap	[2203.3]	{8.3.2}
☐ Vent pipe min 1 1/4in diameter **F13,14**	[2203.4]	{7.5.11}
☐ Vent 1in max penetration into tank **F13,14**	[2203.4]	{8.7.4}
☐ No cross-connections of vent to other piping	[2203.4]	{8.7.12}
☐ Weatherproof vent cap (gooseneck) **F13,14**	[2203.5]	{8.7.6}
☐ Vent must slope to drain to tank, no traps **F13,14**	[2203.4]	{8.7.1&2}
☐ Vent termination min 2ft from building openings & above snow **F13,14**	[2203.5]	{8.7.5}A

Oil Appliances, General	06 IRC	NFPA 31
☐ Appliance & burners must be listed	[1302.1]	{13.2}
☐ Used appliances not OK to install in dwellings	[n/a]	{12.3}

Oil Piping	06 IRC	NFPA 31
☐ Steel pipe or tubing, {brass pipe}, or type L Cu tubing (no aluminum)	[2202.1, 2203,2]	{8.2.2}
☐ No cast-iron fittings	[2202.2]	{8.2.8.5}
☐ Piping min 3/8in outside diameter	[2203.2]	{8.2.6}
☐ Listed flex metal hose OK to reduce vibration	[2202.3]	{8.2.5}
☐ Shutoff req'd between tank & burner	[2204.2]	{8.6.4}
☐ If shutoff on discharge of oil pump, PRV shall bypass or return surplus oil	[2204.2]	{8.8.8}
☐ Pumps must be L&L and installed AMI	[2204.1]	{8.8.6}
☐ Lines w/ heaters must have PRV & return line	[2204.4]	{8.8.9}

FIREPLACE STOVES (SOLID FUEL)

IRC 1414.1 requires fireplace stoves to be listed, labeled, and tested in accordance with UL737, which in turn references the current edition of NFPA 211. Clearance from combustible material and an effective flue draft are the main issues.

Fireplace Stoves & Solid-Fuel Room Heaters IRC — NFPA 211

- ☐ Install per L&L & AMI ____ {12.1}
- ☐ Fire screen req'd per L&L ____ {12.1}
- ☐ Not in alcove or enclosed space <512cu.ft unless so listed ____ {12.2.2}
- ☐ Not OK in garage ____ {12.2.4}
- ☐ Noncombustible floor material 18in beyond stove on all sides **F15** {12.5.1.2}
- ☐ If legs provide >6in clearance under stove, 2in-thick masonry + metal **F15** ____ {12.5.1.2.1}
- ☐ If legs provide 2in–6in clearance, 4in-thick hollow masonry + metal {12.5.1.2.2}
- ☐ If legs provide <2in clearance, floor req'd to be noncombustible _ {12.5.1.2.3}
- ☐ Fuel storage min 36in from appliance ____ {T12.6.1 note a}
- ☐ Clearance to combustibles 36in or per L&L EXC ____ {12.6.1.1}
  Lesser clearances OK w/ approved clearance-reduction system **T3**, **F15,16** ____ {12.6.2.1}

Connectors

- ☐ Must be accessible for inspection, cleaning, & replacement ____ {9.7.11}
- ☐ Single wall min 18in clearance to combustibles EXC **F15** ____ {9.5.1.1}
  Lesser clearance w/ approved clearance-reduction system **T3** ____ {9.5.1.2.1}
- ☐ Not to pass through wall EXC listed pass-through system OR **F15** ____ {9.7.4}
  Provide clearance to combustibles where passing through ____ {9.7.5}
- ☐ Maintain min 1/4in/ft rise from appliance collar to chimney ____ {9.7.7}

Connection to Masonry Fireplace (Stoves & Fireplace Inserts)

- ☐ Connector must extend to flue liner–not just to firebox ____ {12.4.5.1}
- ☐ If entry to flue above smoke chamber, noncombustible seal req'd below entry ____ {12.4.5.1}

Connection to Masonry Fireplace (Stoves & Fireplace Inserts) (cont.) — NFPA 211

- ☐ No dilution of combustion products in flue w/ habitable space air _ {12.4.5.1}
- ☐ Flue not less than size of appliance collar ____ {12.4.5.1}
- ☐ Flue diameter max 2× appliance collar if chimney exposed outside below roof ____ {12.4.5.1}
- ☐ Flue diameter max 3× appliance collar if no part exposed below roof {12.4.5.1}
- ☐ Installation must allow for chimney inspection & cleaning ____ {12.4.5.1}

FIG. 15

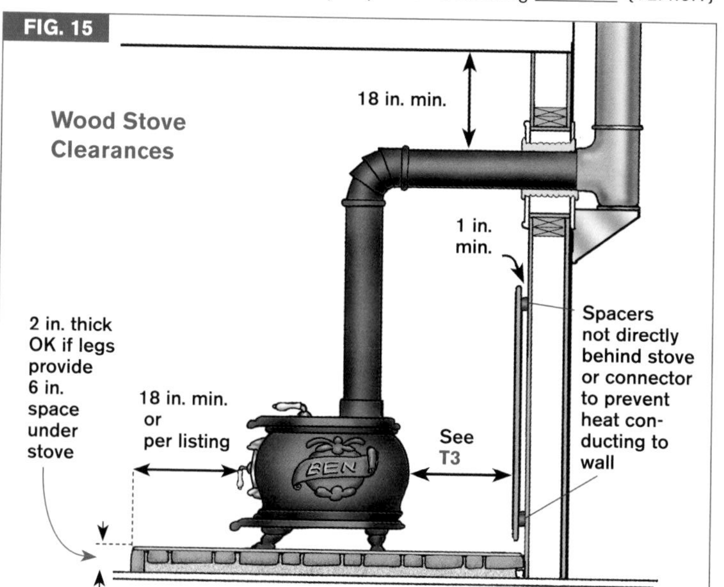

CLEARANCE-REDUCTION SYSTEMS

Clearance-reduction systems are used with solid-fuel, oil-burning, and gas-burning appliances. The information in **T3** is virtually identical in IRC chapter 13, chapter 24, and in NFPA 211. Clearance-reduction systems provide a practical means of installing appliances in spaces where they otherwise might not fit or would take too much space in a room.

General — 06 IRC — 06 UMC

- ☐ Clearance reductions allowed per **T3**[1306.2, 1803.3.4, 2409.2] {12.6.2.1}
- ☐ Solid-fuel appliances not allowed to be reduced to <12in EXC ____ [1306.2.1] {12.6.2.1.2}
  - Appliances listed for <12in & installed AMI ____ [1306.2.1] {12.6.2.1.2}
- ☐ Unlisted appliance 36in sides & rear 48in top clear OR __ [n/a] {12.6.1}
  - Lesser amount OK w/ approved clearance-reduction system [n/a] {12.6.2.1}
- ☐ No spacers directly behind appliance or connector **F15,16** ____ [F1306.2] {12.6.2.1.3}
- ☐ Use **T3** when clearance w/ no protection is 36in ____ [1306.2] {12.6.2.1.1}
- ☐ 3½in masonry w/o ventilated space max wall reduction 33% ____ [n/a] {12.6.2.1.1}
- ☐ ½in insulation board over unvented batts max wall reduction 50% ____ [n/a] {12.6.2.1.1}
- ☐ Other methods max wall reduction 66% ____ [n/a] {12.6.2.1.1}

Recommended Inspections of Existing Chimneys NFPA 211 14.3

Level 1: All readily accessible areas of chimney, structure, & flue *(annually, during routine cleaning & when replacing appliance w/ similar appliance).*

Level 2: Level 1 + video scan of flue. Verify clearances & suitability of flues *(upon resale of property; upon addition or removal of appliances, adding or replacing w/ dissimilar appliance, or after operating malfunction).*

Level 3: Level 1 & 2 + removal of components as necessary to gain access *(when Level 1 or Level 2 cannot identify conditions deemed critical to renewed or continued use; fire or damage investigations).*

FIG. 16

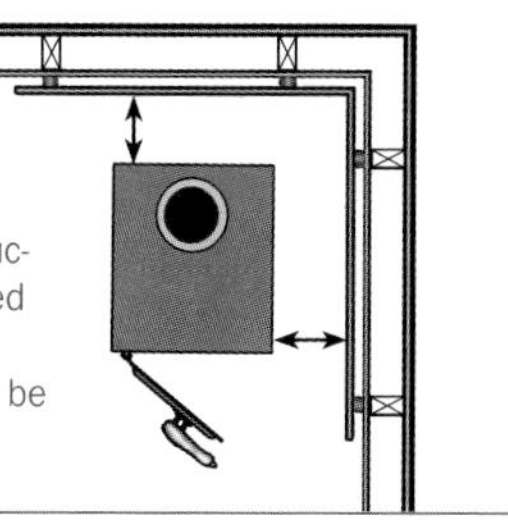

Clearance-Reduction System

The spacers that hold out the clearance-reduction system from the wall must not be located directly behind the appliance or connector. The appliance's distance from the wall must be 12 in. min or in accordance w/ **T3**

TABLE 3 — ALLOWABLE REDUCED CLEARANCES[A] [T1306.2]

Protection Method	Normal 18 in. Clearance Req'd.		Normal 36 in. Clearance Req'd.	
	Wall	Ceiling	Wall	Ceiling
3½-in.-thick masonry w/o ventilated air space	12 in.	n/a	24 in.	n/a
½ in. insulation board over 1-in. glass fiber or mineral wool batts	9 in.	12 in.	18 in.	24 in.
0.024 sheet metal over 1-in. glass fiber or mineral wool over wire & ventilated air space	6 in.	9 in.	12 in.	18 in.
3½-in.-thick masonry w/ ventilated air space	6 in.	n/a	12 in.	n/a
0.024 sheet metal w/ ventilated air space	6 in.	9 in.	12 in.	18 in.
½-in.-thick insulation board w/ ventilated air space	6 in.	9 in.	12 in.	18 in.
1-in. glass fiber or mineral wool batts sandwiched between two sheets 0.024 sheet metal w/ ventilated air space	6 in.	9 in.	12 in.	18 in.

A. Also based on NFPA-211 T9.5.1.2 & T12.6.2.1.

EXHAUST SYSTEMS

Exhaust systems protect indoor air from contamination. They can create a negative pressure that adversely affects the operation of fuel-burning appliances, and this effect must be taken into account in the design of combustion air supplies.

Bathrooms	06 IRC	06 UMC
☐ Openable window min 1.5sq.ft OR	[303.3]	{n/a}
Mechanical 50cfm intermittent or 20cfm continuous	[303.3X]	{n/a}
☐ Air must be exhausted directly to the outside **F17**	[303.3X]	{504.1}
☐ May not discharge to crawl space or attic **F17**	[1506.2][13]	{504.1}
☐ Backdraft damper required **F17**	[n/a]	{504.1}
☐ Outdoor openings screened $^1/_4$in–$^1/_2$in mesh	[303.5]	{n/a}

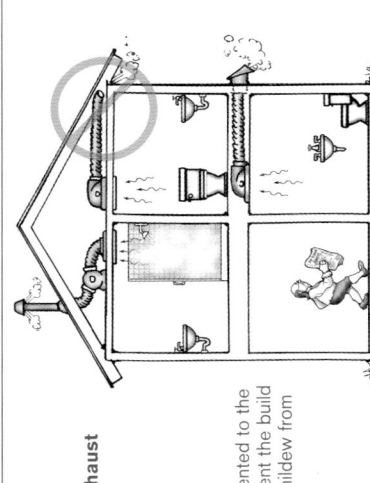

FIG. 17

Bathroom Exhaust Venting

Exhaust air is vented to the outside to prevent the build up of mold or mildew from condensation.

Sauna Heater		
☐ Protect from accidental contact	[1902.1]	{manu}
☐ Install AMI	[1902.2]	{304.1}
☐ Thermostat to limit room temperature to 194°F	[1902.4]	{manu}

Electric Clothes Dryer Ducts	06 IRC	06 UMC
☐ Must be independent of all other systems	[1502.1]	{504.3.1}
☐ Duct diameter per dryer's listing {4in diameter}	[1502.3]	{504.3.2}
☐ Closet req's min 100sq.in makeup air opening in door	[n/a]	{504.3.2}
☐ Must terminate on outside of building **F18**	[1502.2]	{504.3.1}
☐ Must have backdraft damper & no screens **F18**	[1502.2]	{504.3.1}
☐ Must terminate ≥3ft from other building openings	[1502.2][14]	{504.5}
☐ Flex transition ducts not concealed in construction	[1502.4]	{504.3.2.1X}
☐ Transition ducts must be L&L & max 8ft {6ft}	[1502.4]	{504.3.2.1X}
☐ Duct construction [min 0.016in-thick rigid] metal	[1502.5]	{504.3.2.1}
☐ Duct smooth interior, joints in direction of air flow	[1502.5]	{504.3.2.1}
☐ No sheet metal screws or fasteners in interior **F18**	[1502.5]	{504.3.2.1}
☐ [Max length 25ft minus 2$^1/_2$ft per each 45° of bend] **F18**	[1502.6]	{n/a}
☐ {Max length 14ft minus 2ft each 90° turn >2} **F18**	[n/a]	{504.3.2.2}
☐ If make & model of dryer known, longer lengths AMI	[1502.6X1][15]	{504.3.2.2}
☐ Longer lengths OK w/ large radius bends per ASHRAE* calculations	[1502.6X2][16]	{n/a}
☐ Ductless (condensing) dryers OK per L&L	[1502.1X]	{n/a}

The American Society of Heating, Refrigerating and Air-Conditioning Engineers

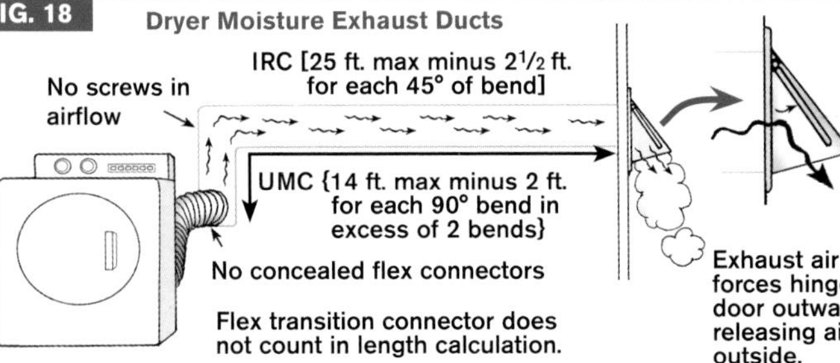

No screws in airflow
IRC [25 ft. max minus 2$^1/_2$ ft. for each 45° of bend]
UMC {14 ft. max minus 2 ft. for each 90° bend in excess of 2 bends}
No concealed flex connectors
Flex transition connector does not count in length calculation.
Exhaust air forces hinged door outward, releasing air outside.

FIG. 18 Dryer Moisture Exhaust Ducts

Gas Clothes Dryers & Ducts

Gas Clothes Dryers & Ducts	06 IRC	06 UMC
☐ Must not connect to vent or chimney	[2439.3]	{905.4A}
☐ Vent rough-in req'd for any dryer space	[2439.5.2]	{905.2A}
☐ Closet install req's [100sq.in] makeup air opening	[2439.4]	{905.3A}
☐ Must terminate on outside of building	[2439.3]	{905.2A}
☐ Must have backdraft damper & no screens	[2439.3]	{504.3.1}
☐ Transition ducts must be metal & L&L max 8ft	[2439.5]	{905.4C}
☐ Flex transition ducts not concealed in construction	[2439.5]	{905.4C}
☐ Duct smooth interior, joints in direction of air flow	[2439.5]	{905.4B&C}
☐ No sheet-metal screws or fasteners in duct interior	[2439.3]	{905.4B&C}
☐ Min 4in diameter duct	[2493.5]	{n/a}
☐ [Max length 25ft minus $2^{1}/_{2}$ft per each 45° of bend]	[2439.5.1]	{manu}
☐ If make & model of dryer known, longer lengths AMI	[2439.5.1X]	{manu}

KITCHEN EXHAUST

Kitchen exhaust systems convey grease as well as moisture and must be constructed of material with smooth inner walls to prevent a buildup of material. With the exception of open-top broilers, kitchen exhaust is an optional item.

Connections for Gas Appliances

Connections for Gas Appliances	06 IRC	06 UMC
☐ Gas shutoff valve & union req'd ahead of connector	[2420.5]	{1312.4}
☐ Range gas connector max 6ft	[2422.1.2.1]	{1312.4}
☐ Flex gas connector req'd for moveable appls	[2422.1.3]	{n/a}

Freestanding Ranges

Freestanding Ranges	06 IRC	06 UMC
☐ Must be listed as household type	[2447.3]	{n/a}
☐ Vert clearance to combustibles 30in **F19** OR	[1901.1]	{916.1B}
Per L&L & AMI **F19** OR	[1901.1]	{916.1B}
24in w/ millboard & metal or w/ metal hood	[n/a]	{916.1B}
☐ Side clearance to combustibles AMI OR	[1901.2]	{916.1A}
6in min side & rear for unlisted appliances	[Ø]	{916.1A}

Built-In Ranges

Built-In Ranges	06 IRC	06 UMC
☐ Install AMI	[1901.2, 2447.1]	{916.2A&C}
☐ Vert clearance to combustibles 30in **F19** OR	[1901.1]	{916.2B}
Per L&L & AMI **F19** OR	[1901.1]	{916.2B}
24in w/ millboard & metal or w/ metal hood	[n/a]	{916.2B}
☐ Must be level	[n/a]	{916.2D}

Open-Top Broilers

Open-Top Broilers	06 IRC	06 UMC
☐ Install AMI	[1505.1]	{920.0}
☐ Metal exhaust hood min $^{1}/_{4}$in clear to combustibles	[1505.1]	{920.0}
☐ Hood at least as wide as broiler unit	[1505.1]	{920.0}
☐ Must go to outdoors & have backdraft damper	[1505.1]	{920.0}
☐ Min 24in from cooking surface & combustibles	[1505.1]	{920.0}

Range Hood

Range Hood	06 IRC	06 UMC
☐ Terminate outside {min 3ft from PL} EXC	[1503.1]	{504.5}
Ductless range hoods AMI	[1503.1X]	{304.1}
☐ Duct req's backdraft damper	[1503.1]	{504.1}
☐ Exterior openings screened $^{1}/_{4}$in–$^{1}/_{2}$in mesh	[303.5]	{n/a}
☐ Ducted hoods min 100cfm intermittent or 25cfm continuous	[1503.3]	{n/a}
☐ Ducts smooth interior & metal EXC	[1503.2]	{504.2}
PVC OK for downdraft under slab, max 1in of PVC coupling above slab	[1503.2X]	{504.2X}

Microwave over Range

Microwave over Range	06 IRC	06 UMC
☐ Must be L&L for install over range **F19**	[1504.1]	{304.1}
☐ Install AMI	[1504.1]	{304.1}

FIG. 19 Range Clearances

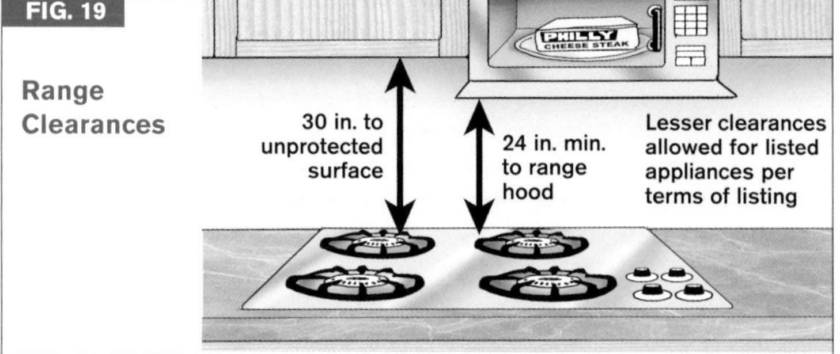

OTHER VENTILATING APPLIANCES

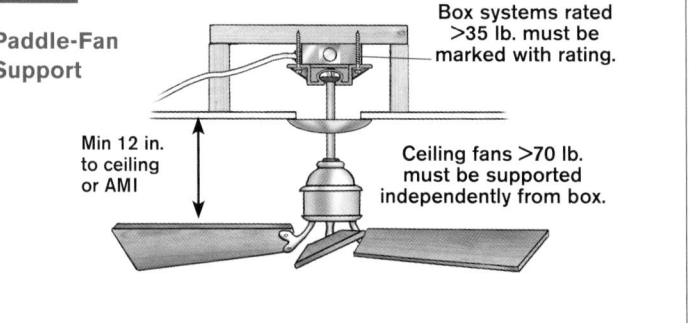

Ventilation appliances for household comfort must be installed in accordance with the terms of their listing, and the effect of vibration must be considered in their supports.

Whole-House Fan	**06 IRC**	**06 UMC**
☐ Consider effect of fan on combustion air **F20**	[1701.2, 2407.4]	{701.1.4}
☐ Lockable breaker or disconnect within sight	[T4001.5]	{309.0}
Heat Recovery Ventilator		
☐ Install per listing & AMI	[1307.1]	{504.4.1}
☐ Exhaust must terminate outdoors	[1501.1][13]	{504.1}
Kiln (Noncommercial)		
☐ Install AMI	[n/a]	{930.2}
☐ Base 2in thick & 12in beyond base of kiln	[n/a]	{930.5.1}
☐ Clearance min 18in to noncombustibles,		
☐ 36in to combustibles or per listing	[n/a]	{930.5.1}
☐ Control side clearance 30in min	[n/a]	{930.5.1}
☐ Provide makeup air equal volume to exhaust	[n/a]	{930.5.4}
☐ Ducted hood req'd	[n/a]	{930.5.2}

FIG. 20 Whole-House Fan

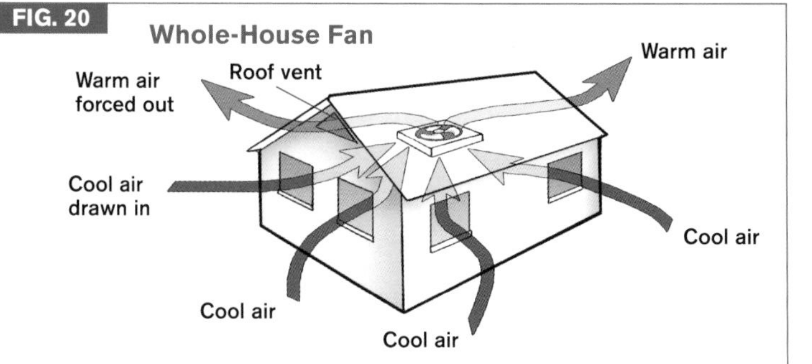

Paddle Fans	**06 IRC**	**05 NEC**
☐ Paddle fan boxes & systems listed for such use	[4001.6]	{422.18}
☐ Listed boxes & box systems max weight 70lb **F21**	[3805.9][17]	{314.27D}[17]
☐ Max weight marked on boxes rated >35lb **F21**	[3805.9][17]	{314.27D}[17]

FIG. 21 Paddle-Fan Support

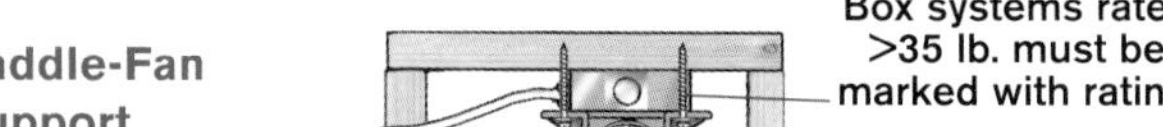

DUCTS

Ducts for distribution of conditioned air must be properly installed to prevent air leakage and resist condensation. Manufactured ducts must meet standards for their flame-spread and smoke-developed index. The Air Conditioning Contractors of America (ACCA) publishes a design manual that is used for determining optimum size of ducts. Their website is www.acca.org.

General

	06 IRC	06 UMC
☐ Design per ACCA Manual D or other approved method	[1601.1]	{601.2}
☐ Vibration isolators max 10in length ______	[1601.2.2]	{602.7}
☐ Metal duct min thickness per **T4** ______	[1601.1.1]	{602.1}

TABLE 4 GAUGE OF METAL DUCTS [1601.1.1(2)] {T6-9}

Type & Exposure	Size (in.)	Min. Thickness (in.)	Equivalent Galvanized Sheet Gauge	Approximate Aluminum B&S Gauge
Round or enclosed rectangular	≤14	0.013	30	26
	>14	0.016	28	24
Exposed rectangular	≤14	0.016	28	24
	>14	0.019	26	22

Use of Building Cavities as Ducts

☐ Gypsum products only OK in ducts if limited to 125°F	[1601.1.1]	{602.1}
☐ Gypsum [only OK for return] {OK for supply or return}	[1601.1.1]	{602.1}
☐ Gypsum not OK if subject to condensation ______	[1601.1.1]	{602.1}
☐ Stud & joist cavities only OK for return ______	[1601.1.1]	{602.2}
☐ Cavities not OK in fire-resistance rated assembly __	[1601.1.1]	{n/a}
☐ Stud wall cavities not OK from one floor level to next	[1601.1.1]	{n/a}
☐ Cavities must be fireblocked from adjacent spaces _	[1601.1.1]	{n/a}

Ducts between House & Garage

	06 IRC	06 UMC
☐ No duct openings in garage ______	[309.1.1]	{n/a}
☐ Min 26gauge sheet steel in garage ______	[309.1.1]	{n/a}
☐ Seal openings around duct penetration in wall/ceiling	[309.1.2]	{n/a}
☐ Protective barrier if exposed to vehicle damage ____	[1307.3.1]	{604.4}

Return Air Sources

☐ Return air permitted to be diluted w/ outdoor air ____	[1602.1]	{n/a}
☐ Outdoor air inlet screen req'd min 1/4in, max 1/2in mesh ______	[1602.3]	{n/a}
☐ Individual room returns OK if ≤ supply to room OR _	[1602.2X]	{n/a}
Must be open to area ≥25% of area served ______	[1602.2]	{n/a}
☐ Adjoining rooms included in return area if connected by permanent air transfer openings ______	[1602.2]	{n/a}
☐ Not OK to draw from closet, bathroom, kitchen, garage, mechanical room, or other dwelling unit______	[1602.2]	{n/a}
☐ Must not adversely affect source of combustion air ___	[1701.4]	{904.7A}
☐ If sole source is room w/ fuel-burning appliance: __	[1602.2X2]	{n/a}

- Return source min 10ft from firebox or draft hood **F22**
- Room size min 100cu.ft/1kBtu of appliances in room
- Supply to room must be at least same amount as return

☐ Sole source min 10ft from fireplace firebox opening	[1602.2X3]	{n/a}
☐ Gas furnace min 2sq.in/kBtu output rating or AMI **T5** _	[2442.2]	{manu}

TABLE 5 GAS FURNACE MINIMUM TOTAL AREA OF SUPPLY & RETURN DUCTS [2442.2]

Furnace Btus:	50	60	70	80	100	125	150
Min. sq. in.	100	120	140	160	200	250	300
Round trunk duct	12 in.	14 in.	14 in.	16 in.	16 in.	18 in.	20 in.

FIG. 22

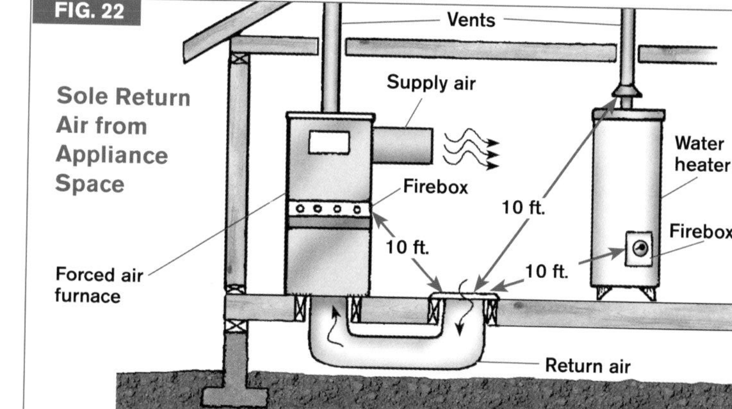

Underfloor-Floor Supply Plenums

	06 IRC	06 UMC
☐ No fuel gas lines or plumbing cleanouts in plenum	[1601.4]	{608.13}
☐ Limited to 2 stories	[n/a]	{608.1}
☐ Min 4mil moisture barrier	[1601.4.1]	{608.12}
☐ Remove loose combustible materials & tightly enclose	[1601.4.1]	{608.2}
☐ Materials min flame-spread rating 200 (1× lumber)	[1601.4.2]	{608.3}
☐ Duct from furnace min 6in below combustible framing	[1601.4.3]	{608.11}
☐ Receptacle req'd max 18in below floor openings	[1601.4.3]	{608.7.1}
☐ Area of receptacle min 3in beyond opening all sides	[1601.4.3]	{608.7.2}
☐ Perimeter of receptacles min 1in vert lip	[1601.4.3]	{608.7.3}
☐ Floor access req'd min 18in × 24 in {24in × 24in}	[1601.4.4]	{608.4}
☐ Furnace controls must start fan at ≤150°F	[1601.4.5]	{608.5}
☐ Furnace high limit on outlet max 200°F	[1601.4.5]	{608.6}

Installation–General

	06 IRC	06 UMC
☐ 18in ground clearance req'd for portion of duct that would obstruct crawl space access	[n/a]	{604.1}[18]
☐ Joints must be airtight	[1601.3.1]	{602.4}
☐ Pressure-sensitive tape must be marked UL 181 **F23**	[1601.3.1]	{602.4}
☐ Round metal duct crimp joints lap min $1\frac{1}{2}$in	[1601.3.1]	{602.4}
☐ Round metal duct crimp joints min 3 screws	[1601.3.1]	{602.4}
☐ Metal duct max support spacing max 10ft	[1601.3.2]	{T6-7}
☐ Connections to flanges mechanically fastened **F23**	[1601.3.1]	{manu}

FIG. 23 Duct Splices

Step 1.
Peel jacket & insulation from core & butt cores together over collar.

Metal collar, min 4in. wide

Band clamps

Step 2.
Apply approved tape & secure w/ band clamps.

UL181 tape

Step 3.
Pull jacket & insulation back together & apply two wraps of tape.

Factory-Made Ducts

	06 IRC	06 UMC
☐ Factory-made ducts L&L & installed AMI **F23–26**	[1601.2]	{602.3}
☐ Factory-made ducts Class O or Class 1 (no Class 2)	[1601.1.1]	{604.3}
☐ 2-story max vert riser on factory-made duct	[n/a]	{604.3}
☐ Ground clearance min 4in	[1601.3.6]	{604.3}
☐ [Not OK] {OK} as liner inside pipe or masonry	[1601.3.5]	{604.3}
☐ Avoid installing flex duct where exposed to sunlight	[n/a]	{602.3}
☐ Flex duct max support spacing AMI {4ft} **F24**	[1601.3.2]	{T6-10}
☐ Flex duct support straps min $1\frac{1}{2}$in wide **F24**	[1601.3.2]	{T6-10}
☐ Mechanical fasteners for flex ducts marked "181B-C"	[1601.3.1][19]	{602.4}

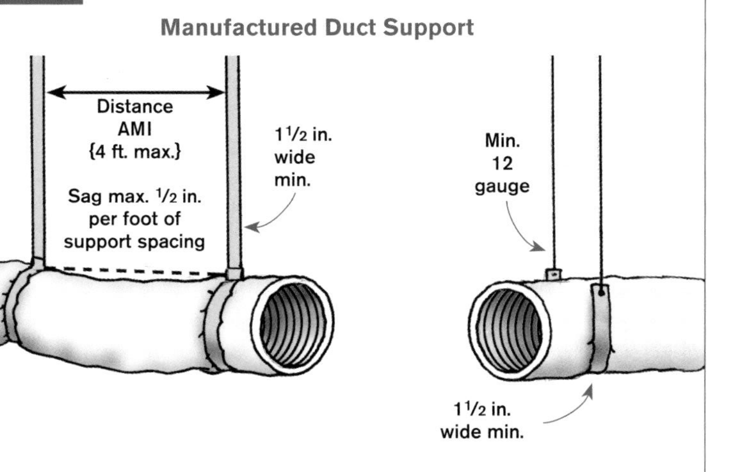

FIG. 24 Manufactured Duct Support

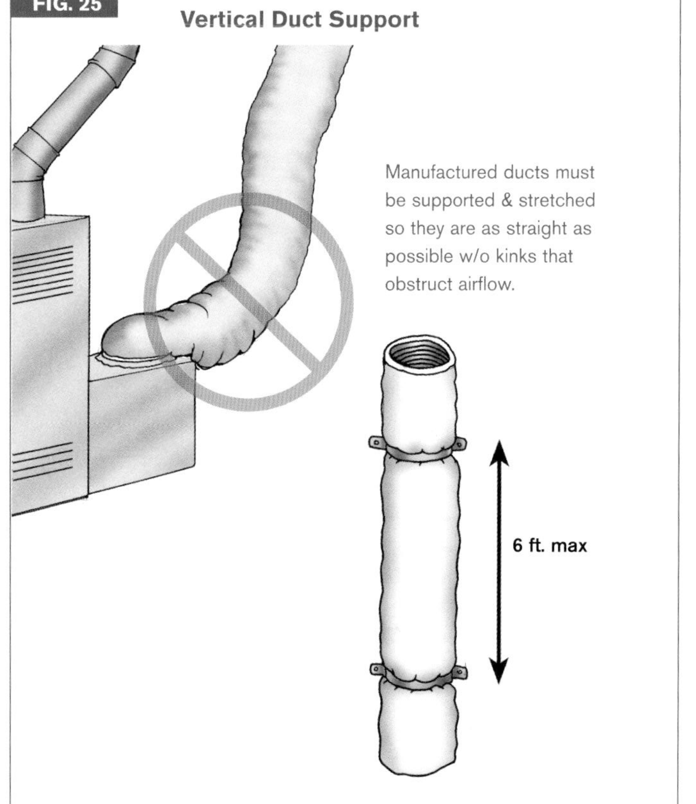

Manufactured ducts must be supported & stretched so they are as straight as possible w/o kinks that obstruct airflow.
6 ft. max

FIG. 25 Vertical Duct Support

FIG. 26 Stretch Manufactured Ducts

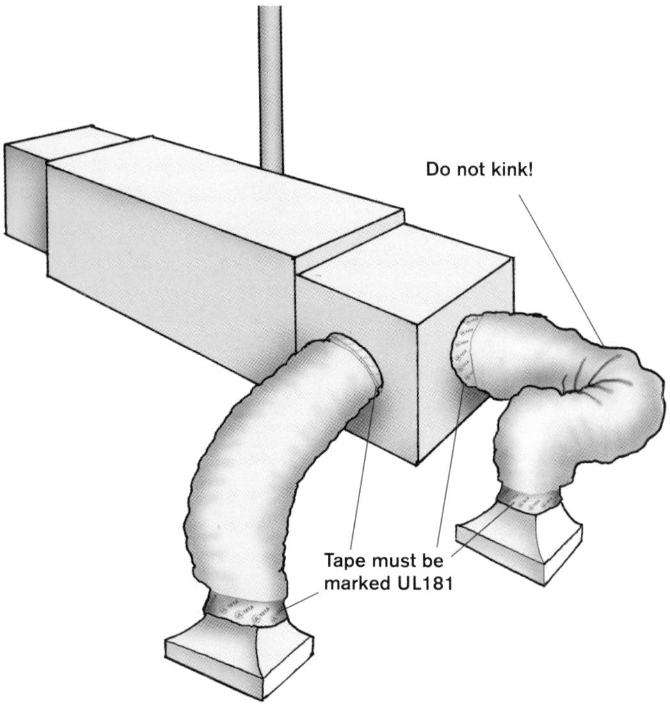

Do not kink!

Tape must be marked UL181

Underground Ducts

	06 IRC	06 UMC
☐ Metal duct concrete encased min 2in thick OR	[1601.1.2]	{604.2}
Protected from corrosion in approved manner & AMI	[1601.1.2]	{n/a}
☐ No flex duct on or underground	[1601.3.5]	{604.3}
☐ Slope to accessible drainage point	[1601.1.2]	{n/a}

Insulation

(The UMC insulation requirements are based on table 6-6 and the heating-degree-day number and the cooling-degree-day number. 1-in fiberglass insulation approximately = R3)

	06 IRC	06 UMC
☐ Coverings, linings, & adhesives max flame-spread index 25, max smoke-developed index 50	[1601.2.1][20]	{605.0}[20]
☐ Vapor retarder req'd on cooling ducts in spaces conducive to condensation	[1601.3.4]	{T6-6}
☐ Min R8 for supply & return, R6 in floor trusses EXC	[1103.2.1]	{T6-6}
Not req'd inside building thermal envelope	[1103.2.1][21]	{605.0XB}
☐ Runs ≤10ft need not exceed R3.5	[n/a]	{605.0X}
☐ Crawl–typical min R3.5	[n/a]	{T6-6}
☐ R-value must be labeled on factory-made ducts	[1601.2.1]	{605.0}

FIG. 27 Loop-System Hydronic

Water shutoff
Backflow prevention device
Pressure regulator
Barometric damper
Tridicator (gauge)
Expansion tank
Electronic zone valves
Out to radiators
Insulation
Pressure-relief valve & drain
Return from radiators
Oil burner
Oil supply
Circulator pump (can be on supply or feed)
Zone purge valves

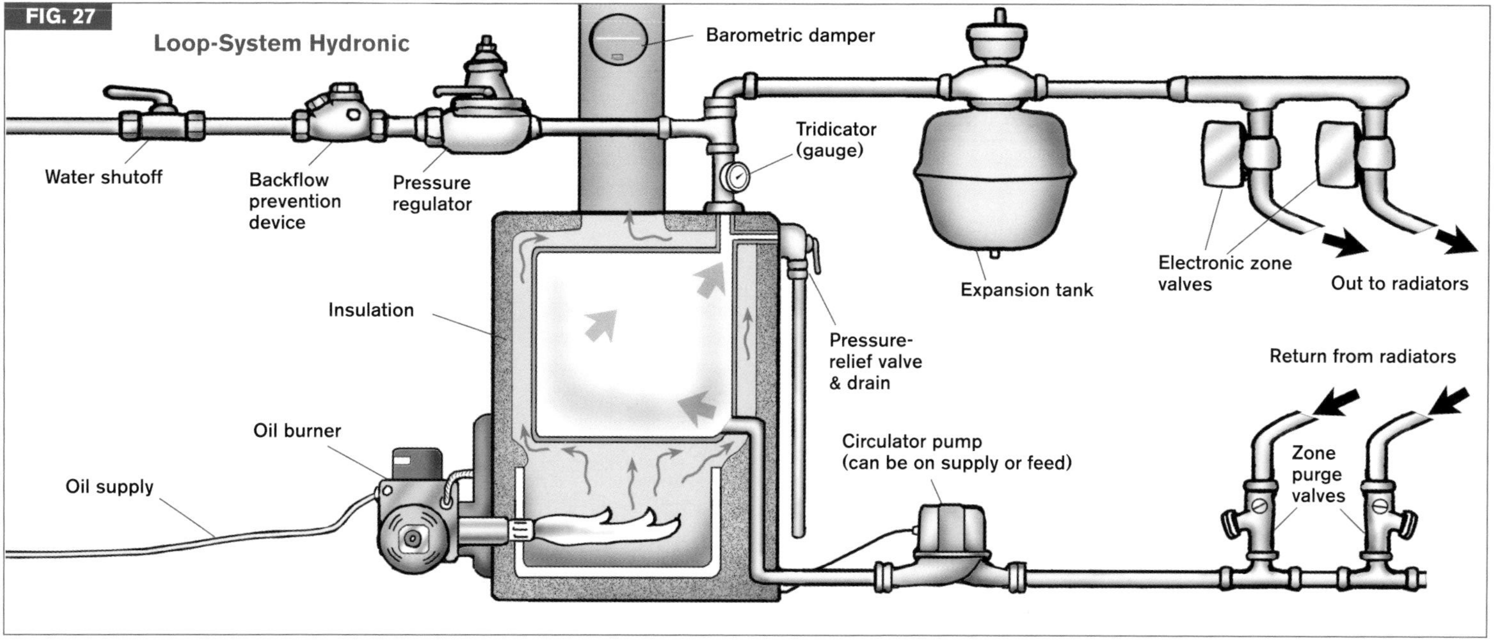

BOILERS

Steam and hot-water heating systems have some advantages compared to forced-air heating systems. There is not the concern over airborne particles of dust, pollen, or mold; and the heat energy is stored in the thermal mass of the water and piping, creating a system that is sometimes more efficient. Hot-water systems can be "zoned" more readily than forced air, providing greater climate control.

Steam & Hot-Water Boilers	**06 IRC**	**06 UMC**
☐ Install AMI	[2001.1]	{304.1}
☐ Installer req'd to supply permanent control diagram & operating instructions	[2001.1]	{1021.0}
☐ Pressure regulator req'd on water feed line **F27,29**	[manu]	{1005.1}
☐ Backflow preventer req'd on potable water feed line	[2902.5.1]	{UPC 603.0}
☐ PRV req'd **F27**	[2002.4]	{1011.2}

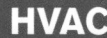

Steam & Hot-Water Boilers (cont.)

	06 IRC	06 UMC
☐ PRV set at max rating of boiler	[2002.4]	{1011.2}
☐ PRV req's drain ≤18in from floor or to open receptor **F27**	[2002.4]	{1007.0}
☐ Low-water cutoff req'd {gas boiler only}	[2002.5]	{904.5}

Hot-Water Boilers

☐ Temperature & pressure gauge req'd **F27**	[2002.2]	{1005.3}
☐ Expansion tank req'd **F27,30**	[2003.1]	{1006.1}
☐ Open tanks req support 2× weight of water in tank to prevent strain on piping	[2003.1]	{1006.2}
☐ Open tanks req vent & overflow drain	[2003.1]	{1006.2}
☐ Pressurized tanks approximately 10% system volume	[2003.1.1]	{1006.4}

Steam Boilers & Piping

☐ Water-gauge glass & pressure gauge req'd	[2002.3]	{1005.3}
☐ Steam piping min 1in clearance to structure EXC	[n/a]	{1201.2.7.8.4}
Reduction to 1/2in OK at penetrations	[n/a]	{T3-1 note6}

FIG. 28 Backflow Preventer

FIG. 29 Pressure Regulator

HYDRONIC PIPING

Hydronic piping systems can distribute heat to radiators or baseboard convectors. The most common use of these systems is in radiant heat panels embedded in floors, where they create a quiet and efficient heating system. Since these panels are embedded in the structure, it is important they be properly installed and tested before covering.

General

	06 IRC	06 UMC
☐ Provide means of system drain down	[2101.2]	{1201.2.7.8.8}
☐ Maintain backflow protection to potable water	[2101.3]	{UPC 603.0}
☐ No contact w/ material causing corrosion or damage	[2101.5]	{1201.2.7.8.5}
☐ Drilling, notching, & protection per building code	[2101.6]	{1201.2.7.4}
☐ Provide for expansion & contraction	[2101.8]	{1201.2.7.5}
☐ Tee fittings–supply side NOT on branch opening	[2101.7]	{n/a}

Exposed Piping

☐ Support piping to avoid strain **T6**	[2101.9]	{1201.2.6}
☐ Pressure test min 100psi water for 15 {30} minutes	[2101.10]	{1201.2.8.3}
☐ Wrap/sleeve pipes through concrete walls or floors	[2603.3]	{1201.2.7.8.1}
☐ Insulate materials over 140°F to protect occupants	[n/a]	{1201.3.5.5.2}

TABLE 6 HYDRONIC PIPING SUPPORT [2101.9]

Material	Max. Horiz. Spacing	Max. Vert. Spacing
CPVC ≤1 in.	3 ft.	5 ft.
CPVC ≤1 1/4 in.	4 ft.	10 ft.
Cu alloy pipe	12 ft.	10 ft.
Cu tubing	6 ft.	10
Steel pipe	12 ft.	15 ft.
Plastic tubing	32 in.	4 ft.

FIG. 30 Expansion Tanks

Expanded water

As water temperature increases & pressure rises, expanded water pushes against the diaphragm.

Rubber diaphragm

Precharged air

Expanded water reaching max. temperature

Air cushion

As water temperature reaches its max., the pressure compresses the air & prevents excessive pressure from building up in the pipes.

Embedded Piping (Radiant Heat)

	06 IRC	06 UMC
☐ Materials–steel pipe, Cu tubing {type L}, PEX-AL-PEX, PB, PEX, [CPVC, or polypropylene]	[2103.1][22]	{1204.1}
☐ Plastic pipe rated min 100psi at 180°F	[2103.1]	{1204.1}
☐ Pressure test 100psi water for 30 minutes	[2103.3]	{1207.0}
☐ Maintain operating pressure on pipe when placing concrete	[n/a]	{1203.2}

HEAT EXCHANGERS

General

☐ Double wall req'd if toxic transfer fluid	[2902.5.2]	{603.4.4.1}[23]
☐ Space between double walls open to atmosphere	[2902.5.2]	{603.4.4.1}
☐ Single wall OK if essentially nontoxic	[2902.5.2]	{603.4.4.1}[23]

Solar

☐ Backflow protection req'd	[2301.5]	{603.0}
☐ Reduced-pressure principle backflow preventer req'd on solar if chemicals used in water	[2902.5.5]	{T6-2}
☐ Backflow prevention devices accessible for servicing	[2902.6]	{603.3.4}

POOL & SPA HEATERS

General

	06 IRC	06 USP*
☐ Must be L&L & installed AMI	[2006.1,2441.1]	{406.1}
☐ Maintain marked clearances	[2006.2]	{405.1.1}
☐ Pressure-limiting device req'd	[n/a][24]	{405.2}
☐ Temperature-relief valve req'd	[2006.3][24]	{405.3}
☐ Bypass valves or integral bypass system req'd	[2006.4]	{408.1}
☐ Piping between heater & check valve metal or AMI	[2006.1]	{408.3}
☐ Flanges or unions in piping within 12in of heater	[2006.1]	{408.5}

2006 Uniform Swimming Pool, Spa, & Hot Tub Code by IAPMO

SOLAR

General

	06 IRC	06 USEC*
☐ Collectors [& thermal storage] must be L&L	[2301.3.1&2]	{701.8}
☐ Roof must be designed for weight of collectors	[2301.2.2]	{305.0}
☐ Roof-mounted collector shall not reduce fire-resistance rating of roof {OK for 1- & 2-family dwellings}	[2301.2.2]	{704.2.1}
☐ PRV [& temperature-relief valves] req'd	[2301.2.3]	{408.2}
☐ No valves can isolate sections from TPRV	[2301.2.3]	{408.2}
☐ Protect system from freezing	[2301.2.5]	{313.5}
☐ Closed loops (heat exchangers) req expansion tank	[2301.2.6]	{602.3}
☐ Valves req'd to allow system isolation from solar loop	[2301.2.8]	{407.2}
☐ Water temperature inside dwelling max 180°F	[2301.2.9]	{n/a}
☐ Tempering mixer req'd to limit potable water temperature	[manu]	{320.4}
☐ Backflow protection req'd	[2301.5]	{403.1}
☐ All roof/wall penetrations flashed & watertight	[2301.2.7]	{313.6}

2006 Uniform Solar Energy Code, by IAPMO

PROPANE STORAGE & DISTRIBUTION

IRC section 2412 and UMC section 1313.0 both defer all rules for LP gas storage to NFPA 58, the Liquefied Petroleum Gas Code, published by the National Fire Protection Association.

LP gas has different characteristics from natural gas. Storage containers are pressurized, and the gas vaporizes upon release into an appliance. LP gas is heavier than air. When an appliance is located in a pit or basement, gas leakage could result in an invisible pool of combustible material awaiting an ignition source, such as an appliance automatically starting. One way to provide safety against such hazards is to have an interlock that shuts off the gas flow if leakage occurs. Another method is to install drains to the outside to prevent LP from accumulating, as in F32.

Tanks — NFPA 58

- ☐ Tank clearances **F31** ____ {Annex I}
- ☐ Not indoors or building is considered hazardous location ____ {T6.20.2.2}
- ☐ Replace containers w/ excessive dents or corrosion ____ {5.2.1.4}

Piping Systems

- ☐ Piping suitable for natural gas also OK for LP ____ {5.8.1.2}
- ☐ Permitted pipe/tubing: black, galvanized, red brass, Cu type L or K, & CSST ____ {5.8.3.1}
- ☐ Cu tubing to appliance per **T7** ____ {15.1}
- ☐ Buried metal pipe min 12in cover ____ {6.8.3.12}[25]
- ☐ Grounding & bonding not req'd ____ {6.20.1.3}
- ☐ Pit or trench under valves or regulators considered hazardous location in terms of allowable electrical equipment ____ {T6.20.2.2}

Underground Plastic Piping

- ☐ Plastic (polyethylene or polyamide) only OK underground ____ {6.8.4.1}
- ☐ Underground plastic min cover 12in; 18in if external damage likely ____ {6.8.4.2}
- ☐ Assembled anodeless riser req'd for transition to above ground ____ {6.8.4.3}
- ☐ Horiz portion of riser min 12in below grade ____ {6.8.4.3}
- ☐ 14AWG tracer wire req'd, one end brought above ground ____ {6.8.4.6}

FIG. 31

Propane Tank Clearances

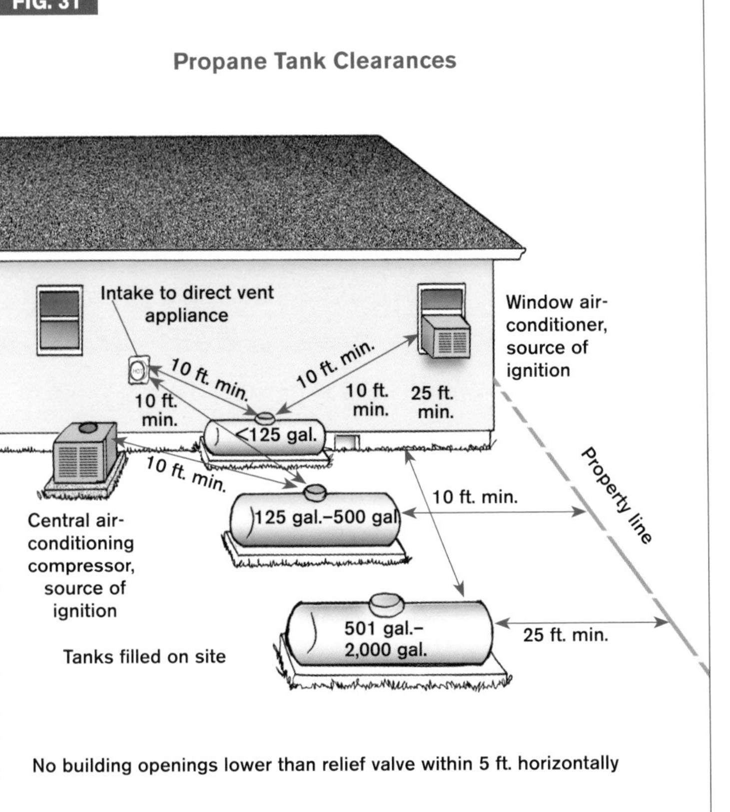

No building openings lower than relief valve within 5 ft. horizontally

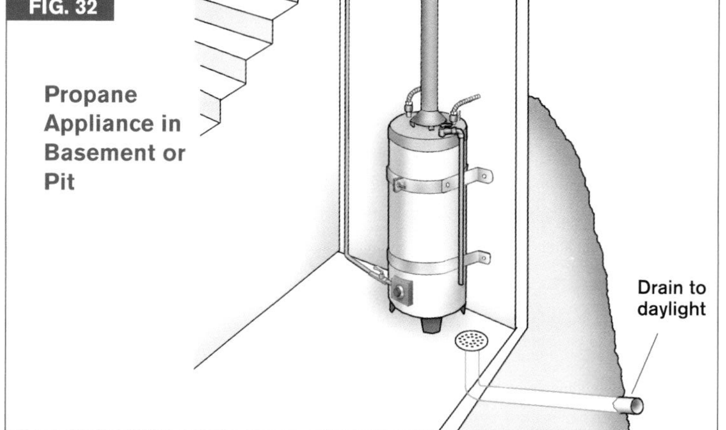

FIG. 32 Propane Appliance in Basement or Pit

- A great amount of energy is available in a propane tank, and care is needed in its handling and use.
- Propane in its natural state is odorless. An odorant, such as ethyl mercaptyn, is added to alert occupants to the presence of propane leaks. However, the odorant may lose its effectiveness, particularly if there is rust inside the tank.
- Propane is explosive at lower concentrations (2%–3% of air) than natural gas.
- Propane has a heating value of about 2,500Btu/cu.ft. (compared to approximately 1,050Btu/cu.ft. for natural gas).
- Propane gas valves should be protected from freezing in cold areas.
- Stationary tanks are built to ASME (American Society of Mechanical Engineers) standards. They are steel. They must have a legible data plate. The container capacities are expressed in maximum amounts of water gallons or pounds.
- Portable tanks are built to DOT (Department of Transportation) standards and may be aluminum or steel. Their date of manufacture is stamped into the body or neck ring. Container capacities are expressed in the amount of propane in pounds.

TABLE 7 COPPER TUBING SIZES FOR PROPANE APPLIANCES {T15.1(J)}

Capacities in kBtu between Regulator & Appliance[A]					
Tubing Length (ft.)	Outside Diameter Copper Tubing, Type L				
	3/8 in.	1/2 in.	5/8 in.	3/4 in.	7/8 in.
10	49	110	206	348	536
20	34	76	141	239	368
30	27	61	114	192	296
40	23	52	97	164	253
50	20	46	86	146	224
60	19	42	78	132	203
80	16	36	67	113	174
100	14	32	59	100	154
125	12	28	52	89	137
150	11	26	48	80	124
200	10	22	41	69	106
250	9	19	36	61	94
300	8	18	33	55	85

A. Based on an 11 in. setting and a 0.5 in. water column drop.

COMBUSTION AIR FOR GAS-BURNING APPLIANCES

Note: IRC Chapter 17 is for oil-burning and solid-fuel appliances & Chapter 24 is for gas-burning appliances.

General	06 IRC	06 UMC
☐ Req'd for natural-draft appliance EXC	[2407.1]	{701.1.1}
Direct-vent appliance AMI &	[2407.1]	{701.1AX1}
Clothes dryers w/ makeup air	[2407.1X]	{701.1.1X2}
☐ Appliances other than Cat. I provide air AMI	[2407.1]	{701.1.2}
☐ Draft hood in same space as appliance	[2407.3]	{701.1.3}
☐ Consider effect of exhaust fans (kitchen, bath)	[2407.4]	{701.1.4}

Indoor Combustion Air

☐ Indoor air source alone only OK if ACH >0.40/hr	[2407.5]	{701.2}
☐ Min volume of appliance space 50cu.ft/kBtu/hr **F33**	[2407.5.1]	{701.2.1}
☐ When air infiltration rate is known, min volumes:	[2407.5.2]	{701.2.1.1}
• Non-fan-assisted appliance = (21cu.ft/ACH)kBtu		
• Fan-assisted appliance = (15cu.ft/ACH)kBtu		
☐ Equations may not be used for infiltration rates >0.60	[2407.5.2]	{701.2.2}
☐ Non-fan-assisted min = 35cu.ft/kBtu	[n/a]	{701.2.2}
☐ Fan-assisted min = 25cu.ft/kBtu	[n/a]	{701.2.2}
☐ Indoor air volume includes rooms directly communicating w/ appliance space **F34**	[2407.5]	{701.2}
☐ Openings connecting indoor spaces req'd to be located in upper & lower 12in of appliance space **F34**	[2407.5]	{701.2}
☐ Openings connecting indoor spaces on same level min 100sq.in each min 1sq.in/kBtu; if on different stories min 2sq.in/kBtu **T8**	[2407.5.3]	{701.3.1}

FIG. 33

All Air from Indoors

Space w/ >.040 ACH sufficient if volume ≥50cu. ft./kBtu.

FIG. 34

Closet Installation with All Air from Indoors

Space w/ volume <50cu. ft./kBtu OK if 2 permanent openings to a space w/ sufficient volume & 0.40 ACH.

Outdoor Combustion Air	06 IRC	06 UMC
☐ Min dimension of openings 3in (7in area **T9**)	[2407.6]	{701.4}
☐ 2 opening method: top & bottom 12in **F41**	[2407.6.1]	{701.4.1}
☐ 2 openings direct to exterior: 1sq.in/4,000Btu/hr **F35**	[2407.6.1]	{701.4.1}
☐ 2 openings to vert ducts: 1sq.in/4,000Btu/hr **F37,38**	[2407.6.1]	{701.4.1}
☐ 2 openings to horiz ducts: 1sq.in/2,000Btu/hr **F36**	[2407.6.1]	{701.4.1}

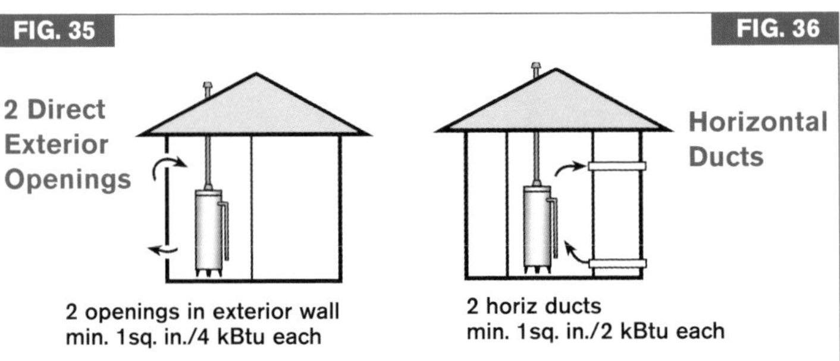

FIG. 37

Vertical Ducts to Outdoors

Two vert ducts
min 1sq. in./4k Btu each

Air moves readily through vert ducts, which are allowed to be the same size as direct exterior openings. These ducts can go outdoors or to a space that is equivalent to outdoors, such as a properly ventillated attic. Ducts that terminate in an attic must be higher than the attic insulation & should not be screened or they could become blocked by insulation.

FIG. 38

Vertical Ducts to Attic

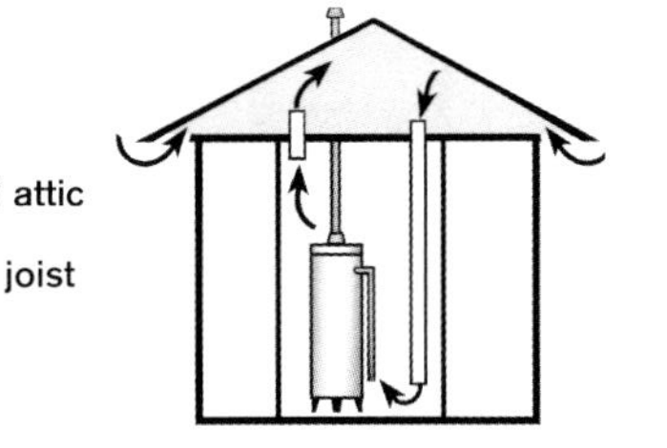

Two openings in ventilated attic
min 1 sq. in./4k Btu each
& sleeved min. 6 in. above joist

Single-Opening Method

	06 IRC	06 UMC
☐ Single opening OK at upper 12in of enclosure, direct or through ducts, min 1sq.in/3kBtu & ≥ cross-sectional area of vent connectors **F39** ______	[2407.6.2]	{701.4.2}
☐ Appliance clearance min 1in sides & back, 6in front _	[2407.6.2]	{701.4.2}

FIG. 39

Single-Opening Method

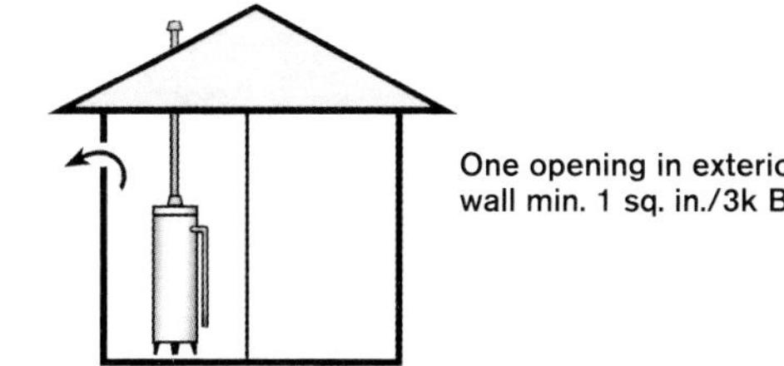

One opening in exterior
wall min. 1 sq. in./3k Btu

A single opening can be an effective combustion air source. Since it is an alternative to direct exterior openings or vert ducts, it req's a larger size than the individual openings using those other methods. A single opening is sometimes the most practical solution in a remodeling project. For example, when an unfinished basement is converted to a habitable space, the appliance space might become enclosed & req. a separate source of combustion air.

Combined Use of Indoor & Outdoor Air	06 IRC	06 UMC
☐ Indoor openings same location & amounts as **F40**	[2407.7.1]	{701.5.1}
☐ Outdoor either 2-opening or single-opening method	[2407.7.2]	[701.5.2}
☐ Outdoor opening size must compensate for insufficient indoor air volume per formula in **F40**	[2407.7]	{701.5.3}

FIG. 40

Combined Indoor & Outdoor Air

Size of exterior opening (**X**) determined by:
A = Actual volume = (Y + Z)
k = Thousands of Btus in appliances
R = Required volume (cu. ft.) = 50 × k
Min. opening size = same as for outdoor air source × [1 minus (A/R)]

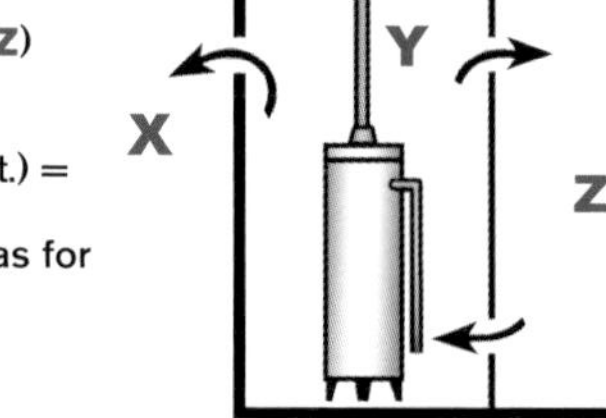

Example: Appliance Btus = 120,000. If using outside air source only, opening (X) must be min 40 sq. in. (120,000 ÷ 3,000)
If Y = 1,000 & Z = 2,000, then A = 3,000. R = 6,000 (50 × 120)
$X = 40 \times [1 - (A/R)] = 40 \times [1 - 1/2] = 40 \times 1/2 = 20$ sq. in.

Mechanical Combustion Air Supply	06 IRC	06 UMC
☐ Min. supplied rate from outdoors 0.35 cu.ft/minute	[2407.9]	{701.7}
☐ If exhaust also supplied, makeup air must compensate	[2407.9.1]	{701.8.1}
☐ Appliance interlock req'd to burners	[2407.9.2]	{701.8.2}
☐ If mechanical source is venting system for building, combustion air demands are in addition to ventilation requirements	[2407.9.3]	{701.8.3}

Openings & Ducts		
☐ Use known net clear opening area of louvers & grilles	[2407.10]	{701.9A}
☐ Assume 25% clear if wood, 75% if metal	[2407.10]	{701.9A}
☐ Screen mesh size min 1/4in	[2407.10]	{701.9B}[26]
☐ Duct min cross-sectional dimension 3in	[2407.6]	{701.4}
☐ Ducts galvanized steel or equivalent corrosion resistance, strength and rigidity EXC	[2407.11][27]	{701.10}[27]
☐ Joist/stud space OK as combustion air duct if no more than 1 req'd fireblock removed	[2407.11X]	{701.10X}
☐ Ducts must end in unobstructed space	[2407.11]	{701.10}
☐ Ducts may serve only one enclosure	[2407.11]	{701.10}
☐ Single duct cannot serve both upper & lower openings; maintain duct separation to source of combustion air	[2407.11]	{701.10}
☐ No screens on duct ending in attic **F38,40**	[2407.11]	{701.10}
☐ Horiz ducts to upper part of enclosure may not slope down to source (upper opening not OK from below) **F42**	[2407.11]	{701.10}
☐ Space around chimney liner not OK to supply air	[2407.11]	{701.10}
☐ Exterior openings min 12in above adjoining grade	[2407.11]	{701.10}
☐ No dampers in combustion air ducts	[n/a]	{701.11}
☐ No corrosive fumes at source of combustion air	[2407.12]	{n/a}

FIG. 41

Crawl Space & Attic Openings

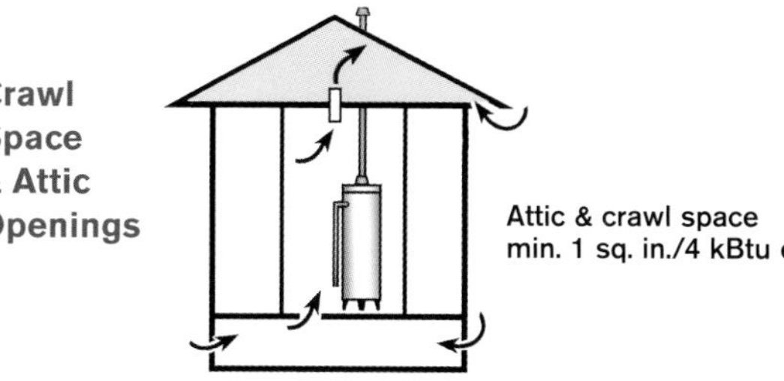

The depressurizing effect of flue gases passing through the vent pipe helps draw combustion air through the openings. The upper opening also provides ventilation for the appliance space, & it provides a secondary outlet in the event of a blocked flue. Because it relies on convection, ducts from this upper opening must be horiz or rise upward; lighter gases will not travel downward through the duct in **F42.**

FIG. 42

Crawl Space Cannot Be the Only Source

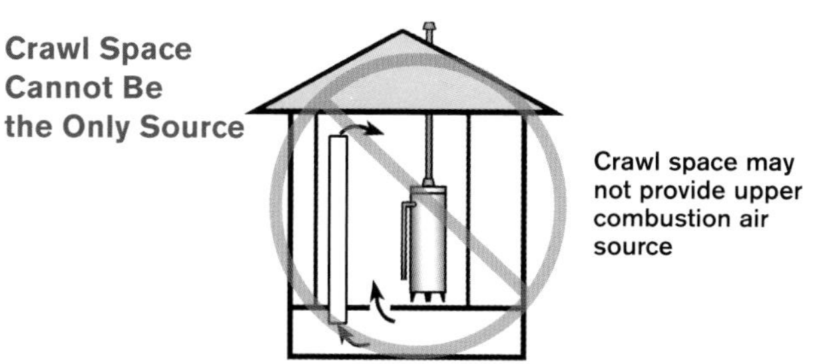

TABLE 8 COMBUSTION AIR-OPENING SIZES

	Inside Air[A]		Outside Air Openings	
Btu	Opening size[B]	Cu. ft. min. (sq. ft.[C])	1 in./4 kBtu/hr.	1 in./2 kBtu/hr.
30k	100 sq. in.	1,500 (188)	15 sq. in.	7.5 sq. in.
40k	100 sq. in.	2,000 (250)	20 sq. in.	10 sq. in.
50k	100 sq. in.	2,500 (313)	25 sq. in.	12.5 sq. in.
60k	100 sq. in.	3,000 (375)	30 sq. in.	15 sq. in.
80k	100 sq. in.	4,000 (500)	40 sq. in.	20 sq. in.
100k	100 sq. in.	5,000 (625)	50 sq. in.	25 sq. in.
125k	125 sq. in.	6,250 (781)	62.5 sq. in.	31.3 sq. in.
150k	150 sq. in.	7,500 (938)	75 sq. in.	37.5 sq. in.

A. For construction w/ known air infiltration rate >0.40/hr.
B. Req'd opening between confined space (<50 cu. ft. per kBtus) & unconfined space.
C. Square footage for an 8 ft. ceiling.

TABLE 9 AREA OF A CIRCLE

Diameter (in.)	Area (sq. in.)	Diameter (in.)	Area (sq. in.)
3	7	9	63.6
4	12.6	10	78.5
5	20	12	113
6	28	14	154
7	38.5	16	201
8	50.3		

NATURAL GAS PIPING

The required size of gas piping depends on the appliance demand, gas pressure, Btus per cubic foot of gas, and the length of the run. Modern corrugated stainless-steel tubing (CSST) systems gas can be supplied to a central manifold with a regulator, and low pressure lines will run from the manifold to each appliance. Fuel gas piping must be protected from damage and leaks and must be tested before being put into use.

Electrical — 06 IRC — 06 UMC

- ☐ Gas pipe not OK as grounding electrode in earth ____ [2410.1] {1311.13B}
- ☐ Above-ground gas piping bonded to create fault return path ____ [2411.1] {1311.13A}
- ☐ Piping considered bonded when connected to appliance w/ equipment grounding conductor ____ [2411.1] {1311.13A}

General

- ☐ LP gas storage and piping per NFPA 58 (**p. 152**) ____ [2412.2] {1313.0}
- ☐ Pipe (exc Fe) must be marked "Gas" in black letters on yellow label 5ft. on center–exc for pipes in same room as equipment ____ [2412.5] {n/a}
- ☐ Meters in multiple installations must be permanently identified (metal tag) to indicate the premises served ____ [2412.7] {1309.6.5}

Pipe Size

- ☐ Size per max demand based on appliance ratings ____ [2413.2] {1309.4.2}
- ☐ If specific appliance not known, use **T10** ____ [2413.2] {1309.4.2}
- ☐ Size per tables or manufacturer **T11** ____ [2413.3] {1309.4.3}

TABLE 10 TYPICAL GAS APPLIANCE DEMAND [T2413.2] {T13-1}

Appliance	Typical kBtu/hr.	Actual kBtu/hr.	Typical cu.ft./hr.[A]	Actual cu.ft./hr.
FAU or hydronic boiler	100		91	
Space & water heating units	120		109	
Instantaneous WH 2 gpm	143		130	
Instantaneous WH 4 gpm	285		259	
Storage tank WH 30–40 gal.	35		32	
Storage tank WH 50 gal.	50		45	
Built-in oven	25		22	
Built-in cooktop	40		36	
Freestanding range	65		59	
Barbecue	40		36	
Clothes dryer	35		32	
Direct-vent fireplace	40		36	
Gas log	80		73	
Total cu. ft./hr. max. gas demand				

A. based on 1100 Btu/cu. ft.–consult local gas provider for actual values

TABLE 11 **PROCEDURES FOR SIZING GAS PIPE [2413.4.1,2] {1317.1}**

1. Determine Btu/cu .ft. from local gas provider.
2. Determine cu. ft./hr. demand for each appliance.
3. Sketch layout w/ piping lengths to each appliance. **F43**
4. Determine total cu. ft./hr. demand on each pipe section.
5. Determine length to most remote appliance.
6A. (Longest-length method) Use **T12** row for that length for all appliances.
6B. (Branch-length method) Use same **T12** row for all sections in series w/ most remote appliance.
For other branches, use actual length of each branch.

The longest-length method is more conservative & compensates for pressure losses throughout the system. The branch-length method has less leeway & consideration should be given to the lengths of pipe fittings. The IRC, UMC, & NFPA 54 accept both methods. The UMC allows the branch-length method only up to a total demand of 250 cu. ft./hr.

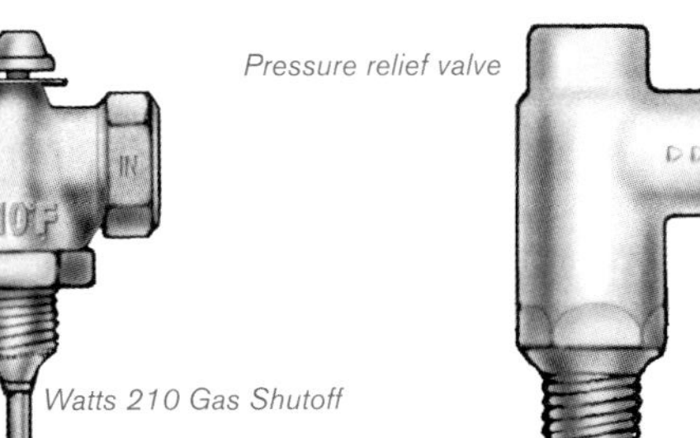

Watts 210 Gas Shutoff

Pressure relief valve

FIG. 43

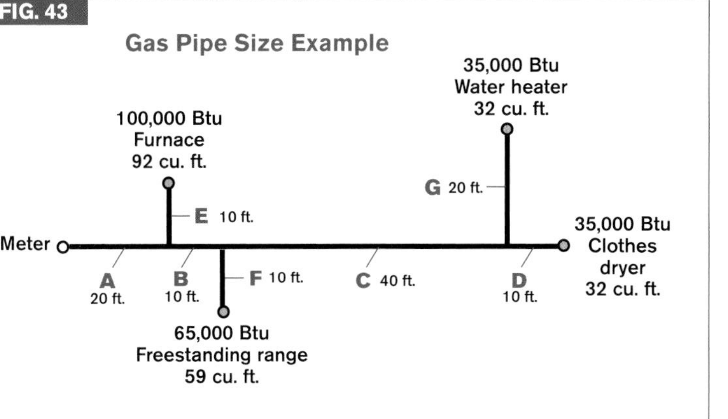

GAS PIPE SIZE EXAMPLE FILL-IN

Pipe Section	Total cu. ft./hr.	Longest Length	Longest-Length Method	Actual Lengths	Branch-Length Method
A	214	90 ft.	1 1/4 in.	90 ft.	1 1/4 in.
B	123	90 ft.	1 in.	90 ft.	1 in.
C	64	90 ft.	3/4 in.	90 ft.	3/4 in.
D	32	90 ft.	1/2 in.	90 ft.	1/2 in.
E	91	90 ft.	3/4 in.	30 ft.	1/2 in.
F	59	90 ft.	3/4 in.	40 ft.	1/2 in.
G	32	90 ft.	1/2 in.	80 ft.	1/2 in.

TABLE 12 CU. FT. CAPACITY OF SCHEDULE 40 METALLIC GAS PIPE[A] [T2413.4(1)] {13-8}

Length	Pipe Size					
	1/2 in.	3/4 in.	1 in.	1 1/4 in.	1 1/2 in.	2 in.
10	172	360	678	1,390	2,090	4,020
20	118	247	466	957	1,430	2,760
30	95	199	374	768	1,150	2,220
40	81	170	320	657	985	1,900
50	72	151	284	583	873	1,680
60	65	137	257	528	791	1,520
70	60	126	237	486	728	1,400
80	56	117	220	452	677	1,300
90	52	110	207	424	635	1,220
100	50	104	195	400	600	1,160
125	44	92	173	355	532	1,120
150	40	83	157	322	482	928
175	37	77	144	296	443	854
200	34	71	134	275	412	794

A. Based on inlet pressure <2 psi, pressure drop 0.5 in. water column, specific gravity 0.60.

Pipe & Tubing Materials — 06 IRC — 06 UMC

- ☐ Cast-iron pipe not acceptable ____ [2414.4.1] {1309.5.2.1}
- ☐ Steel (galvanized or black pipe) OK min schedule 40 [2414.4.2] {1309.5.2.2}
- ☐ Cu & brass tubing OK if gas <0.3 grains hydrogen sulfide per 100cu.ft (check w/ supplier) ____ [2414.5.2] {1309.5.2.3}
- ☐ Cu tubing type K or L ____ [2414.5.2] {1309.5.2.3}
- ☐ CSST OK **F46** ____ [2414.5.3] {1309.5.3.4}
- ☐ Plastic only OK underground & outside ____ [2414.6] {1309.5.4}

Joints, Fittings, & Coatings — 06 IRC — 06 UMC

- ☐ Pipe must be de-burred, brushed, & chip & scale blown [2414.7] {1309.5.5}
- ☐ Defects in pipe may not be repaired, must be replaced [2414.7] {1309.5.5}
- ☐ Use coated metal pipe & fittings if in corrosive environment ____ [2414.8] {1309.5.6}
- ☐ Metal joints threaded, flanged, brazed, or welded ____ [2414.10.1] {1309.5.8.1}
- ☐ Tubing joints w/ approved fittings or brazing >1,000°F ____ [2414.10.2] {1309.5.8.2}
- ☐ Flared joints nonferrous only & suitable for conditions ____ [2412.10.3] {1309.5.8.3}
- ☐ Plastic heat-fusion joints AMI ____ [2414.11] {1309.5.9B}

Pipe Installation

- ☐ Not OK in circulating air ducts, vents, or clothes chutes [2415.1] {1311.2.5}
- ☐ Pipe may not pass through one townhouse to another [2415.1][28] {n/a}
- ☐ No piping concealed in solid partitions (concrete block walls) ____ [2415.2] {1311.3.3}
- ☐ No unions or bushings in concealed locations ____ [2415.3] {1311.3.2}
- ☐ Piping through foundation walls req's sleeve ____ [2415.4] {1311.1.5}
- ☐ Seal space between pipe & sleeve ____ [2415.4] {1311.1.5}
- ☐ Pipe other than black or steel req's steel nail plate protection if pipe closer than 1 1/2in of face of framing ____ [2415.5] {n/a}
- ☐ Tubing in hollow walls req's protection along entire length OR Install in single runs w/o rigid attachment & provide strike-plate protection at framing **F47** ____ [manu] {1311.3.4}
- ☐ Exterior pipes above ground securely supported & protected ____ [2415.7] {1311.2.1}
- ☐ Exterior pipes above ground min 3 1/2in elevation ____ [2415.7] {n/a}
- ☐ Exterior pipes above roof min 3 1/2in elevation ____ [2415.7] {n/a}
- ☐ Cap unused gas outlets ____ [2415.12] {1311.7.2}

Underground Piping	06 IRC	06 UMC
☐ Fe pipe req's protection–not just zinc coating	[2415.8]	{1311.1.3}
☐ Fe wrapping must be factory applied {approved} EXC	[2415.8.2	{1311.1.3}
Field wrapping OK where stripped for threading	[2415.8.2X]	{1211.1.3}
☐ Min cover depth 12in	[2415.9]	{1311.1.2A}
☐ Where 12in not possible, bridge or install in conduit	[n/a]	{1311.1.2A}
☐ Min cover 18in unless damage unlikely at 12in depth	[n/a]	{1311.1.2A}
☐ 8in depth OK to individual lights, grills, etc, if not exposed to damage	[2415.9.1]	{Ø}
☐ Pipes in trenches on firm continuous bearing	[2415.10]	{1311.1.2B}
☐ Piping prohibited underground beneath building EXC	[2415.11]	{1311.1.6}
In conduit sealed in building & vented on exterior **F44**	[2415.11]	{1311.1.6}
☐ Interior end of conduit sealed **F44**	[2415.11]	{1311.1.6}
☐ Conduit min 4in outside building & sealed **F44**	[2415.11]	{1311.1.6}
☐ Conduit vented above grade outside **F44**	[2415.11]	{1311.1.6}
☐ Yellow insulated min 18AWG {14AWG} Cu tracer req'd w/ buried plastic pipe, req'd to terminate above ground at each end	[2415.14.3]	{1311.1.7B}

Inspections		
☐ Complete rough inspections before cover or concealment	[109.1.2]	{1314.3.1}
☐ Leave all joints exposed until tested	[2417.3]	{1314.3.1}
☐ Test pressure min $1\frac{1}{2} \times$ working pressure	[2417.4]	{1314.4.2}
☐ Test pressure min 3psig	[2417.4]	{1314.4.2}
☐ Test min time 10minutes	[2417.4]	{1314.4.3}
☐ Test medium air, nitrogen, or CO_2 (not oxygen)	[2417.2]	{1314.1.7}
☐ Test gauge scale not >5× test pressure	[2417.4.1	{1314.4.1}
☐ Turn off {cap} all outlets before pressure test	[2417.3.4]	{1314.3.4}
☐ Inspect for open fittings or valves before turning on gas	[2417.6.2]	{1314.6.2}

FIG. 44

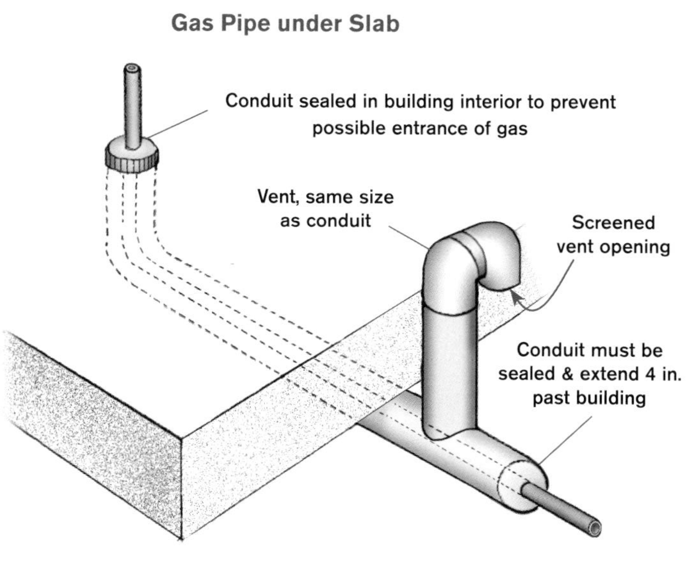

Gas Pipe under Slab

Conduit sealed in building interior to prevent possible entrance of gas

Vent, same size as conduit

Screened vent opening

Conduit must be sealed & extend 4 in. past building

Inspections (cont.)	06 IRC	06 UMC
☐ Check for leakage after turning on gas	[2417.6.3][29]	{1314.6.3}[29]
☐ Soapy water or gas detector OK for leak detection	[2417.5.1]	{1314.5.2}
☐ No matches for leak detection	[2417.5.1]	{1314.5.2}
☐ Purge appliances before placing in operation	[2417.7.4]	{1314.7.4}

Piping Support

	06 IRC	06 UMC
☐ Methods include hooks, bands, straps, brackets, & hangers	[2418.2]	{1311.2.6A}
☐ Support intervals per **T13**	[2424.1]	{1311.2.6B}

TABLE 13 — GAS PIPING SUPPORT INTERVALS [T2424.1] {13-3}

Steel pipe nominal size (in.)	Support spacing (ft.)	Nominal size of smooth-wall tubing (in. outer diameter)	Support spacing (ft.)
1/2	6	1/2	4
3/4 or 1	8	5/8 or 3/4	6
1 1/4 or larger (horizontal)	10	7/8 or 1 (horizontal)	8
1 1/4 or larger (vertical)	Every floor level	1 or larger (vertical)	Every floor level

FIG. 45

Drips & Sloped Piping

	06 IRC	06 UMC
☐ Slope piping min 1/4in/15ft for other than dry gas	[2419.1]	{1311.2.4}
☐ Install drips wherever condensate could collect	[2419.2]	{1311.6.1}
☐ Install trap to prevent water from running back to meter	[2419.2]	{1311.6.1}
☐ Drips must be located where readily accessible	[2419.3]	{1311.6.2}
☐ Sediment trap req'd downstream of shutoff valve & as close as practical to appliance inlet EXC **F45**	[2419.4]	{1312.7}
Ranges, clothes dryers, gas lights, & outdoor grills	[2419.4]	{1312.7}

Valves & Shutoffs

	06 IRC	06 UMC
☐ Valve req'd ahead of meter {regulator}	[2420.2]	{1311.9.1}
☐ Emergency shutoff req'd outside building	[n/a]	{1311.9.3}
☐ Meter valve must be accessible	[2420.1.3]	{n/a}
☐ In common systems serving >1 building each building req's exterior shutoff	[2420.3]	{1311.9B}
☐ Each appliance req's accessible shutoff valve **F45**	[2420.5]	{1312.4}
☐ Union req'd downstream of valve	[2420.5]	{1312.4}
☐ Valve must be within 6ft of appliance EXC	[2420.5]	{1312.4}
Decorative fireplace valves may be remote if identified	[2420.5X]	{na/}
☐ Shutoff valves inside fireplaces AMI	[2420.5.1]	{1312.4}

Connectors

	06 IRC	06 UMC
☐ Shutoff valve upstream of connector **F45**	[2422.1.2.4]	{1312.4}
☐ Rigid metal connectors OK	[2422.1]	{1312.1}
☐ CSST installed AMI OK	[2422.1]	{1312.1}
☐ L&L indoor connectors installed AMI	[2422.1]	{1312.1}
☐ L&L outdoor connectors installed AMI	[2422.1][30]	{303.1}
☐ {Semi-rigid metal tubing OK}	[Ø]	{1312.1}
☐ Quick-disconnect devices installed AMI	[2422.1]	{1312.5}

Connectors (cont.)

	06 IRC	06 UMC
☐ Max length 3ft exc dryers & ranges 6ft EXC	[2422.1.2.1]	{n/a}
Rigid piping OK >3ft & <6ft	[2422.1.2.1X][31]	{1312.4}
☐ May not be ganged together	[2422.1.2.1]	{manu}
☐ Must be entirely in same room as appliance	[2422.1.2.3]	{1312.1}
☐ May not pass through appliance wall EXC	[2422.1.2.3]	{manu}
OK to pass through fireplace insert equipped w/ protective grommet or sleeve & installed AMI	[2422.1.2.3X]	{manu}
☐ Nonmetallic hose connectors only OK outdoors	[n/a]	{1312.2B}

FVIR Water Heater

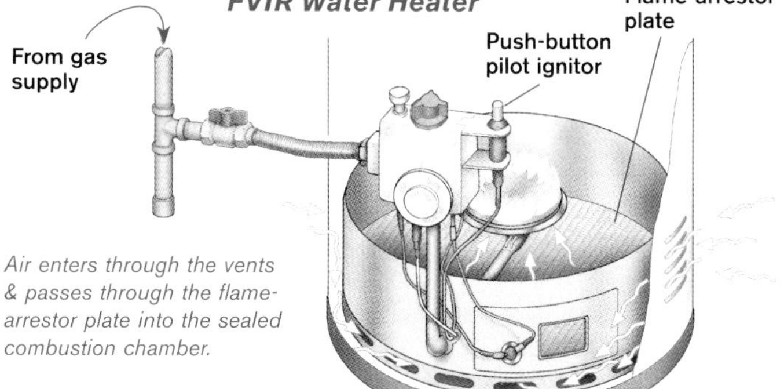

Air enters through the vents & passes through the flame-arrestor plate into the sealed combustion chamber.

Since July 2003, all 30 gal.–50 gal. gas-fired residential water heaters must conform to a new standard that mandates FVIR. This addresses the problem caused by improper storage of gasoline or other flammable liquids near gas-fired water heaters. The new standard includes testing for lint, dust, & oil (LDO) on arrestor plate designs. Since the chamber is sealed, the arrestor plate must be self-cleaning.

Tankless Water Heater

Instantaneous, or on-demand, water heaters are becoming very popular. They save the energy that would otherwise go toward keeping the tank warm, & when properly sized they can supply unlimited amounts of hot water. These units often have a Cat. III stainless-steel vent. They require much larger gas supplies than conventional water heaters.

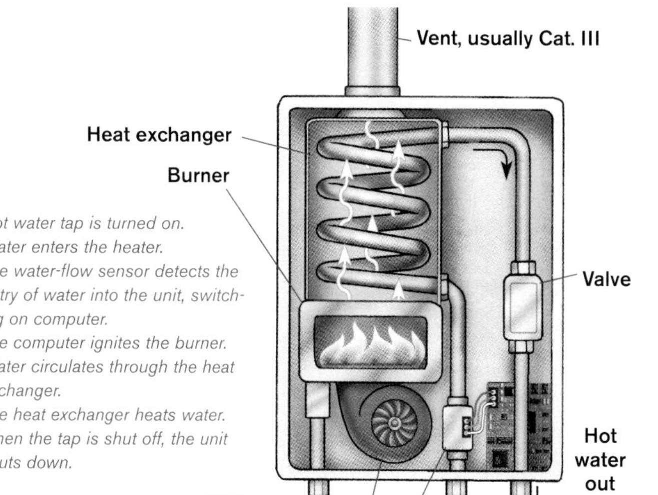

1. Hot water tap is turned on.
2. Water enters the heater.
3. The water-flow sensor detects the entry of water into the unit, switching on computer.
4. The computer ignites the burner.
5. Water circulates through the heat exchanger.
6. The heat exchanger heats water.
7. When the tap is shut off, the unit shuts down.

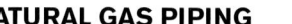

CSST (CORRUGATED STAINLESS-STEEL TUBING)

CSST is becoming a very popular material for residential gas piping. It should be installed only by workers who have successfully completed a training program offered by the CSST manufacturer. Each manufacturer has slightly different requirements for its products. CSST systems can be a traditional "series" design, a hybrid design using pipe and tubing, or a manifold system. Many CSST systems are supplied at "medium" pressure (>0.5 psig) to a central regulator and manifold, then run to individual appliances at low pressure.

Pipe Installation — 06 IRC

- ☐ Support per manu tables ____ (manu)
- ☐ Bending radius per manu tables ____ (manu)
- ☐ No direct burial–routing through conduit OK ____ (manu)
- ☐ Striker plates per manu **F47** ____ (manu)
- ☐ Avoid kinking, twisting, or contact w/ sharp objects ____ (manu)
- ☐ Protect where passing through sheet metal ____ (manu)
- ☐ Regulators in vented area or w/ vent limiters ____ (manu)

Medium Pressure (MP) Regulators F46

- ☐ MP regulators must be accessible ____ [2421.1]
- ☐ MP regulators req tee between shutoff & regulator ____ [2421.1]
- ☐ Capped tee fitting req'd downstream to allow pressure readings ____ [2421.1]
- ☐ Vented regulators must be vented to outdoors EXC ____ [2421.3]
  Outdoor vent not req'd w/ approved vent-limiting device ____ [2421.3X]
- ☐ Vent piping must run independently to outdoors ____ [2421.3.1][32]

CSST is regulated by American National Standards Institute ANSI/IAS LC 1-1997/CSA 6.26-M97 Fuel Gas Piping Systems Using Corrugated Stainless Steel Tubing (CSST). The standard requires that a contractor be certified before installing CSST.

FIG. 46 CSST Manifold

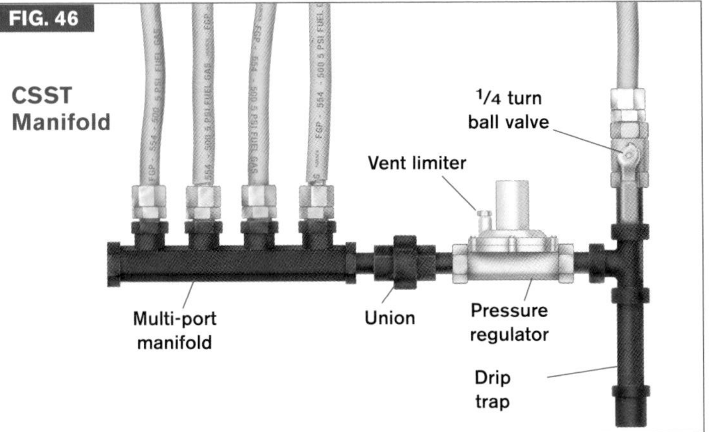

FIG. 47 CSST Strike Plates

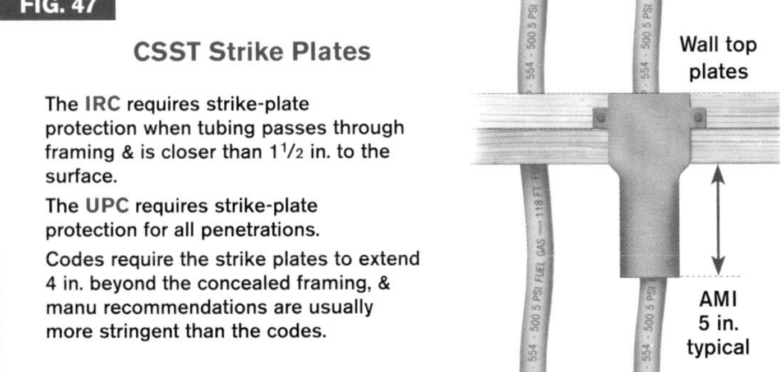

The **IRC** requires strike-plate protection when tubing passes through framing & is closer than 1 1/2 in. to the surface.

The **UPC** requires strike-plate protection for all penetrations.

Codes require the strike plates to extend 4 in. beyond the concealed framing, & manu recommendations are usually more stringent than the codes.

GAS APPLIANCE VENTING

The majority of gas appliances use a Category I vent, by which flue gases that are lighter than the outside atmosphere establish a flue draft to the exterior by gravity. These appliances often have a connector between the appliance and vent or chimney. High-efficiency appliances have flue gases that cool to the dew point and must be forced through a sealed vent to the exterior.

General	06 IRC	06 UMC
☐ Choose vent material based on appliance category	[2427.1]	{802.1}
☐ Cat. I induced draft is "gravity" vent	[2427.1]	{802.1}
☐ Select type of venting system from **T14** & AMI	[2427.4]	{802.4.1}
☐ Properly support all vents AMI	[2426.6]	{802.6.5}
☐ All vents L&L exc single wall or plastic	[2426.1]	{n/a}
☐ Condensate drain also req'd for Cat. I or II if local experience shows need	[2427.8&9]	{802.9.2}
☐ Sheet-metal shield to 2in above attic insulation **F49**	[2426.4]	{n/a}
☐ Protect w/ steel strike plate extending 4in beyond framing plate when inside wall	[2426.7][33]	{n/a}

Vent Terminations	06 IRC	06 UMC
☐ Roof penetration req's flashing	[2427.6.3]	{802.6.1}
☐ Must have listed cap	[2427.6.3]	{802.6.1}
☐ Decorative shrouds only if L&L & AMI	[2427.6.4.1]	{802.6.2.4}
☐ Chimneys min 3ft above roof	[2427.5.3]	{802.5.2.1}
☐ Chimneys min 2ft higher than building within 10ft	[2427.5.3]	{802.5.2.1}
☐ Single wall min 5ft vert above flue collar	[2427.7.3]	{802.7.3.1}
☐ Single wall min 2ft above roof	[2427.7.3]	{802.7.3.2}
☐ Single wall min 2ft higher than building within 10ft	[2427.7.3]	{802.7.3.2}
☐ Type B ≤12in diameter min 1ft above roof up to 6/12 slope **T15**	[2427.6.4]	{802.6.2}
☐ Type B min 2ft above vert walls within 8ft **F48**	[2427.6.4]	{802.6.2}

TABLE 14 Type of Venting System to Be Used [T2427.4] {T8-1}

Appliances	Type of Vent	IRC	UMC
Listed Cat. I Listed appliances w/ draft hoods Appliances listed for B vent	Type B gas vent	2427.6	802.6
	Chimney	2427.5	802.5
	Single-wall metal pipe	2427.7	802.7
	Listed chimney lining for gas	2427.5.2	802.5.1.3
	Special vent listed for appliance	2427.4.2	802.4.3
Listed vented wall furnaces	Type BW	2427.6	802.6 928.0
Cat. II, III & IV appliances	As specified by manu.	2427.4.1 2427.4.2	802.4.2 802.4.3
Unlisted appliances	Chimney	2427.5	802.5
Decorative appliances in vented fireplaces	Chimney	2427.5	802.5
Direct-vent appliances	As specified by manu.	2427.2.1	802.2.5
Appliances w/ integral vent	As specified by manu.	2427.2.2	802.2.6

Vent Terminations (cont.)	06 IRC	06 UMC
☐ Type B or L min 5ft vert above flue collar (except wall furnace)	[2427.6.5]	{802.6.2.1}
☐ Wall furnace min 12ft from bottom of furnace **F55**	[2427.6.5]	{802.6.2.2}
☐ Direct vent per **T16**	[2427.8]	{802.8.3}

FIG. 48

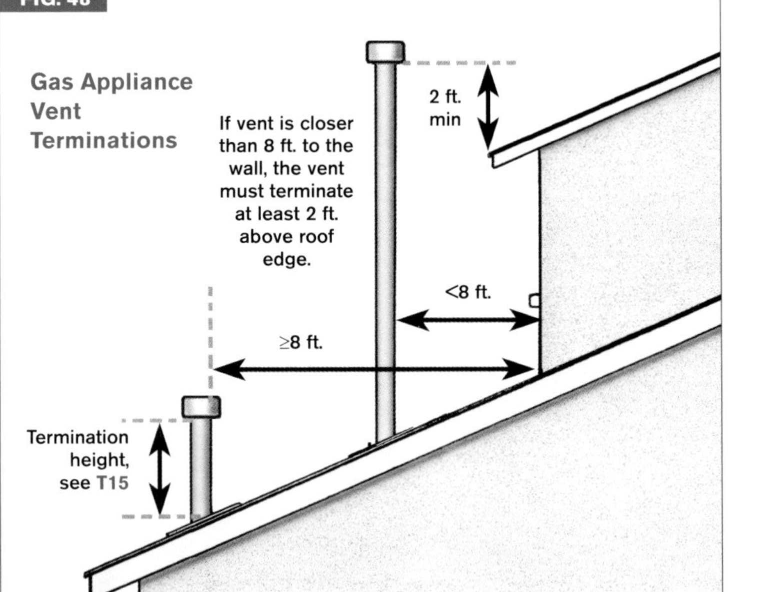
Gas Appliance
Vent
Terminations
If vent is closer than 8 ft. to the wall, the vent must terminate at least 2 ft. above roof edge.
2 ft. min
<8 ft.
≥8 ft.
Termination height, see T15

Single-Wall Vents

	06 IRC	06 UMC
☐ No single wall outdoors in cold climates	[2427.7.2][34]	{802.7.2}[34]
☐ Must go direct to exterior	[2427.7.4]	{802.7.4.1}
☐ May not originate in unoccupied attic	[2427.7.6]	{802.7.4.2}
☐ May not pass through attic, floor, or inside wall	[2427.7.6]	{802.7.4.2}
☐ Roof penetration req's thimble 18in above & 6in below	[2427.7.5]	{802.7.4.5}
☐ Min 6in clear to combustibles	[2427.7.7]	{802.7.4.4}

TABLE 15 TERMINATION OF B AND BW VENTS IRC F2427.6.5] AND {UMC F 8-2}

Roof Slope	Height above Roof
up to 6/12	1ft.
>6/12 to 7/12	1 ft. 3 in.
>7/12 to 8/12	1 ft. 6 in.
>8/12 to 9/12	2 ft.
>9/12 to 10/12	2 ft. 6 in.
>10/12 to 11/12	3 ft. 3 in.
>11/12 to 12/12	4 ft.
>12/12 to 14/12	5 ft.
>14/12 to 16/12	6 ft.
>16/12 to 18/12	7 ft.
>18/12 to 20/12	7 ft. 6 in.
>20/12 to 21/12	8 ft.

Vent Connectors for Category I Appliances

	06 IRC	06 UMC
☐ No connection to chimney serving solid fuel	[2427.5.6.1]	{802.5.5.1}
☐ Must be as straight as practical	[2427.10.6]	{802.10.6}
☐ Min $^1/_4$in/ft slope toward appliance **F50**	[2427.10.8]	{802.10.8}
☐ No dips or sags	[2427.10.8]	{802.10.8}
☐ Must be as short as practical	[2427.10.9]	{802.10.9.1}
☐ Provide adequate support	[2427.10.10]	{802.10.10}
☐ Entire connector must be accessible	[2427.10.12]	{802.10.12}
☐ Attach to appliance w/ screws or AMI	[2427.10.7]	{802.10.7}
☐ Size increases only at appliance outlet connection	[2427.10.3.5]	{802.10.3.5}

Single-Wall Connectors for Category I Appliances

06 IRC 06 UMC

- ☐ Not in attic, crawl space, or other unconditioned space EXC ______ [2427.10.2.2] {802.10.2.2}
  OK [in unconditioned basement or garage] within exterior walls if local 99% winter design temp ≥5°F ______ [2427.10.2.2X]{802.10.2.2X}
- ☐ Uninsulated horiz connector max 75% of vert **F50** ______ [2427.10.9] {802.10.9.2}
- ☐ Min 6in clearance to combustibles ______ [2427.7.7] {802.7.4.4}
- ☐ No passing through interior wall, floor, or ceiling ______ [2427.10.14][35]{802.10.14.1}
- ☐ Max horizontal length 75% of vert height ______ [2427.10.9] {802.10.9.2}

Type B Double-Wall Vent Connectors

- ☐ Min clearance to combustibles per L&L (typical 1in) [2427.10.5] {802.10.5}
- ☐ Max horizontal length 100% of vertical height ______ [2427.10.9] {802.10.9.3}

Connectors to Masonry Chimneys

- ☐ Inspection req'd before connecting to existing chimney [2427.5.5] {802.5.4}
- ☐ Approved liner req'd ______ [2427.5.5.1] {802.5.4.2}
- ☐ Enter above extreme bottom of chimney ______ [2427.10.11] {802.10.11}
- ☐ No connection to chimney serving solid fuel ______ [2427.10.13] {802.10.13}
- ☐ Enter not less than 12in from chimney bottom ______ [2425.9] {n/a}
- ☐ Secure connector to prevent blocking chimney ______ [2427.10.11] {802.10.11}
- ☐ No decorative shrouds unless L&L & PMI ______ [2427.5.3] {802.5.2.4}

FIG. 49

Vent Offsets

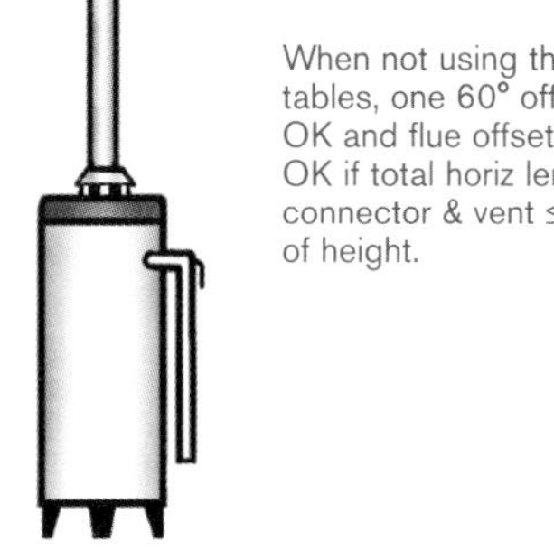

Appliance manufacturers supply venting tables giving max Btu capacity for different sizes, lengths, & offsets.

When the offset includes more than two 90° elbows, subtract 5% from the table capacity for each elbow up to 45° & 10% for each elbow up to 90°.

When not using the tables, one 60° offset OK and flue offsets ≤45° OK if total horiz length of connector & vent ≤75% of height.

Multiple Appliances Vented in Common	06 IRC	06 UMC
☐ Use venting tables supplied w/ appliance if draft-hood appliance & induced-draft connected to common vent	[2427.10.3.1]	{802.10.3.1}
☐ Join multiple connectors as high as possible per available headroom & clearance **F50**	[2427.10.3.4]	{802.10.3.4}
☐ Connect smaller appliance above larger as possible per available headroom & clearance **F50**	[2427.10.4]	{802.10.4.1}
☐ Use tables for size exc 2 draft-hood equipped appliances OK to size for 100% of larger + 50% of smaller	[2427.10.3.4]	{802.10.3.4}
☐ Reduce connector table capacity 5% each elbow up to 45° & 10% each elbow >45° up to 90° **F50**	[2428.3.7]	{803.2.7}

Vent Size (Area) for Appliances with Draft Hoods

☐ Chimneys min size same as flue collar	[2427.5.4]	{802.5.3}
☐ If 2 appliances, 100% of larger + 50% of smaller	[2427.5.4]	{802.5.3}
☐ Chimneys max size 7× area of smallest flue collar	[2427.5.4]	{802.5.3}
☐ Vents min size same as flue collar	[2427.6.9.1]	{802.6.3.1}
☐ If 2 appliances, 100% of larger + 50% of smaller	[2427.6.9.1]	{802.6.3.1}
☐ Vents max size 7× area of smallest flue collar	[2427.6.9.1]	{802.6.3.1}
☐ Offsets 45° max except one of 60° OK **F49**	[2427.6.9.2]	{802.6.1.1}

Vent Size Cat. I Appliances–Induced or Draft Hoods

☐ Size per tables (supplied w/ appliance) **F50**	[2427.6.9]	{802.6.3}
☐ Connector not >2 sizes larger than flue collar	[2428.2.11]	{803.1.10}
☐ No elbows if using "zero lateral length" part of table	[2428.2.3]	{803.1.2}
☐ Vent tables w/ lateral length allow for 2°–90° elbows	[2428.2.3]	{803.1.2}
☐ Reduce table capacity 5% each elbow up to 45° & 10% each elbow >45° up to 90°	[2428.2.3]	{803.1.2}
☐ Reductions for elbows in common vents as above	[2428.3.6]	{803.2.6}
☐ Reductions for common vent connectors as above	[2428.3.7]	{803.2.7}

FIG. 50 Common Venting Example

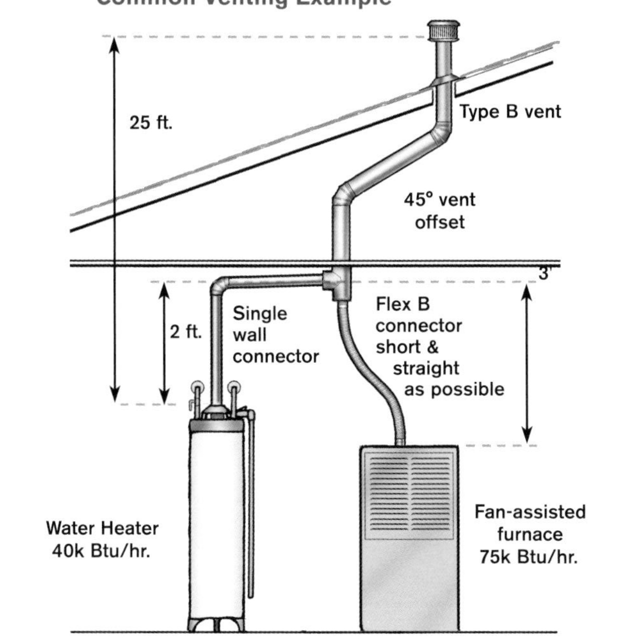

The venting table's capacity must be reduced by 5% for each offset up to 45°, so the offsets above the ceiling require a total 10% reduction. The venting tables allow for two 90° elbows in the connector, so no adjustment is needed for the two at the water-heater connector. In this actual example, the tables tell us that both connectors would need to be 4 in. in diameter & the common vent would need to be 5 in.

GAS APPLIANCES–LOCATION

The IRC requires appliances to be listed and labeled unless accepted under the provisions for alternate materials and methods. The UMC accepts unlisted appliances when acceptable to the building official. In either case, appliances are provided with nameplate instructions that must be followed in their location and installation.

General	**06 IRC**	**06 UMC**
☐ Boiler or FAU not in bedroom or bath [or storage closet] unless direct vent	[2406.2]	{904.1}
☐ If bedroom or bath unconfined space: vented room heaters, wall furnaces, gas fireplace heaters, & decorative appliances OK	[2406.2]	{902.0B}
☐ Not in space that opens only to bedroom, bath, or storage closet unless space used for no other purpose & solid self-closing weatherstripped door & combustion air from exterior	[2406.2][36]	{n/a}
☐ Ignition source ≥18in above floor in garages **F5** EXC	[2408.2]	{308.0}
Flammable vapor ignition resistant appliances	[2408.2X]	{n/a}
☐ Elevate [6ft] or guard (bollards) garage appliances **F5**	[2408.3]	{308.0}
☐ Equipment at grade req's pad above adjoining grade	[2408.4]	{n/a}
☐ Suspended equipment min 6in above grade	[2408.4]	{n/a}

Boilers & FAUs in Closets & Alcoves

☐ FAUs in alcoves or closets L&L for alcove	[2409.3.2]	{904.2B}
☐ Boilers in alcoves or closets L&L for alcove	[2409.4.2]	{904.2B}
☐ FAUs & boilers clearance AMI EXC	[2409.3.2, 2409.4.2]	{904.2A&B}
Clearance reduction OK if room large in comparison w/ size of equipment (see glossary)	[2409.3.3, 2409.4.3]	{904.2A&B}
☐ No clearance reduction in alcoves	[2409.3.2, 2409.4.2]	{904.2B}

GAS FORCED-AIR FURNACES (FAUs)

High-efficiency Category IV condensing furnaces are becoming increasingly popular. Their installation is very different than conventional Category I appliances, and it is important to follow the installation instructions for the specific furnace.

General	**06 IRC**	**06 UMC**
☐ Electrical receptacle req'd within 25ft	[3801.11]	{309.0}
☐ Individual circuit req'd	[3603.1]	{NEC422.12}
☐ Automatic outlet air temperature limit 250°F	[manu]	{306.0}
☐ Air filter req'd	[2442.1]	{312.1}

Condensing Furnace (Cat. IV) Venting F51

☐ Size venting AMI	[2426.5]	{802.6.3.2}
☐ Install vent & support AMI	[2426.5]	{802.6.1.1}
☐ Positive-pressure systems req'd to be gas tight	[2427.3.3]	{802.3.4.3}
☐ No mixing of natural to forced draft connectors or vents	[2427.3.3]	[802.3.4.4}
☐ Burner interlock req'd to forced-vent fan	[2427.3.3]	[802.3.4.5}

Condensing Furnace (Cat. IV) Vent Termination

☐ Furnaces w/ combustion air piping terminating in same location as vent piping are considered direct vent – see **T16** on **p. 172**	[2427.8X1]	{802.8.1X1}
☐ Min 3ft above forced-air inlet within 10ft	[2427.8]	{802.8.1}
☐ Min 4ft below, 4ft horiz, or 1 ft above building openings	[2427.8]	{802.8.2}
☐ Min 12in above grade	[2427.8]	{802.8.2}
☐ No vent termination where vapor would be a nuisance	[2427.8]	{802.8.4}

Condensate Disposal

☐ Provide means to collect & dispose of condensate	[2427.9]	{310.1}
☐ Auxiliary drain pan req'd if condensate stoppage could damage any building component EXC	[2404.10][37]	{n/a}
Automatic cutout installed in drain system	[2404.10X][37]	{n/a}

Condensate Disposal (cont.)

	06 IRC	06 UMC
☐ Not drained to public way or nuisance location	[2427.9]	{310.1}
☐ Drainpipe min $^3/_4$in w/ $^1/_8$in/ft slope or AMI	[manu]	{310.1}
☐ May drain to indirect waste **F10**	[manu]	{310.1}

FIG. 51

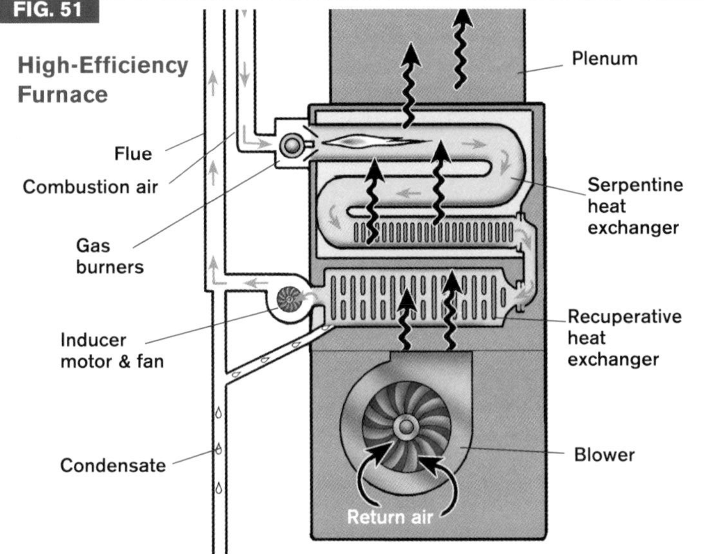

High-efficiency furnaces extract so much heat from the fuel gas, the flue gases are cooled below their dew point. As a result, water must be drained from the venting system. Since this water is a result of condensation, high-efficiency furnaces are also known as condensing furnaces.

GAS APPLIANCES IN FIREPLACES

Due to concerns over air pollution and energy loss, many fireplaces today are installed primarily for appearance rather than for burning wood. Gas appliances in fireplaces must have a sufficient source of combustion air.

Vented Decorative Gas Appliances

	06 IRC	06 UMC
☐ Not allowed in bedroom if room considered confined space	[2406.2]	{907.1}
☐ Must be L&L & installed AMI	[2432.1]	{907.2}
☐ Maintain open vent (block damper in open position)	[manu]	{manu}
☐ Fireplace screen req'd	[manu]	{907.3}
☐ Appliance w/ pilot or ignition system req's pilot safety	[2432.2]	{306.0}
☐ Shutoff inside firepit OK if AMI	[2420.5.1]	{1312.4}

Log Lighters

☐ Req'd valve ≤6ft of fireplace & in same room	[2420.5]	{1312.4}
☐ Hard gas pipe inside firepit (no flex)	[2422.1.1]	{1312.1}

Unvented Gas Log Heaters

☐ Not in factory-built fireplace unless L&L to UL127 & AMI	[1004.4]	{n/a}

DIRECT-VENT GAS APPLIANCES

Direct-vent appliances draw their source of combustion air from the same area where they vent combustion gases. This arrangement has the advantage of placing equal pressure on the inlet and outlet of the firebox, which is sealed and has no open flame on the building interior. Because these appliances do not use the interior air for combustion, they can be located in rooms that are considered confined spaces.

Direct-Vent Gas Fireplaces F52

☐ OK in bedroom or bath	[2406.2]	{908.1X}
☐ Must be L&L & AMI	[2427.2.1]	{908.2D}
☐ CSST or flex req's grommet through appliance wall	[2422.1.2.3X]	{1312.1}
☐ Vent termination clearances **T16**	[2427.8]	{802.8.3}

FIG. 52

Direct-Vent Gas Fireplace

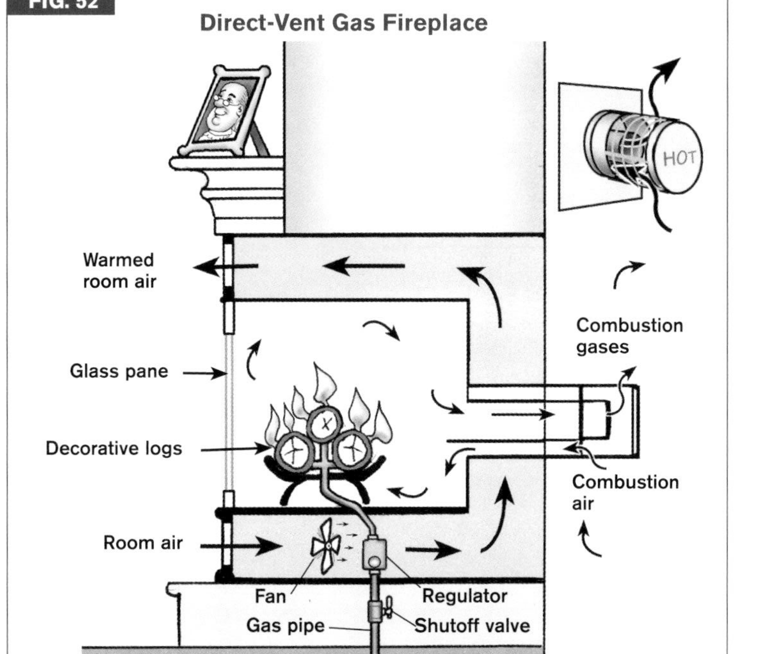

A direct-vent fireplace can vent horizontally out a sidewall or vertically to the roof. With a completely enclosed chamber, it draws in outside air for combustion & expels gases to the outside. The front glass enclosure allows radiant heat to pass into the room. It heats a room w/o robbing it of oxygen or of the heated air it provides, & keeps it free of fumes & combustible materials, such as embers or ash.

Direct-Vent Wall Gas Heaters

	06 IRC	06 UMC
☐ Install AMI	[2427.2.1, 2429.1]	{304.1}
☐ Indoor combustion air not req'd	[2407.1]	{928.1D}

FIG. 53

Direct-Vent Wall Furnace

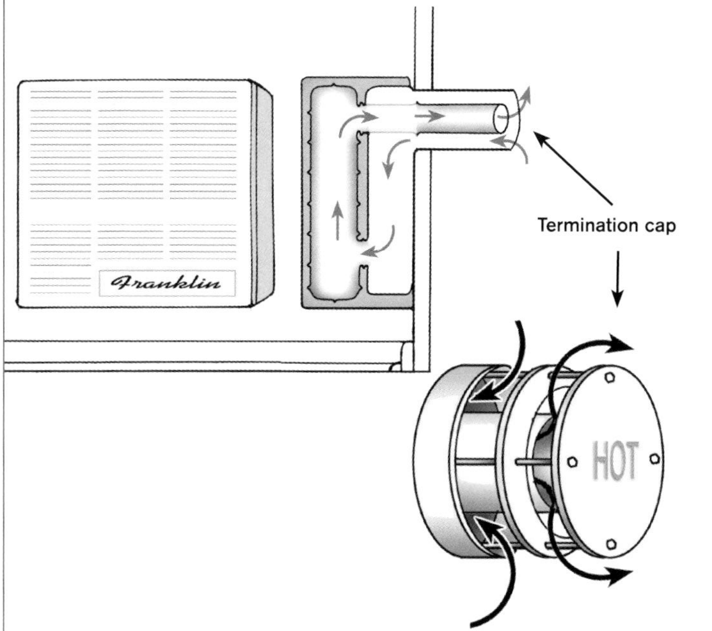

Franklin

Termination cap

HOT

Direct-Vent Appliance Vent Termination — **06 IRC** — **06 UMC**

- ☐ OK to terminate through wall **F52,53** ______ [2427.8] {802.8.3}
- ☐ Termination clearances **T16** ______ [2427.8] {802.8.3}
- ☐ Bottom of vent terminal min 12in above grade ______ [2427.8] {802.8.3}

TABLE 16 — DIRECT-VENT GAS APPLIANCES TERMINATION CLEARANCE FROM BUILDING OPENINGS [2427.8] & {802.8.3}

Appliance Rating	IRC	UMC
≤10k Btu	6 in.	6 in.
>10k Btu–50k Btu	9 in.	9 in.
>50k Btu	12 in.	12 in.

GAS WALL FURNACES

Gas-fired wall furnaces can be located only in rooms that are large enough to meet the combustion air requirements of the appliance. Because they need indoor air for combustion, they are typically found only in older buildings with high air-infiltration rates. If a building is upgraded in terms of energy compliance, it might not be possible to use wall furnaces as the heat source. Vent installation on wall furnaces is especially important, in that clearances inside the wall are less than the normal minimums for B vents. When a wall furnace is installed in an existing building, one side of the wall above the furnace should be opened for inspection of vent clearances.

Clearances

- ☐ From sidewall–install AMI (typical 6in min) **F54** ______ [2436.3] {928.1&2}
- ☐ From door swing [12in] {AMI} **F54** ______ [2436.4] {928.1&2}
- ☐ Do not rely on doorstops to maintain clearance ______ [2436.4] {928.2}
- ☐ Clearance below structural projections AMI (typical 18in) [2436.3] {928.1&2}

FIG. 54 Wall Furnace Clearances

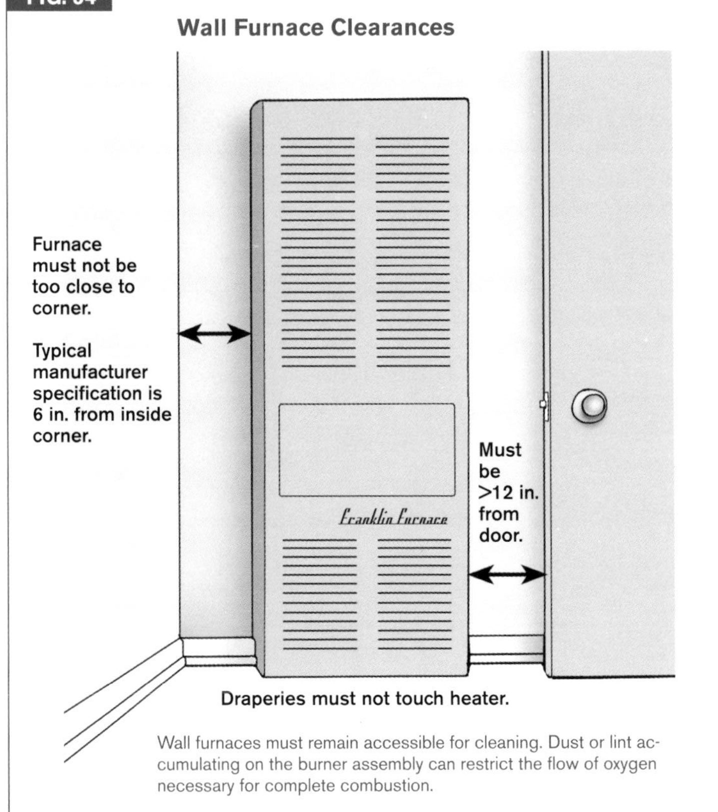
Furnace must not be too close to corner.
Typical manufacturer specification is 6 in. from inside corner.
Franklin Furnace
Must be >12 in. from door.
Draperies must not touch heater.

Wall furnaces must remain accessible for cleaning. Dust or lint accumulating on the burner assembly can restrict the flow of oxygen necessary for complete combustion.

Furnace Installation	06 IRC	06 UMC
☐ Fan assist only if L&L & AMI	[2436.1]	{304.1}
☐ No ducts attached to wall furnaces	[2436.5]	{928.1A}
☐ Must remain accessible for cleaning	[2436.6]	{928.1E}
☐ Header plate at top of furnace AMI	[2436.1]	{928.1C}
☐ Stud bay depth AMI	[2436.1]	{928.1A}
☐ Unlisted furnaces not OK {in combustible construction}	[2404.3]	{928.1B}
Vent Installation		
☐ Cut top & floor plates flush to stud	[2427.6.1]	{928.1C}
☐ Furnace stud space vented by ceiling plate spacers AMI **F56**	[2427.6.1]	{928.1C}
☐ Subsequent ceiling stud plates firestopped AMI **F57**	[2427.6.1]	{928.1C}
☐ Single-story systems OK only in single story or top of multistory	[2427.6.1]	{928.1C}
☐ Multistory systems OK in single story or multistory	[2427.6.1]	{928.1C}
☐ Sleeve around vent in insulated assembly **F57**	[2426.4]	{n/a}
☐ Sleeve min 2in above insulation in insulated attic **F57**	[2426.4]	{n/a}
☐ Vent min height 12ft above bottom of furnace **F55**	[2427.6.5]	{802.6.2.2

FIG. 55

Flue Termination

Min. 12 ft.

FIG. 56

Wall Furnace Flue

BW vent
Each floor above ceiling line requires firestop spacers.
Stud space above furnace vented by spacer supplied with furnace.
Ceiling line above wall furnace

FIG. 57

BW Vent Sleeve

2 in. min.
Insulation sleeves should be at least 2 in. higher than insulation.
Header plate supplied w/ heater is the first fire block.

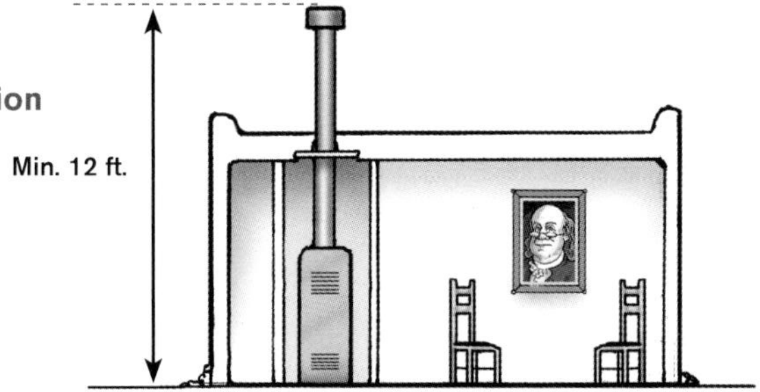

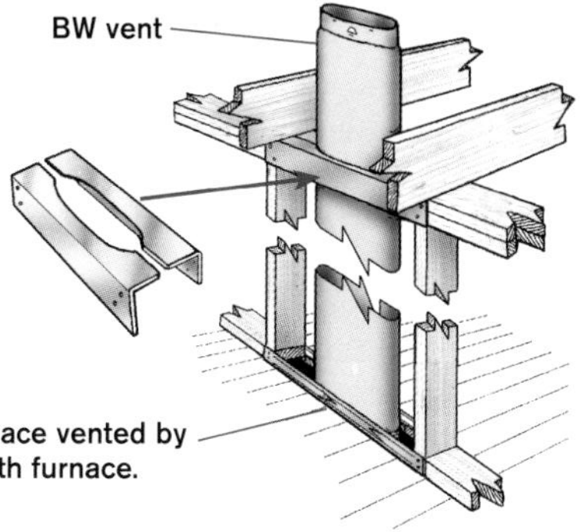

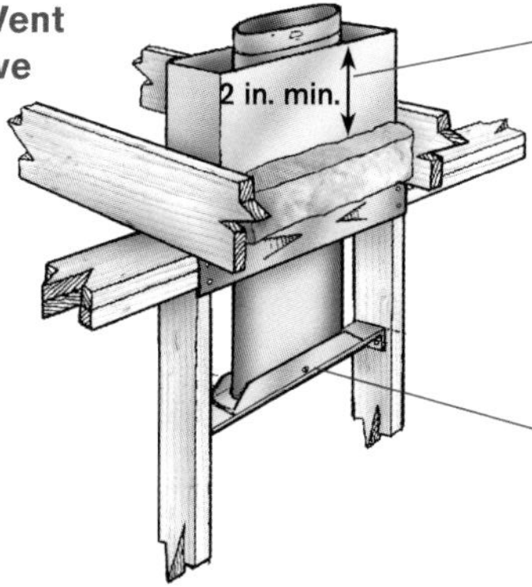

GAS FLOOR FURNACES

In addition to these specific rules for gas-burning floor furnaces, the access and passageway illumination requirements on **p. 130** also apply. Grills for these furnaces can be hot, and care must be taken to protect young children from those grills.

Underfloor Area	**06 IRC**	**06 UMC**
☐ Must be L&L & installed AMI	[2437.1]	{912.1A}
☐ Unlisted furnaces only in noncombustible floors	[Ø]	{912.1B}
☐ Appliance must fit through access opening	[1305.1.3]	{305.0}
☐ [Access opening min 22in × 30in] {In foundation wall, min 18in × 24in–trap door min 24in × 24in}	[1305.1.4]	{912.8}
☐ May not be in concrete slab on grade	[2437.2]	{912.1A}
☐ May not project into habitable space below	[2437.5&6]	{912.11&12}
Excavation Clearances		
☐ 6in to ground (2in if factory sealed) **F58**	[2437.4]	{912.7}
☐ 12in side clearance; 18in on control side **F58**	[2437.4]	{912.7}
Above Floor		
☐ Not in doorway, landing, passageway, or exit way	[2437.2]	{912.4A}
☐ Flat-register type min 6in from wall **F59**	[2437.2]	{912.4B}
☐ Min 12in from door swing or draperies **F58,59**	[2437.2]	{912.4C}
☐ Two adjoining sides must have 18in clearance **F60**	[2437.2]	{912.4B}
☐ Wall-register type min 6in to inner corner **F59**	[2437.2]	{912.4B}
☐ Thermostat must be in same room	[2437.2]	{912.1C}

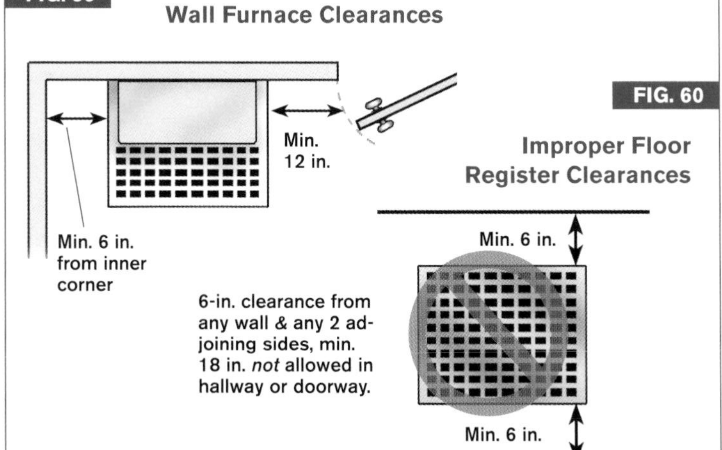

ROOM HEATERS

Room heaters that are not direct vent must be supplied with sufficient combustion air from the interior and are typically found only in older buildings with high air-infiltration rates. Because these must connect to a venting system, they should be secured to prevent displacement of the vent. New provisions in each code allow a limited use of unvented heaters in bedrooms and baths. These should be provided with a nontamperable oxygen depletion sensor.

Room Heaters	**06 IRC**	**06 UMC**
☐ Secure to floor	[1307.2]	{304.4}
☐ Install AMI	[2446.1]	{304.1}
☐ Flame safeguard (pilot safety) req'd	[2446.1]	{306.0}
☐ Unlisted circulating type min clearances 12in all around	[Ø]	{924.3A}
☐ Unlisted radiating type min clearances 18in all around	[Ø]	{924.3B}

Unvented Heaters		
☐ Unvented heater may NOT be sole heat source	[2445.2]	{924.1}[38]
☐ Unvented heater ≤6kBtu OK in bath	[2406.2][39]	{924.1.1X1}[40]
☐ Unvented heater ≤10kBtu OK in bedroom	[2406.2][39]	{924.1.1X2}[40]
☐ Rooms must meet unconfined space requirements	[2406.2]	{924.1.1X}[40]
☐ Adjacent spaces w/ permanent large openings (doorway or archway) part of room volume	[2445.5]	{701.2}
☐ Unvented heater req's oxygen-depletion sensor	[2445.6]	{924.1.1X}[40]

A roaring fire heats only via radiant heat & can remove heated air from the house, resulting in a net loss of energy. For this reason, building codes require a source of combustion air to the firepit. When using a fire, it is a good idea to turn down the thermostat. Glass doors help prevent air loss, & fireplace inserts help recover heat from the fire & radiate it outward to the room.

GLOSSARY

Accessible (for inspection): Capable of being exposed without damage to the building or component structure or finishes. May require removal of access doors or limited fasteners.

Air-conditioning: The process of heating, cooling, humidifying, dehumidifying, filtering, or otherwise treating air in a building. Most non-trade persons relate this term to cooling only.

Air handler: A blower or fan enclosed in a metal box used for the purpose of distributing supply air to a room, space, or area.

Alcove: A room or space that does not have at least 12 times the volume of a furnace or other air-handling device or 16 times the volume of a boiler. When the ceiling is greater than 8 ft., the volume is calculated based on an 8-ft. height. See "Room large in comparison to size of equipment."

Backflow preventer: A device or means to prevent backflow. Most common is the reduced-pressure zone type. This valve consists of two independently acting check valves, internally pressure-forced to the normally open position. These checks are separated by an intermediate chamber that is equipped with automatic-relief means. Should there be a reversal of flow, the downstream liquid will drain instead of placing backpressure on the supply liquid.

Building thermal envelope: The basement walls, exterior walls, floor, roof, and any other building element that encloses conditioned spaces.

Check valve: A device used to prevent the flow of liquids in a direction not intended in the design of the system. Check valves are not backflow preventers.

Chimney: A primarily vertical structure containing one or more flues, for the purpose of carrying gaseous products of combustion and air from an appliance to the outside atmosphere.

Combustible material: A material of or surfaced with wood, compressed paper, plant fiber, plastics, or other material that can ignite or burn, whether flame proofed or not or whether plastered or unplastered.

Combustion air: Air required for complete combustion of a fuel, including theoretical air and excess air. This may be interpreted to include dilution air, which is air introduced into the draft hood and is mixed with flue gases, and ventilation air, which cools appliances.

Confined space: A room or space having a volume less than 50 cu. ft. for each 1,000Btu input rating of all fuel-burning appliances in the room or space.

Decorative gas appliance for vented fireplaces: A vented appliance designed for installation within the fire chamber of a fireplace, wherein the primary function lies in the aesthetic effect of the gas flames.

Decorative shroud: A partial enclosure for aesthetic purposes that is installed at the termination of a venting system that surrounds or conceals the chimney or vent cap.

Dilution air: Air that combines with flue gases at the draft hood of an appliance. See "Combustion air."

Direct-vent appliances: Appliances that are constructed and installed so that all air for combustion is derived from the outside atmosphere and all flue gases are discharged directly to the outside atmosphere, usually by a flue pipe inside the combustion air pipe.

Draft: The flow of gases or air through a chimney or flue caused by pressure differences. An induced-draft appliance has a fan to overcome the resistance of the combustion chamber while still delivering flue gas lighter than atmosphere to the vent. A forced-draft appliance delivers flue gas under positive pressure. Natural draft is caused by the height of the chimney and the difference in temperature of hot gases and outside atmosphere.

Draft diverter/draft hood: A nonadjustable device integral to an appliance or made part of the connector that provides for the ready escape of flue gases from the appliance in the event of insufficient draft, backdraft, or stoppage. A draft diverter (typical on water heaters) prevents backdraft from entering the appliance and neutralizes the stack effect on the operation of the appliance.

Draft regulator: A device that functions to maintain a desired draft in the appliance by automatically reducing the draft to the desired value. These are usually adjustable, such as the barometric damper on an oil-furnace flue. Most barometric dampers are free to move in either direction and protect against both excessive draft (that could allow the flame to lift) and backdraft.

Duct: A continuous passageway for the transmission of air (usually forced) made of factory-built components.

Energy-recovery ventilator: Same as "heat-recovery ventilator" with a heat exchanger core designed for more moderate climates.

Evaporative cooler: A device used for reducing the sensible heat of air for cooling by the process of evaporation of water into an airstream. Also known as a "swamp cooler."

Factory-built fireplace: A fireplace composed of listed factory-built components assembled in accordance with the terms of the listing to form the completed fireplace. The appliance must be suitable for solid fuel and be equipped with a listed and properly installed chimney.

Fan-assisted appliance: An appliance equipped with an integral mechanical means to either draw or force products of combustion through the combustion chamber and/or heat exchanger.

Fireblock: Building materials installed to resist the free passage of flame to other areas through concealed spaces of the building.

Fireplace stove: A freestanding solid-fuel burning device designed to be operated with the firebox door either open or closed.

Firestop: Materials installed to resist free passage of flame through an assembly, typically around a duct, vent, or chimney passing through a ceiling assembly.

Flame safeguard: A device that will automatically shut off the fuel supply to a main burner or group of burners when the means of ignition of those burners becomes disabled and when flame failure occurs.

Flue: A term often substitute for vent; or a passageway intended to carry hot gases through a chimney.

Flue collar: The outlet of an appliance designed for the attachment of a draft hood, vent connector, or venting system.

Flue gases: Products of combustion plus excess air in appliance flues or heat exchangers.

Forced draft: A vent system using a fan or other mechanical means to expel flue gases under positive static vent pressure.

Furnace: A device that is completely self-contained and designed to supply heated air to spaces remote from or adjacent to the furnace location. A central furnace uses ducts to supply heat to spaces.

Gas connector: Tubing or piping that connects the gas supply piping to the appliance.

Ground: *n:* The solid surface of the earth; firm or dry land, or soil. *v:* To connect an electrical conductor with a fault current path, which becomes part of the circuit.

H

Habitable room: A room used for living, sleeping, eating, or cooking. Bathrooms, closets, halls, storage spaces, and laundry rooms are not considered habitable rooms.

Heat pump: A refrigeration system that extracts heat from one substance and transfers it to another portion of the same substance or to a second substance at a higher temperature for a beneficial purpose.

Heat-recovery ventilator: A house ventilation fan that expels air. As this air is expelled, the outside vent simultaneously draws in fresh air, warming it with heat recovered from the indoor air being expelled. These are most popular in colder climates.

HSPF: Heating Season Performance Factor is the measure of a system's efficiency in heating mode. The higher the number, the more efficient the system.

I

Induced draft: A portion of the vent system using a fan or other mechanical means to cause the removal of gases under non-positive static vent pressure.

Induced draft burner: A burner that depends upon a draft that is induced by a fan that is integral to the appliance and is downstream from the burner.

L

Labeled: Devices, equipment, or materials affixed with a seal or other identifying mark to attest that the product complies with specific standards of a testing laboratory, inspection agency, or other organization concerned with product evaluation. See "Listed."

Liquefied petroleum (LP) Gas: LP or propane gas is composed primarily of propane, propylene, butanes, or butylenes or mixtures thereof that is gaseous under normal atmospheric conditions but is capable of being liquefied under moderate pressure at normal temperatures. LP gas is typically stored in tanks on-site. LP gas is heavier than air.

Listed: Equipment that is shown on a list published by an approved testing agency determining that the equipment complies with nationally recognized standards when installed in accordance with the manufacturer's installation instructions. See "Labeled."

Log lighter, gas-fired: A manually operated solid-fuel ignition device for installation in a vented solid-fuel burning fireplace. These are intended to help initiate a fire in a fireplace.

Low-pressure hot-water heating boiler: A boiler furnishing hot water at pressures not exceeding 160 psi or temperatures not exceeding 250°F.

Low-pressure steam-heating boiler: A steam boiler that operates at pressures not exceeding 15 psi.

Luminaire: A complete lighting fixture, including the lamp(s), mounting assembly, and cover.

M

Makeup air: Air provided to replace air being exhausted.

MP regulator: A line pressure regulator that reduces gas pressure from the range of 0.5 psig or greater to a lower level. These are often found in CSST systems with a central manifold.

N

Natural-draft burner: A burner that depends on draft induced by the natural draft created by the height and/or temperature/pressure differences of the flue.

Natural gas: Gas, usually odorized methane, supplied to the site by municipal supply system and metered at the site. Natural gas is lighter than air.

Non-combustible material: A material of which no part will ignite and burn when subjected to fire, or material having a structural base of non-combustible material with a surfacing material not over 1/8 in. thick and a flame-spread index not higher than 50. This does not apply to surface-finish materials; the entire material must be noncombustible.

O

Ordinary tightness: Buildings of ordinary tightness are those that do not meet the standards for "unusually tight construction."

Plenum: A chamber, other than the occupied space being conditioned or the ductwork, that forms part of the air-circulation system.

Power vent: See "Forced draft."

Pressure-relief valve (PRV): A device designed to protect against high pressure and to function as a relief mechanism.

R

Room heater (liquid or gas fuel): A room heater installed in the space to be heated and not intended for duct connection.

Room heater (solid fuel): A solid-fuel-burning appliance designed to be operated with the fire chamber door closed.

Room large in comparison to size of equipment: A room having at least 12 times the volume of a furnace or other air-handling device, or 16 times the volume of a boiler. When the ceiling is greater than 8 ft., the volume is calculated based on an 8 ft. height. See "Alcove."

Temperature and pressure-relief valve (TPRV): A device designed to protect against high pressure or temperature and to function as a relief for either.

Unconfined space: A room or space having at least 50 cu. ft. for each 1,000 Btu of the fuel-burning appliances contained in the room or space.

Unlisted: An appliance not shown to comply with nationally recognized standards by an approved testing agency. The appliance could have nameplate instructions.

Unusually tight construction: Construction with walls and ceilings having a vapor retarder of 1 perm or less with sealed or gasketed openings, weather-stripping on openable windows and doors, and caulking or sealant at joints. Buildings of unusually tight construction are required by many energy codes and have an average air infiltration rate <0.35 ACH.

V

Vent: A pipe or other conduit composed of factory-made components, with a passageway for conveying combustion products and air to the atmosphere, listed and labeled for use with a specific type or class of appliance. Special gas vents are listed and labeled for use with Category II, III, and IV gas appliances. B vents are listed and labeled for use with appliances with draft hoods and other Category I appliances listed for use with B vents. Type BW vents are listed and labeled for use with wall furnaces. Type L vents are rated for use with oil-burning appliances and may be used for gas appliances that require type B vents.

Vent connector: A device that connects an appliance to a chimney, flue, or vent.

Vented gas appliance categories:

Category I: An appliance with nonpositive vent static pressure and with a gas vent temperature that avoids excessive condensate production in the vent.

Category II: An appliance with nonpositive vent static pressure and a vent gas temperature that is capable of causing excessive condensate production in the vent.

Category III: An appliance with a positive vent static pressure and with a vent gas temperature that avoids excessive condensate production in the vent.

Category IV: An appliance with a positive vent static pressure and a vent gas temperature capable of causing excessive condensate production in the vent.

Wood stove: See "Fireplace stove" and "Room heater (solid fuel)."

HVAC CODE CHANGES IN 2006

1. 2003 UMC did not req protection against flood hazards.
2. 2003 IRC did not req work space to be level or 30in wide.
3. 2003 UMC did not address appliances in attics.
4. Allowance for 50ft distance is new in the 2006 IRC.
5. Allowance for unlimited distance is new in the 2006 IRC.
6. New requirement for equipment pad under all outdoor appliances.
7. The circuits were req'd to be sized to 125% of their load. By considering them to be continuous loads, the net effect on circuit size is the same.
8. 25amp circuits were not included in the 2003 IRC.
9. New option for an interlocked detector in lieu of a pan or secondary.
10. The requirement for an internal blockage detector is new.
11. The exemption for service receptacles at swamp coolers is new in 2006.
12. New requirement to remove exterior piping of abandoned tanks.
13. Bath fans must now be ducted to or directly connected to the outside of the building.
14. Dryer ducts must now terminate at least 3ft in any direction from other building openings.
15. Booster fans no longer allowed unless make & model of dryer are known & booster installed AMI.
16. ASHRAE calculations now OK for large-radius bends.
17. The rules for box support on paddle fans have been reformatted & moved to the section on boxes, w/ a reference in the appliances section of the electrical codes.
18. New UMC rule to create 18in access under ducts that restrict crawl access.
19. Mechanical fasteners for flex nonmetallic ducts must now be L&L.
20. Adhesives must now also meet the flame & smoke requirements.
21. In 2003, ducts req'd R-5 inside the building but outside the conditioned space. In 2006, ducts inside the building thermal envelope do not req insulation.
22. PEX & PP piping new in the 2006 IRC.
23. 2006 UPC includes new section on indirect-fired water heaters.
24. 2003 IRC req'd a TPRV, not just a temperature limiting valve.
25. New rule for min 12in cover over buried propane lines.
26. The min size of screen mesh is new in the 2006 UMC.
27. Ducts of other materials must now have equivalent strength & rigidity as galvanized steel.
28. New rule that piping may not pass through one townhouse unit to another.
29. The word test has been replaced w/ check to clarify that another pressure test is not being req'd.
30. There are now separate ASTM standards for indoor & outdoor connectors.

31. The max length of rigid piping connectors has been extended to 6 ft.

32. A new rule req's MP regulator vent piping to run independently to outdoors.

33. The rule requiring strike-plate protection is new. Modern vent systems are sometimes plastic, although this rule also includes steel vent piping.

34. The definition of cold has been revised to mean a 99% winter design temperature below 32°F.

35. In 2003, the restriction on connectors was to interior walls & fire-resistance rated walls. It now applies to all walls.

36. This section has been reformatted to eliminate the "exceptions" language.

37. New requirement for auxiliary pan under condensing furnaces where condensate blockage could cause building damage.

38. The 2003 UMC prohibited all unvented fuel-burning room heaters. The 2006 UMC prohibits them from being the primary heat source.

39. The 2003 IRC specified the oxygen-depletion sensor in the text, & the 2006 IRC refers to the same rule in 2445.6.

40. These installations req specific advance approval from the AHJ.

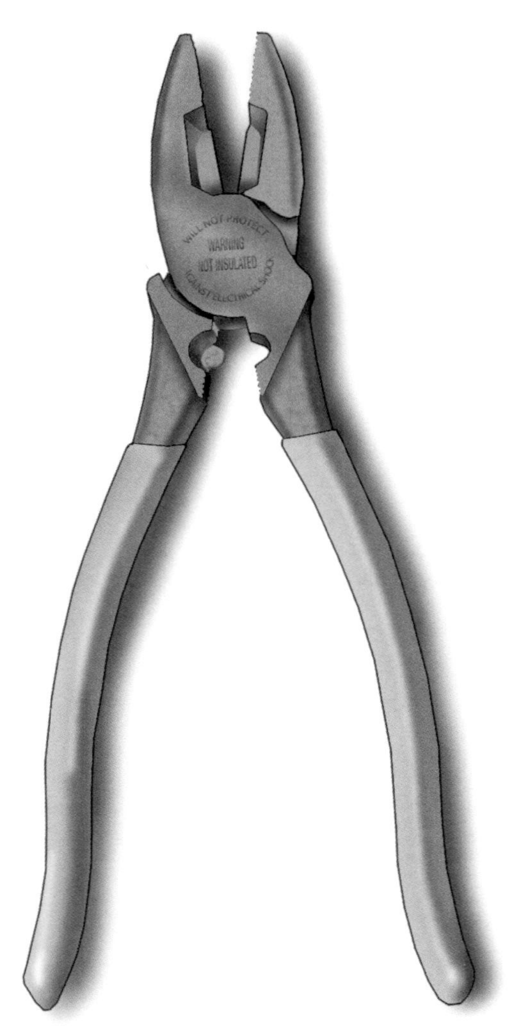

Side Cutters

Code ✓Check® Electrical Fourth Edition

BY REDWOOD KARDON, DOUGLAS HANSEN, AND MICHAEL CASEY

KEY TO THIS SECTION

Code references are followed by two bracketed numbers, ex:

☐ **Code reference** T1,F1 ________ **[123.4]** **{123.4}**

Code numbers on left, in straight brackets, ex: [123.4], refer to 2002 NEC.

Code numbers on right, in braces, ex: {123.4}, refer to 2005 NEC.

T1 refers to Code Check Electrical table 1.

F1 refers to Code Check Electrical figure 1.

[manu] = Typically required by manufacturer's installation instructions.

Adoption and enforcement varies greatly, so check with your local jurisdiction.

An X after a code number refers to an exception in the code.

EXC = When placed at end of text line signals an exception in following line.

OR = When placed at end of text line signals an alternative in following line.

[n/a] = Not addressed by the 2002 NEC.

{n/a} = Not addressed by the 2005 NEC.

[Ø] = Prohibited by the 2002 NEC.

Codes ending in numbers separated by commas refer to multiple code sections, ex: [210.8A2,5X2] = 2002 sections 210.8(A)(2) & 210.5(A)Exception2

A colored code citation followed by a superscript number indicates a change in the code. Ex: [210.8B3][9] refers to a code change in the IRC, listed as #9 in the Code Change Summary on pp 237–238.

EXAMPLES

Example of a code with figure number (from page 201)

☐ Terminal bar for EGCs req'd to be provided **F26,27** [408.20] {408.40}

Panels require a terminal bar for the equipment grounding conductors. Found in 408.20 of the 2002 NEC and 408.40 of the 2005 NEC and is also shown in figures 26 and 27.

Example of a code change (from page 201)

☐ OCPDs readily accessible {& max height 6ft 7in} _ [240.24A] {240.24A}[22]

The explanation on p. 237 states that the 02 code only limited the height of breakers used as switches, and in 05 it applies to all breakers.

Example using "EXC" (from page 198)

☐ Each bldg or structure requires GES EXC **F24** ___ [250.32A] {250.32A}
Bldg w/ only 1 branch ckt with EGC ________ [250.32AX] {250.32AX}

The basic rule requires the disconnecting means to be rated as service equipment with the exception of outbuildings that have only 1 circuit that can be controlled by a snap switch. This example is shown in figure 24.

Example using "OR" (from page 206)

☐ Separate 20A ckt for bath receps only OR ______ [210.11C3] {210.11C3}
Dedicated 20A ckt to each bathroom ________ [210.11C3X] {210.11C3X}

A circuit dedicated to the bathroom receptacles is required, or each bathroom can have its own dedicated circuit that supplies all the equipment in that bathroom.

ABBREVIATIONS

A = amperage, amps (ex.– a 15A breaker)
AC = alternating current or air-conditioning
AC = armored cable, aka "BX"
addl = additional
AFCI = arc-fault circuit interrupter
AHJ = Authority Having Jurisdiction
AL = aluminum
appl(s) = appliance(s)
AWG = American wire gauge
bldg = building
BX® = trade name for AC cable
c&p = cord & plug
CATV = cable television
ckt(s) = circuit(s)
cu = cubic, ex: cu.in–cubic inches
Cu = copper
DB = direct burial
DC = direct current
DW = dishwasher
EGC = equipment grounding conductor
EMT = electrical metallic tubing
ENT = electrical nonmetallic tubing, aka smurf tubing
exc = except
ext = exterior
fixt = lighting fixture(s) (now called luminaire or luminaires)
FLA = full load amps (motor nameplate current rating)
FLC = full load current
FMC = flexible metal conduit, aka greenfield
ft = foot, feet
GEC = grounding electrode conductor
GES = grounding electrode system
GFCI = ground-fault circuit interrupter
horiz = horizontal
hr = hour, hours
HVAC = heating, ventilation & air conditioning
Hz = hertz (frequency, e.g., 60 cycles per second)
IMC = intermediate metal conduit
in = inch, inches
incl = includes, including
kcmil = 1,000 circular mill units (conductor size)
KO = knockouts F9,10
KVA = kilovolt amperes = 1,000 × volts × amps
L&A = lighting & appliance (panelboard)
L&L = listed & labeled, listing & labeling
lb = pound(s)
LCDI = leakage current detection & interruption
LFMC = liquidtight flexible metal conduit, a.k.a. sealtight
LFNMC = liquidtight flexible nonmetallic conduit
lum = luminaire(s) (lighting fixture)
manu = manufacturer
MC = metal clad cable
NM = nonmetallic sheathed cable (Romex®)
OCPD = overcurrent protection device (breaker or fuse)
PMI = per manufacturer's instructions
PV = photovoltaic
PVC = rigid nonmetallic conduit (RNMC)
recep(s) = receptacle outlet(s)
refrig = refrigerator
req = require, requiring
req's = requires
req'd = required
RMC = rigid metal conduit
RNMC = rigid nonmetallic conduit (PVC)
SE = service entrance cable
SFD = single-family dwelling
specs = specifications
sq = square
temp = temperature
UF = underground feeder cable
USE = underground service entrance cable
V = voltage, volts
VA = volt amps
vert = vertical
W = watts
w/ = with
w/o = without

MODEL CODES & ORGANIZATIONS

CSA = Canadian Standards Association; www.cssinfo.com/info/csa.html
IAEI = International Association of Electrical Inspectors; www.iaei.org
IRC = International Residential Code; www.iccsafe.org
NEC = National Electrical Code, published by the NFPA
NEMA = National Electrical Manufacturers Association; www.nema.org
NFPA = National Fire Protection Association; www.nfpa.org
NTRL = Nationally recognized testing laboratory, such as UL or CSA
UL = Underwriters Laboratory; www.ul.com

CONTENTS OF CODE CHECK ELECTRICAL

FIG. 1

The Electrical System

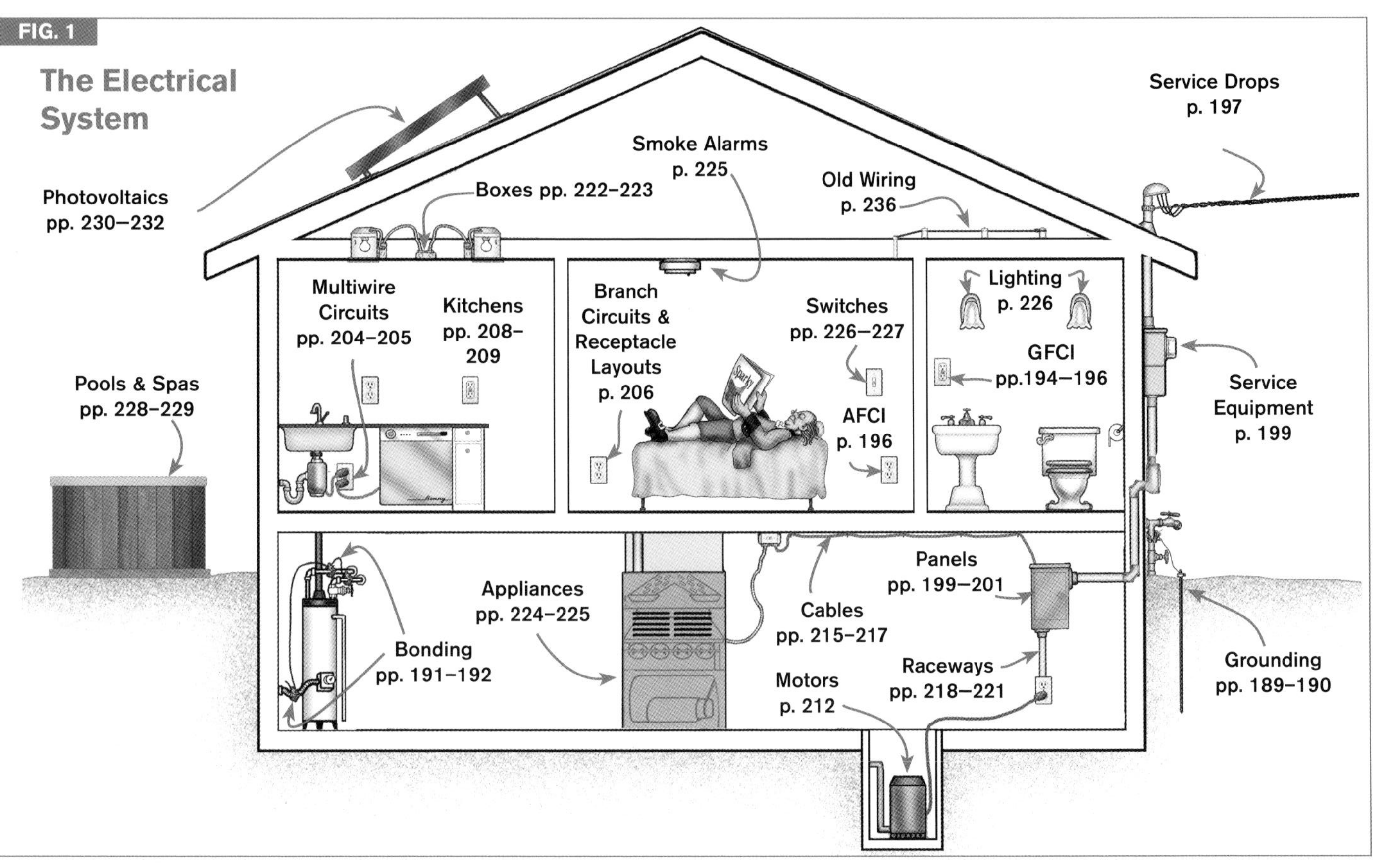
Photovoltaics pp. 230–232
Pools & Spas pp. 228–229
Boxes pp. 222–223
Smoke Alarms p. 225
Old Wiring p. 236
Service Drops p. 197
Multiwire Circuits pp. 204–205
Kitchens pp. 208–209
Branch Circuits & Receptacle Layouts p. 206
Switches pp. 226–227
AFCI p. 196
Lighting p. 226
GFCI pp.194–196
Service Equipment p. 199
Appliances pp. 224–225
Bonding pp. 191–192
Cables pp. 215–217
Panels pp. 199–201
Raceways pp. 218–221
Motors p. 212
Grounding pp. 189–190

GLOSSARY OF ELECTRICAL TERMS

Acccessible: Not permanently concealed or enclosed by building construction.

AFCI, branch/feeder type: An AFCI meeting the standard for interrupting parallel arcs if 75 amps of current are available at the device.

AFCI, combination type: An AFCI meeting the standard for interrupting both series and parallel arcs and requiring less than 75 amps available current to facilitate operation.

Alternating current: Current that flows in one direction and then in the other in regular cycles, referred to as frequency or Hertz.

Apparent power: Available power from a transformer measured in VA.

Appliance: Equipment such as an air-conditioner that uses electrical power.

Branch circuit: The circuit conductors between the final OCPD (breaker or fuse) and the outlet or outlets it supplies.

Branch circuit, general purpose: A branch circuit that supplies a number of outlets for lighting and appliances.

Branch circuit, individual: A branch circuit supplying only one piece of equipment.

Branch circuit, multiwire (residential): A branch circuit consisting of two hot conductors having a 240volt potential between them and a grounded conductor having 120volt difference between it and each hot conductor. F29–33

Branch circuit, small appliance: A branch circuit supplying portable (can be unplugged and moved without tools) household kitchen appliances

Controller: A device to start and stop motors.

Devices: Equipment that carries but does not use electricity. Examples are receptacles, switches, and circuit breakers.

Equipment grounding conductor: A non-current-carrying conductor that provides an alternative path for equipment faults. F14

Feeder: Conductors supplying panelboards other than service panels.

Gooseneck: A curve at the top of a service entrance cable designed to prevent water from entering the open end of the cable.

Grounded conductor: A current-carrying conductor that is connected to earth and that may be a neutral. F5

Ground fault: Current traveling on an unintended path such as an equipment grounding conductor or equipment enclosure.

Hertz: A measure of the frequency of alternating current. In North America the standard is 60 Hertz.

In sight: Visible without obstructions and within 50 feet.

Lighting & appliance panel: An electrical panel for which more than 10% of the circuits are rated at or less than 30 amps and are supplied with neutrals.

Load: The electrical demand in watts or hp of a piece of electrical equipment.

Luminaire: The term now used to describe lighting fixtures.

Open conductors: Individual conductors not contained within a raceway or cable sheathing.

Panelboards: The *guts* of an electrical panel–the assembly of busbars, terminal bars, etc. designed to be placed in a *cabinet*. What is commonly called an electrical panel is, by NEC terms, a panelboard mounted in a cabinet.

Power: A product of volts and amps that can be expressed as either watts (true power) or VA (apparent power). F3

Service: The conductors and equipment providing a connection to the utility.

Service drop: The overhead conductors supplied by the utility.

Service entrance conductors: The conductors on the customer's premises that convey power to the service equipment.

Service equipment: The equipment at which the power conductors entering the building can be switched off to disconnect the premise's wiring from the utility power source.

Service lateral: Underground service entrance conductors.

Service point: The point where the service drop and service entrance meet–it is the handoff between the utility and the customer.

Snap switch: A typical wall switch, including 3-way and 4-way switches.

Ufer: A concrete encased grounding electrode, named after the developer of the system, Herbert Ufer.

Unit switch: A switch that is an integral part of an appliance.

OHM'S LAW

FIG. 2

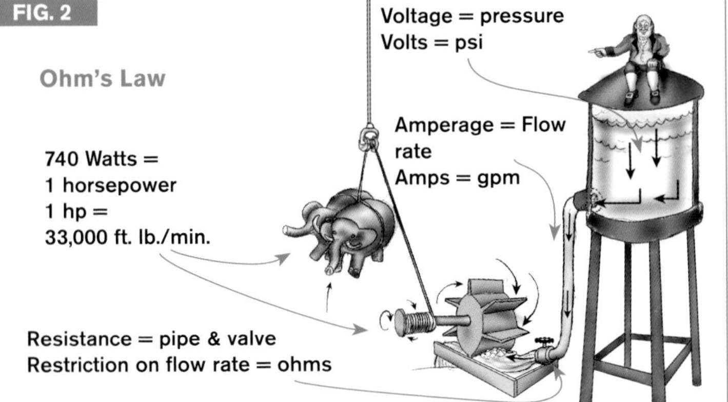

Water flow is often used to describe electricity. In the water analogy, water is the medium to transfer force. In an electrical circuit, the medium is the "free" electrons of conductive metals like copper or aluminum. Water transmits force through the water in a pipe; in an electrical circuit, the force is transmitted over the electrons of a conductor. In F2, we use water to turn a waterwheel. The weight of the water in the tower exerts pressure on the outflow pipe. As the water flows out and under the paddles of the waterwheel, most of the water's kinetic energy (objects in motion) has been converted into rotational force (torque) at the axle of our illustrative contraption. Pressure is said to be dropped across the load. To maintain pressure at the source, new energy must be added–somebody has to scoop up the water, climb the tower, and keep the water level constant. In an electrical system, it is the utility that must maintain this pressure, known as voltage.

The force exerted on the axle in F2 depends on three things: The pressure (volts) in the pipe, the gallons per minute of water flow (amps), and the restrictive effect of the pipe and valve diameter (ohms). There is a simple formula to express this relationship of pressure, flow rate, and resistance. The formula is called Ohm's law after the German scientist George Simon Ohm who first described it. It reads:

I = E/R, or in lay terms,
Amps = Volts/Ohms

E = **electromotive force** = **V** = **voltage**
I = **intensity** = **A** = **current** (measured in amperes)
R = **resistance in ohms**

We can use the energy of the spinning axle to do work (convert energy). The job in this example is to lift 33,000 lb. 1 ft. in 1 min. One healthy draft horse can do it, so we'll call this motor a 1-horsepower motor (1 hp = 33,000ft. lb./min.). Power is the rate at which work is done. More power means more energy is being transferred from one form to another. The rate at which electricity is converted to another form of energy (heat, light, motion) is measured in watts. The power formula is the most commonly used formula for electrical work. It is essential to understanding its application. That formula is expressed as follows:

P (power in watts) = I (amps) × E (volts)

This formula is also useful to utility companies. They can move large amounts of power through small low-current wires by distributing it at high voltage, then lowering the voltage with transformers near the point of use by the customer.

FIG. 3 Ohm's Law & Power Formula

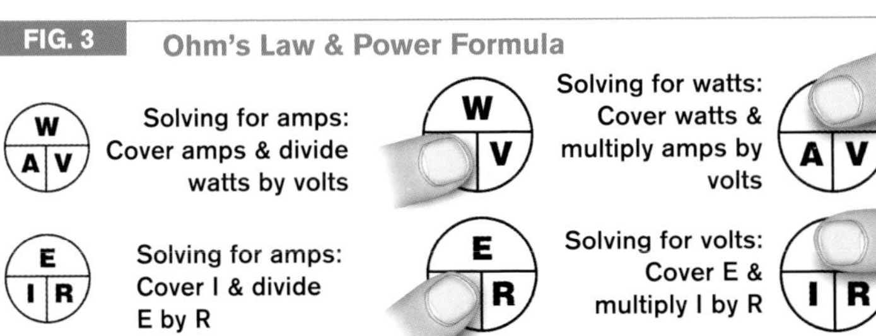

GROUNDING

Benjamin Franklin's dangerous kite experiment proved that lightning was electricity and could be discharged over a conductor and into the ground. Ben's lightning rod and conductor provided the lightning an alternative path to the ground.

A century and a half later Franklin's discovery was applied to the newly emerging electrical infrastructure. Transformers were especially vulnerable to the enormously high voltages of lightning and were grounded to help protect them.

FIG. 4

Ben's Big Idea

The transformer in your neighborhood F5 is a powerful, pulsing (60-Hertz) magnetic field. The high-voltage utility distribution wires transfer energy to the relatively low-voltage (240V) wires connecting to our homes. This 240V secondary coil has a center tap conductor that is grounded, and the voltage between it and either of the "live" ungrounded conductors is 120V. It is important to remember that this intentionally grounded conductor (the "neutral") is designed to carry current all the time. It is a "live" wire.

Once this grounded conductor comes to the building, we connect it again to the earth and to any metal enclosures up to and including those that contain our electrical disconnects and overcurrent protection.

TABLE 1	GEC SIZING (NEC T250.66)	
Cu Service Wire	**AL Service Wire**	**GEC Cu**
2 or <	1/0 or <	8
1 or 1/0	2/0 or 3/0	6
2/0 or 3/0	4/0 or 250 kcmil	4
4/0–350 kcmil	250–500 kcmil	2
>350–600 kcmil	500–900 kcmil	1/0

FIG. 5

Power Transformer

Grounding Electrode System

	2002	2005
☐ Metal water pipe if ≥10ft in contact with soil F6	[250.52A1]	{250.52A1}
☐ Bond around water meters, filters, etc	[250.52A1]	{250.52A1}
☐ Water pipe cannot be only electrode	[250.53D2]	{250.53D2}
☐ Cu rods min 2in diameter and L&L	[250.52.A5]	{250.52A5}
☐ Rods min 8ft in contact with soil	[250.52.A5]	{250.52A5}
☐ Drive rods vertical & fully below grade EXC F6–8	[250.53G]	{250.53G}
If bedrock encountered, rod may be buried horiz 22 ft deep or driven at 45° angle	[250.53G]	{250.53G}
☐ If water pipe & rod are only electrodes, rod max resistance 25 ohms *or* install 2nd rod min 6ft from 1st & bond two rods together	[250.56]	{250.56}
☐ Ufer electrode = 20ft #4 rebar near bottom of footing OR 20ft 4AWG Cu wire near bottom of footing F6	[250.52A3]	{250.52A3}
☐ Ufer must be used if present during construction	[250.50]	{250.50}[1]
☐ Ufer not req'd in in existing bldg if concrete would have to be disturbed to gain access	[n/a]	{250.50X}[1]
☐ Ordinary tie-wires OK for bonding Ufer sections	[250.52A3]	{250.52A3}
☐ Ufer must have min 2in min concrete encasement	[250.52A3]	{250.52A3}
☐ Metal well casings unless bonded to metal pipe	[250.52A7]	{250.52A7}[2]
☐ Metal bldg frame if bonded to other electrodes or if ≥10ft in contact with earth or encased in concrete in earth	[250.52A2]	{250.52A2}[3]
☐ Underground gas pipe not OK as electrode	[250.52B1]	{250.52B1}

FIG. 6

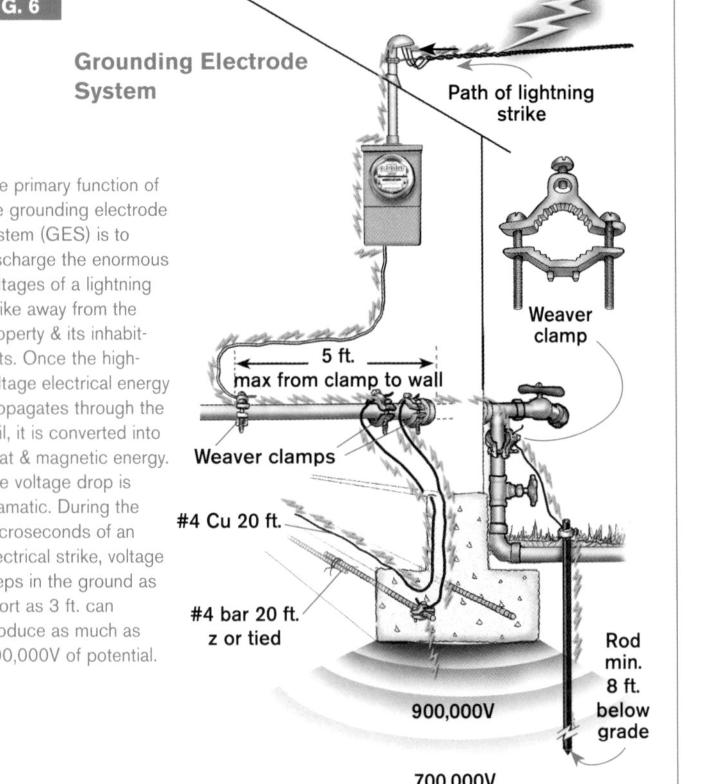

Grounding Electrode System

The primary function of the grounding electrode system (GES) is to discharge the enormous voltages of a lightning strike away from the property & its inhabitants. Once the high-voltage electrical energy propagates through the soil, it is converted into heat & magnetic energy. The voltage drop is dramatic. During the microseconds of an electrical strike, voltage steps in the ground as short as 3 ft. can produce as much as 100,000V of potential.

Grounding Electrode Conductor

	2002	2005
☐ GEC must connect to neutral at or before service	[250.24A1]	{250.24A1}
☐ Size per service conductor size **T1**	[250.66]	{250.66}
☐ 6AWG Cu largest size GEC needed if ending at rod	[250.66A]	{250.66A}
☐ 4AWG Cu largest size GEC needed if ending at Ufer	[250.66B]	{250.66B}
☐ 8AWG must be protected by raceway or armor	[250.64B]	{250.64B}
☐ 6AWG OK unprotected if not subject to damage & following bldg contour **F7**	[250.64B]	{250.64B}
☐ Bond each end of metal raceway enclosing GEC **F8**	[250.64E]	{250.64E}
☐ No splices between service & GES	[250.64C]	{250.64C}
☐ GEC can connect to any part of GES	[250.64F][4]	{250.64F}
☐ Connections to metal water pipe that is part of GES must be within first 5ft of water pipe inside bldg **F6**	[250.52A1]	{250.52A1}

GEC Connections

	2002	2005
☐ Buried clamps must be L&L for direct burial **F6**	[250.70]	{250.70}
☐ Cu water tubing clamps L&L for Cu tubing **F6**	[250.70]	{250.70}
☐ Ufer clamps L&L for rebar & encasement **F6**	[250.70]	{250.70}

Note: Rebar can be brought through the top of a foundation in a protected location, such as the garage, to provide an accessible location for the Ufer clamp.

	2002	2005
☐ Strap-type clamps suitable only for phone systems	[250.70]	{250.70}
☐ Max 1 conductor per clamp unless listed for more	[250.70]	{250.70}
☐ Connections must be accessible EXC	[250.68A]	{250.68A}
Buried or encased connections **F6**	[250.70]	{250.70}

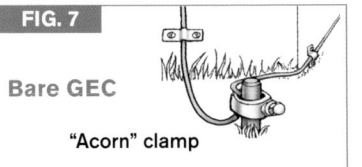

FIG. 7

Bare GEC

"Acorn" clamp

FIG. 8

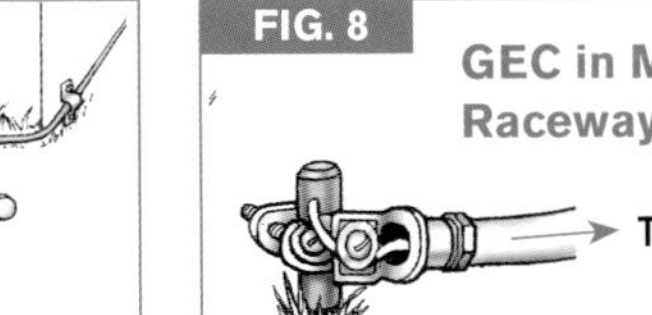

GEC in Metal Raceway

BONDING

Bonding ensures electrical continuity to prevent differences of voltage potential between conductive components. On the line side (ahead of the main disconnect), it provides a path back to the transformer for faults on service conductors and to limit voltage potential to other systems, such as telephones or cable TV. On the load side (after the main overcurrent protection), it provides a path back to the transformer to clear faults and protect against shocks.

Bonding Methods	2002	2005
☐ Use listed materials or exothermic welding	[250.8]	{250.8}
☐ No sheet-metal screws to conductors {or lugs}	[250.8]	{250.8}[5]
☐ Clean nonconductive coatings from contact surfaces	[250.12]	{250.12}

Line-Side Bonding		
☐ Bond all service equipment & conduits & GEC enclosures	[250.92A]	{250.92A}
☐ Threaded fittings OK for bonding service conduit	[250.92B2]	{250.92B2}
☐ Meyers hub OK for bonding service conduit **F10**	[250.92B2]	{250.92B2}
☐ Standard locknuts alone not sufficient on line side of service **F10**	[250.92B]	{250.92B}
☐ Jumpers req'd around concentric KOs on line-side of service **F9** OR Grounding locknuts OK if no remaining concentrics **F10**	[250.92B4]	{250.92B4}
☐ Service neutral can bond line side equipment	[250.142A]	{250.142A}
☐ Size line side bonding jumpers per **T1**	[250.102C]	{250.102C}

Intersystem Bonding		
☐ Min 6AWG Cu bond to CATV or phone electrodes **F11**	[800.40D]	{800.100D}
☐ Bond lightning protection system to GEC **F11**	[250.106]	{250.106}
☐ Provide access for intersystem bonding & GECs on ext of service equipment **F11**	[250.94]	{250.94}[6]

FIG. 9 Fittings with Concentric Rings

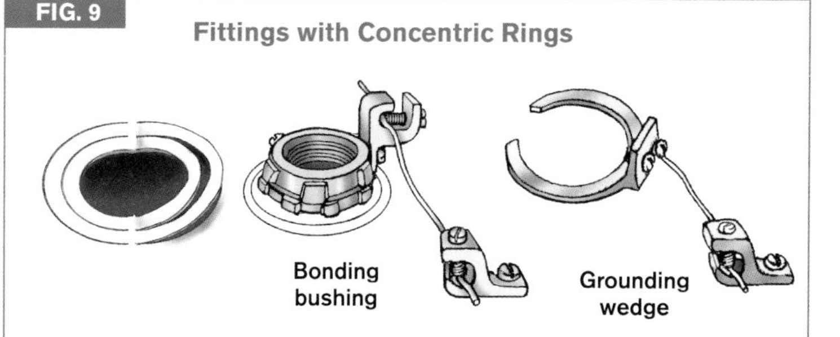

FIG. 10 Fittings with Clean Holes

Load Side Bonding	2002	2005
☐ Bond all metal piping, hot, cold, & gas **F12**	[250.104A,B]	{250.104A,B}
☐ Size water pipe bonding per **T1**	[250.104A1]	{250.104A1}
☐ Size gas pipe bonding per **T2**	[250.104B]	{250.104B}
☐ Bond metal well casings to EGC of pump motor	[250.112M]	{250.112M}

FIG. 11

Intersystem Electrode Bonding

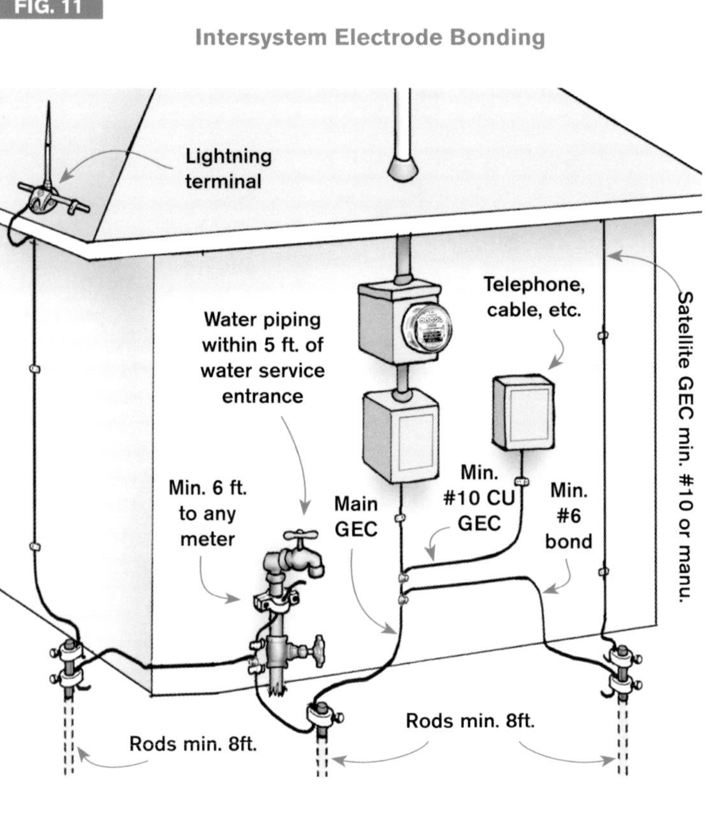

EQUIPMENT GROUNDING CONDUCTORS

Equipment Grounding Conductor (EGC)	**2002**	**2005**
☐ Size eqpmt grounding conductors per T2	[250.122]	{250.122}
☐ EGCs can be bare, covered, or insulated	[250.119]	{250.119}
☐ EGC insulation green or green w/ yellow stripes	[250.119]	{250.119}
☐ EGC >6AWG OK to strip bare or use green tape for entire exposed length inside panels or boxes	[250.119A]	{250.119A}
☐ RMC, IMC, or EMT OK as EGC	[250.118]	{250.118}
☐ FMC & LFMC OK as EGC for non-motor ckts in combined lengths to 6ft if proper fittings *(see **p. 219**)*	[250.118]	{250.118}
☐ Remove paint from contact surfaces of conduit fittings & ensure good contact	[250.96]	{250.96}
☐ EGCs must be run w/ other conductors of ckt EXC	[300.3B]	{300.3B}
Replacing nongrounding receps *(see **pp. 234–235**)*	[250.130C]	{250.130C}

TABLE 2	EQUIPMENT GROUNDING CONDUCTORS (EGCs) NEC T250.122	
Size in Amps of Breaker or Fuse Protecting Circuit	AWG size of Copper EGC	AWG size of Aluminum EGC
15	14	12
20	12	10
30	10	8
100	8	6

FIG. 12

Bonding the Piping System

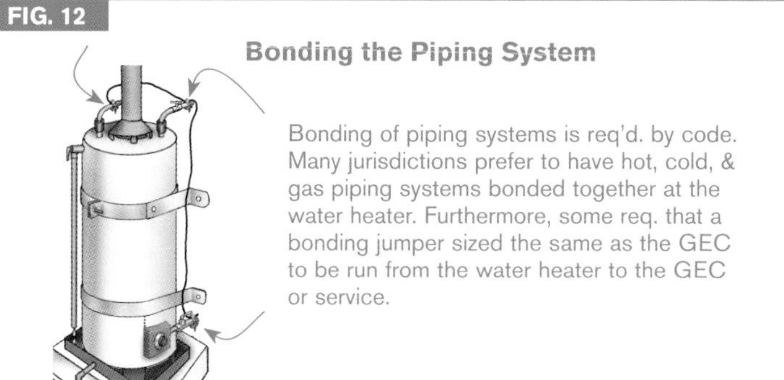

Bonding of piping systems is req'd. by code. Many jurisdictions prefer to have hot, cold, & gas piping systems bonded together at the water heater. Furthermore, some req. that a bonding jumper sized the same as the GEC to be run from the water heater to the GEC or service.

FIG. 13

An Equipment Fault

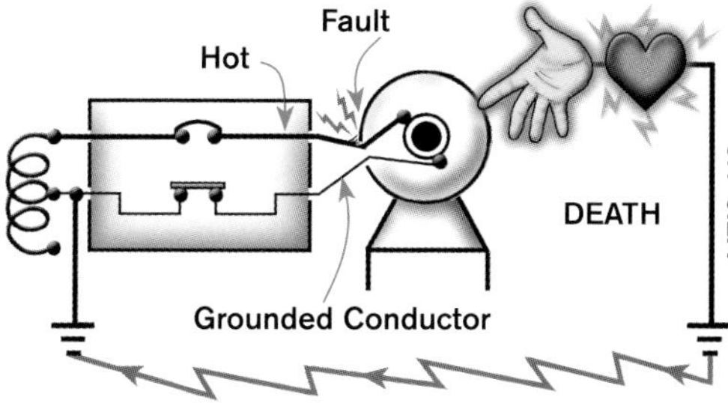

Hot

Fault

Grounded Conductor

DEATH

0.1-0.2A

Unlike the grounded conductors, EGCs are not designed to carry current under normal operation. They are part of a safety backup system that is designed to protect life & property from accidental eqpmt. faults. Here, the motor has a deadly potential to ground (the earth). The current is looking to return to its source at the transformer. A person contacting a grounded surface–ex. plumbing, earth, etc.–could provide a parallel return path.

FIG. 14

Clearing a Fault

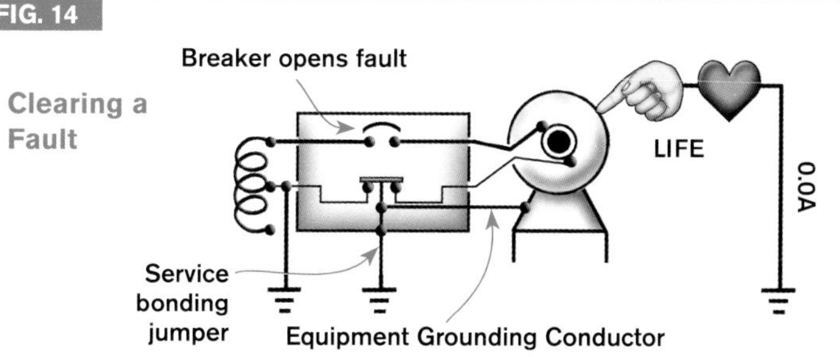

An EGC has been added to the motor, & it serves to carry the fault current & trip the breaker. Even if the fault was insufficient to trip the breaker, the EGC would provide electrical equipotential that would help protect anyone touching the motor.

FIG. 15

Futility of Using Earth as the Equipment Ground

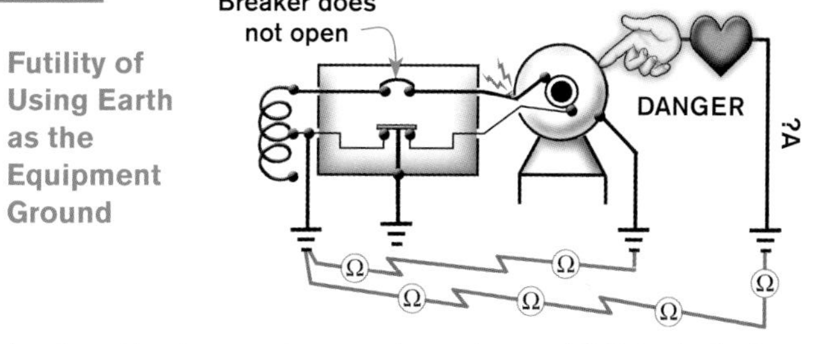

An attempt has been made to use the earth as an EGC. Ironically, the earth might be a good enough conductor to carry the amount of current necessary to shock or electrocute a person, but it is not a good enough conductor to trip the breaker. You can't rely on the earth to clear faults.

GFCI (GROUND-FAULT CIRCUIT INTERRUPTER)

FIG. 16

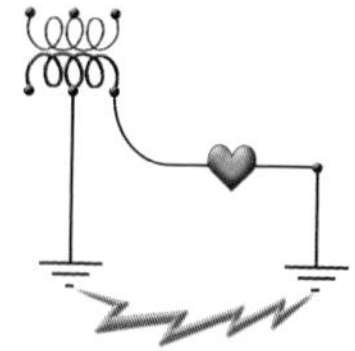

To Ground or Not to Ground

Ground-fault circuit interrupters (GFCIs) detect the escape of electrical current outside the intended circuit. An example is the current returning to the transformer through a person, as represented in **F16**. GFCIs save lives by limiting the duration of the fault.

How does a GFCI work its magic? In **F17**, equal currents are flowing to and from the load. When any electrical current flows, it generates a magnetic field. The magnetic fields generated by the flow of electrons in these two conductors are of opposite polarity (north and south, leaving and returning). The forces are equal and opposite, and their magnetic fields cancel each other. The circuit passes through a coil of wire inside the GFCI, and the GFCI accounts for the electrons on each conductor. As long as the currents are balanced, the GFCI allows current on the circuit.

FIG. 17

GFCI

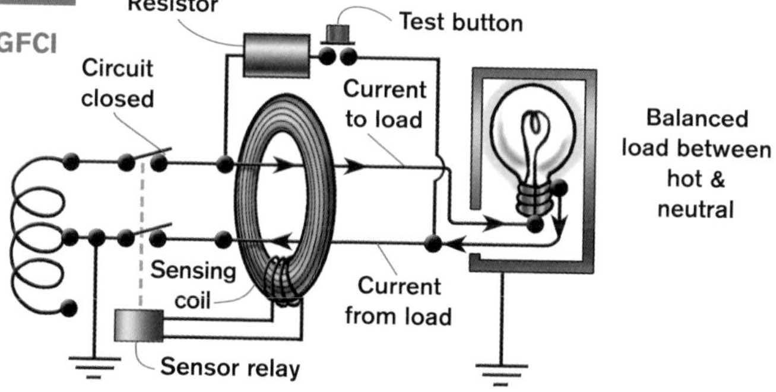

Resistor
Test button
Circuit closed
Current to load
Balanced load between hot & neutral
Sensing coil
Current from load
Sensor relay

During a ground fault, such as the flow of current through a person, the circuit becomes unbalanced **F18**. A magnetic field is generated on the coil around the circuit wires. Solid-state circuitry connected to the sensing coil activates a trip mechanism, and the circuit is interrupted.

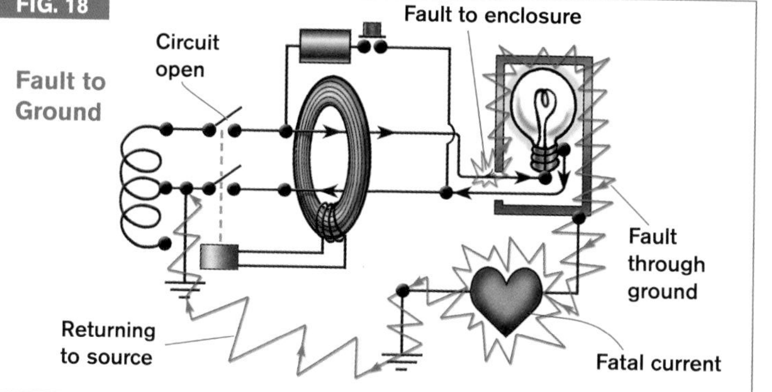

FIG. 18

Fault to Ground

A GFCI also detects improper connections of the neutral (grounded conductor) to ground. A second "injector" coil **F20** surrounds the monitored circuit and induces a small current. Should the neutral have a downstream connection to ground, current will escape outside the circuit, and the sensor coil circuit will be activated as described above.

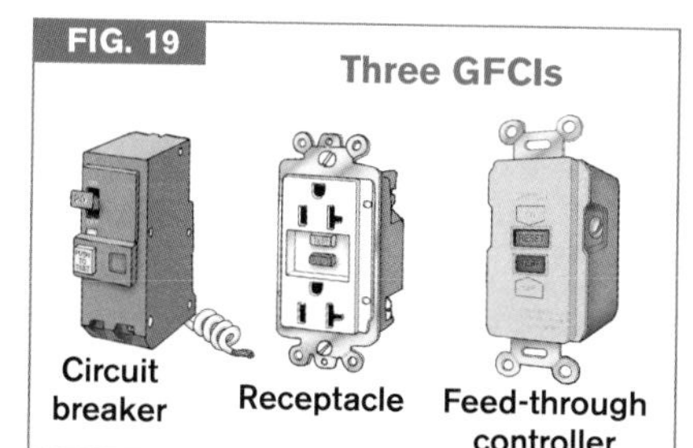

FIG. 19

Three GFCIs

FIG. 20

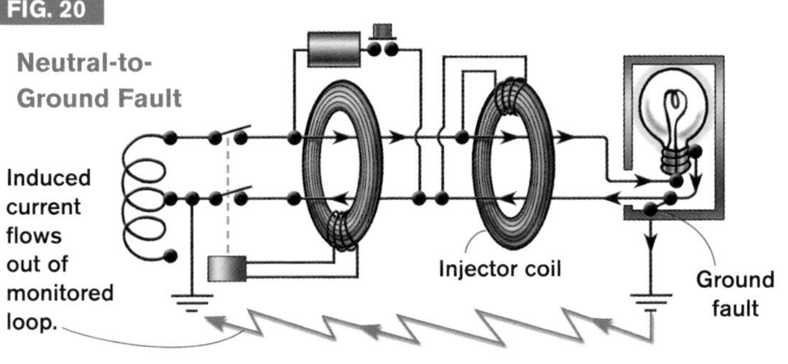

FIG. 21

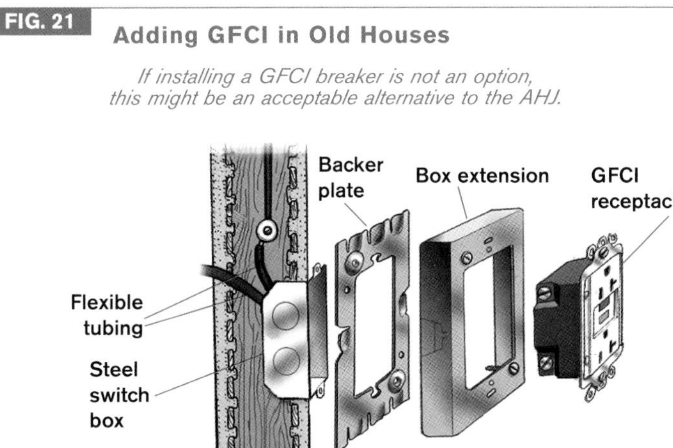

GFCIs take more space inside a box than a conventional receptacle. When adding GFCIs to old houses with shallow boxes, it might be necessary to first add an extension box, as in F21.

> A GFCI will operate properly without an equipment ground. The receptacle should be labeled "no equipment ground" & any downstream protected receptacles should also have that label as well as a label stating that they are GFCI protected. Labels are not required for properly grounded GFCI-protected receptacles.

Required GFCI Protection Locations	2002	2005
☐ All receps serving kitchen counters F37	[210.8A6]	{210.8A6}
☐ All bathroom receps	[210.8A1,B1]	{210.8A1, B1}
☐ All residential ext receps EXC	[210.8A3]	{210.8A3}
De-icing equipment receps w/o ready access	[210.8A3,X]	{210.8A3X}
☐ Receps ≤6ft of {laundry, util or} wet-bar sinks	[210.8A7]	{210.8A7}[7]
☐ Commercial ext areas accessible to public	[n/a]	{210.8B4}[8]
☐ Commercial kitchen receps {in areas w/ permanent food-preparation facilities}	[210.8B3][9]	{210.8B2}[10]
☐ Commercial 120V receps on rooftops	[210.8B2]	{210.8B3}
☐ Outdoor 120V receps for servicing HVAC equipment	[n/a]	{210.8B5}[11]
☐ All garage & unfinished basement receps EXC	[210.8A2,5]	{210.8A2,5}
Receps that are not readily accessible OR	[210.8A2,5X1]	{210.8A2,5X1}
Single receps for appls not easily moved (sump, freezer, clothes washer) or duplex recep for 2 such appls	[210.8A2,5X2]	{210.8A2,5X2}
☐ All 125V, 15A, 20A, & 30A receps for temporary power	[527.6A]	{590.6A}
☐ Receps in boathouses	[210.8A8]	{210.8A8}

Pool, Spa, & Hydromassage Tub GFCI Requirements

	2002	2005
☐ Receps within 20ft of pools & outdoor hot tubs unless cord would have to pass through window or door	[680.22A5,6]	{680.22A5,6}
☐ Receps for pool pump motors any distance from pool	[680.22A5][12]	{680.22A5}
☐ Receps providing spa or hot tub power	[680.43A3]	{680.43A3}
☐ Receps >5ft & <10ft of indoor hot tubs	[680.43A2]	{680.43A2}
☐ Pool cover motor & controller	[680.27B2]	{680.27B2}
☐ Hydromassage (whirlpool) tubs	[680.71]	{680.71}
☐ Underwater pool lights >15V **F78**	[680.23A3]	{680.23A3}
☐ Light fixt <10ft horiz unless >5ft vert from water	[680.22B4]	{680.22B4}
☐ Existing fixt OK <5ft horiz if >5ft vert from water	[680.22B4]	{680.22B3}
☐ GFCI protect all receps ≤10ft from inside wall of spa	[680.43A2]	{680.43A2}
☐ GFCI protect all outlets that supply spa equipment exc Listed package spa w/ integral GFCI *or* Combination pool & spa or hot tub	[680.44]	{680.44}

UL 943–the standard of safety for GFCIs–was revised in 2003, requiring GFCIs to have greater resistance to corrosion, surge currents, and high voltages. The revised standard also requires a reverse line-load test that prevents them from resetting if they are miswired. The contacts on these newer GFCIs ensure proper resetting and prevent some miswiring that could appear from manipulation of the controls on the older GFCIs. In addition, all manufacturer's installation instructions for GFCIs are now standardized for consistency. These instructions require specific methods for checking GFCI operation after installation to ensure that devices are properly wired. As a result, these proven life-savers have become more reliable than ever.

AFCI (ARC FAULT CIRCUIT INTERRUPTER)

Arc-fault circuit interrupters are intended to provide fire protection by opening the circuit if an arcing fault is detected. Though they look similar to GFCI breakers, they do not provide protection against shock hazards at the same sensitivity as a GFCI.

	2002	2005
☐ AFCI protect all 15A & 20A circuits w/ any outlet (recep, lum, or smoke detector) in bedrooms	[210.12B][13]	{210.12B}
☐ AFCI must be "combination type" after January 1, 2008	[n/a]	{210.12B}[14]
☐ Recep device type OK if ≤6ft from panel & wiring between panel & device is metal-clad cable or raceway	[n/a]	{210.12BX}[15]

FIG. 22

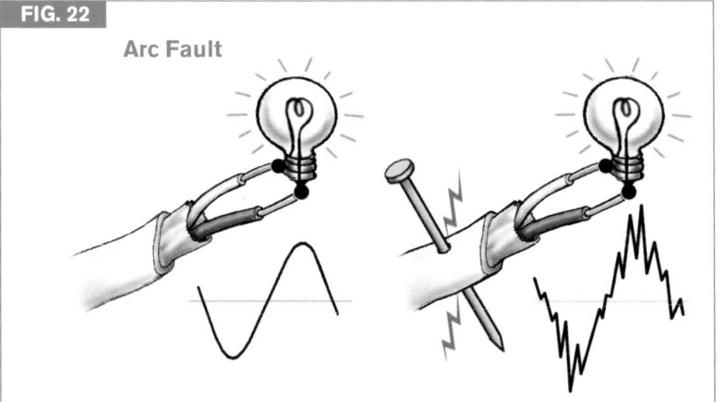

FIG. 23 Service Drop Clearances

SERVICE DROPS

The utility company does not necessarily follow the rules in the NEC. Check with your local jurisdiction and utility to find out what rules apply.

Clearances above Ground	2002	2005
☐ Area accessible only to pedestrians–10ft vert **F23**	[230.24B1]	{230.24B1}
☐ General above ground–12ft vert **F23**	[230.24B2]	{230.24B2}
☐ Above driveway–12ft **F23**	[230.24B2]	{230.24B2}
☐ Above roadway–18ft **F23**	[230.24B4]	{230.24B4}
☐ Any direction from pool water–22 1/2ft	[680.8A][16]	{680.8A}[17]
☐ Trees may not support overhead conductors	[230.10][18]	{230.10}

Clearances above Roof	2002	2005
☐ <4-in-12 slope–min 8ft	F23 [230.24A]	{230.24A}
☐ 4-in-12 slope or greater–min 3ft EXC	[230.24AX2]	{230.24AX2}
18in OK over eave if mast ≤4ft horiz to edge **F25**	[230.24AX3]	{230.24AX3}

Clearances from Openings		
☐ Below or to sides of openable window–3ft **F23**	[230.9A]	{230.9A}
☐ Above decks & balconies–10ft out to 3ft horiz **F23**	[230.9B]	{230.9B}

The NEC does not have a requirement for minimum clearance of open conductors above a window. Check to see if your local utility has a requirement.

UNDERGROUND

	2002	2005
☐ Burial depth & cover per **T3**	[300.5]	{300.5}
☐ Protect USE cable where exposed **F46**	[300.5D1]	{300.5D1}
☐ Warning ribbon in trench 12in above lateral	[300.5D3]	{300.5D3}

SERVICE ENTRANCE CONDUCTORS

Unlike overhead service drop conductors, the service entrance conductors are part of the building. The handoff from the utility to the customer is referred to as the "service point," and usually it is the splice point at the drip loop. For underground service laterals, the service point is specified by the utility.

General

	2002	2005
☐ SFD min wire size 4AWG CU or 2AWG AL **T9**	[T310.15B6]	{T310.15B6}
☐ Identify (white tape) neutral at both ends	[200.6B]	{200.6B}
☐ Protect SE cables where subject to damage w/ metal conduit or RNMC-80 or EMT	[230.50A]	{230.50A}
☐ Secure SE every 30in & 12in from terminations	[230.51A]	{230.51A}
☐ Raintight service head req'd for raceways	[230.54A]	{230.54A}
☐ Raintight service head or taped gooseneck OK in SE cable	[230.54B]	{230.54B}
☐ Install drip loop in conductors	[230.54F]	{230.54F}

Service Riser/Lateral

	2002	2005
☐ Clamp RMC within 3ft of service box	[344.30A]	{344.30A}
☐ Plumbing pipe or fittings not permitted	[110.8]	{110.8}
☐ Min size mast per utility (typically 1 1/4–2in RMC)	[utility]	{utility}
☐ Brace riser to utility or local specs	[230.28]	{230.28}
☐ No unsupported couplings above roof	[local]	{local}
☐ Only service conductors may be supported by mast	[230.28]	{230.28}
☐ Service lateral buried proper depth **T3**	[300.5A]	{300.5A}
☐ Check w/ utility re: other systems in trench	[utility]	{utility}

TWO BUILDINGS

When more than one building is supplied by a service, care must be taken to avoid objectionable currents on the grounding paths between the buildings.

	2002	2005
☐ Each bldg or structure req's GES EXC **F24**	[250.32A]	{250.32A}
Bldg w/ only 1 branch ckt w/ EGC	[250.32A,X]	{250.32A,X}
☐ Multiwire ckt considered one ckt for above rule	[n/a]	{250.32A,X}[19]
☐ Each bldg req's disconnect at bldg **F24**	[225.31]	{225.31}
☐ Disconnect must be rated as service equipment EXC	[225.36]	{225.36}
Garages or outbuildings snap switch or 3-way OK	[225.36X]	{225.36X}
☐ EGC (4-wire feeder) req'd between bldgs EXC	[250.32B1]	{250.32B1}
3-wire feed OK if no other metal path between bldgs	[250.32B1,2]	{250.32B1,2}
☐ Isolate neutral from EGC in feeder & subpanel	[250.32B1]	{250.32B1}
☐ Provide proper cover for buried cable or conduit **T3**	[300.5]	{300.5}

FIG. 24 Grounding at Separate Buildings

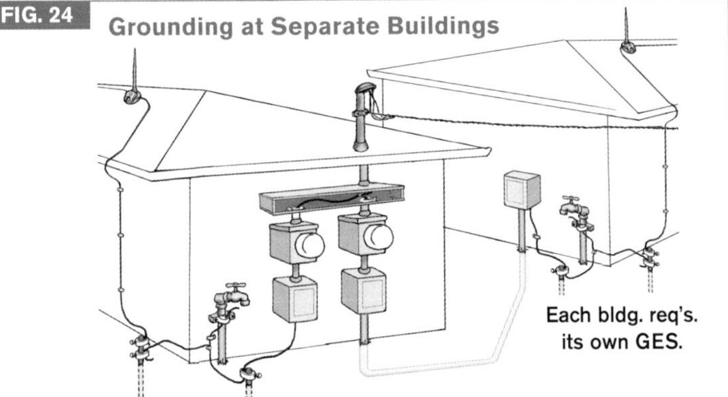
Each bldg. req's. its own GES.

TABLE 3	MINIMUM COVER REQUIREMENTS IN TRENCH				
Cover	UF Cable	Rigid Metal	PVC	GFCI 20A Circuit	30V
General	24 in.	6 in.	18 in.	12 in.	6 in.
2-in. concrete	18 in.	6 in.	12 in.	6 in.	6 in.
Under bldg.	n/a	0	0	n/a	n/a
4-in. slab no vehicles	18 in.	4 in.	4 in.	6 in.	6 in.
Street	24 in.	24 in.	24 in.	24 in.	24 in.
Driveway	18 in.	18 in.	18 in.	12 in.	18 in.

FIG. 25

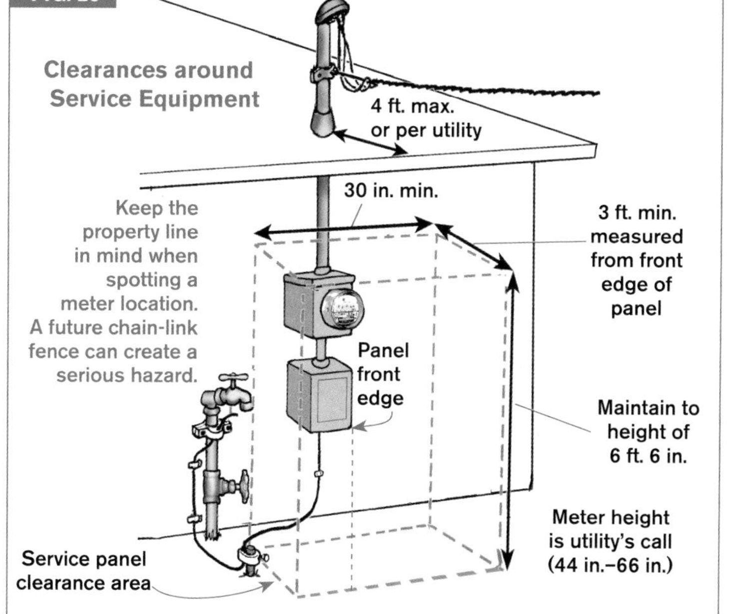

METERS & SERVICE PANELS

An overcurrent protective device (OCPD), such as a fuse or circuit breaker, is specifically designed to protect electrical circuits against the hazardous effects of overcurrents. Service equipment encloses the disconnecting means to shut off the power to a building. A meter is not necessarily service equipment.

	2002	2005
☐ Max 6 disconnects to shut off power panelboard EXC	[230.71]	{230.71}
Max 2 OCPDs for L&A panels ________	[408.16A]	{408.36A}
☐ Provide working space **F25** ________	[110.26]	{110.26}
☐ Occupants to have access to all OCPDs EXC____	[240.24B]	{240.24B}
Multi-unit bldg OK for management-only access to mains {& to other OCPDs in commercial guest suites} _	[240.24BX]	{240.24BX}[20]

	2002	2005
☐ Illumination req'd for indoor equipment ________	[110.26D]	{110.26D}
☐ Max height of breakers [used as switches] 6ft 7in __	[404.8A]	{240.24A}[21]
☐ Enclosure labeled suitable for service equipment ___	[230.66]	{230.66}
☐ Verify location & hookup fees w/ util company ______	[utility]	{utility}
☐ Breakers correct brand per panel labeling________	[110.3B]	{110.3B}
☐ Antioxidant on AL conductors PMI________	[110.3B,110.14]	{110.3B,110.14}

FIG. 26

Service Panel

This is a common style of panelboard. When used as a service, the white neutral & the bare or green equipment grounds *must* be bonded together *and* to the enclosure. This connection is essential to the safe operation of the grounding system.

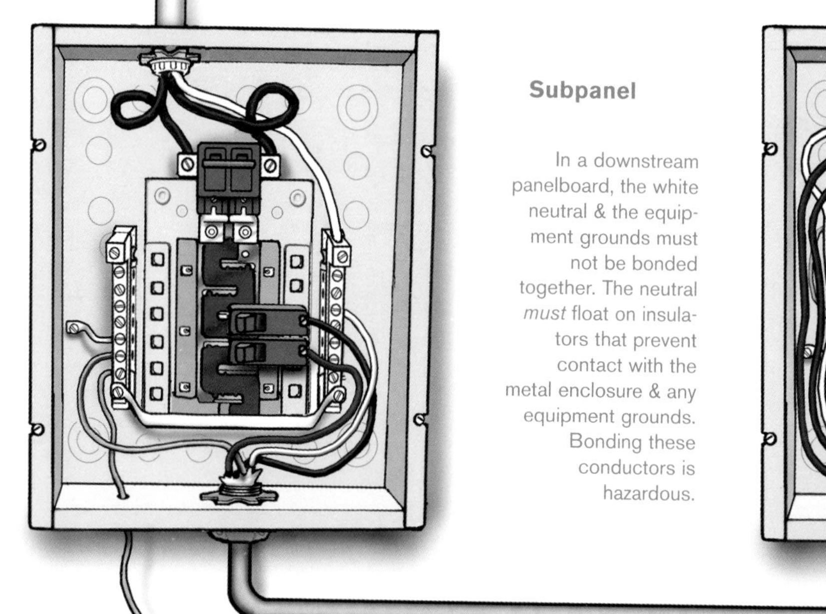

FIG. 27

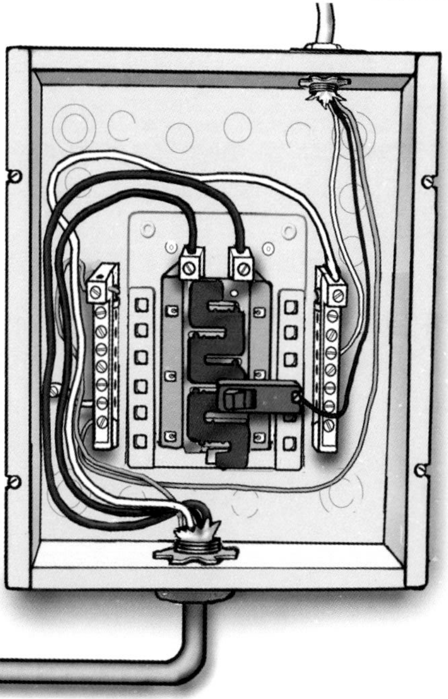

Subpanel

In a downstream panelboard, the white neutral & the equipment grounds must not be bonded together. The neutral *must* float on insulators that prevent contact with the metal enclosure & any equipment grounds. Bonding these conductors is hazardous.

The neutral conductor is bonded to service enclosures. Once the neutral conductors leave the service panel, they must be isolated from contact with equipment and enclosures, otherwise current could return on EGCs, enclosures, and piping. Subpanel feeders require four conductors–two ungrounded (hot) conductors, a neutral, and an EGC (which can be metal conduit).

PANELBOARDS & PANEL ENCLOSURES

Panels and their enclosures must be installed and used in accordance with all instructions and specifications from the manufacturer.

Location & Enclosures	2002	2005
☐ Front working clearance min 30in wide by 36in deep **F25**	[110.26A]	{110.26A}
☐ Wet location enclosures req'd to be weatherproof	[312.2A]	{312.2A}
☐ Surface-mounted wet location boxes 1/4 space from wall	[312.2A]	{312.2A}
☐ Open KOs must be filled (not taped)	[110.12A]	{110.12A}
☐ No OCPDs in clothes closet or bathroom	[240.24D,E]	{240.24D,E}
☐ OCPDs readily accessible {& max height 6ft 7in}	[240.24A]	{240.24A}[22]
☐ Max panel setback in noncombustible wall (drywall) 1/4in	[312.3]	{312.3}
☐ Max panel setback in combustible wall (wood) 0in	[312.3]	{312.3}
☐ Max plaster gap at side of flush mount panel 1/4in	[n/a]	{312.4}[23]

OCPDs & Wiring		
☐ Secure cables entering panel **F28**	[312.5C]	{312.5C}
☐ Breakers correct brand per panel labeling	[110.3B]	{110.3B}
☐ Isolate neutral from EGCs & GES exc at service **F26,27**	[408.20]	{408.40}
☐ Modifications to split neutral bar PMI **F26,27**	[manu]	{manu}
☐ Terminal bar for EGCs req'd to be provided **F26,27**	[408.20]	{408.40}
☐ Purpose of breakers & fuses legibly marked	[408.4]	{408.4}
☐ Ckt identification must be clear, evident, & specific	[n/a]	{408.4}[24]
☐ Approved handle ties OK for 240V ckts w/single pole breakers **F70**	[240.20B2]	{240.20B2}
☐ Handle tie for multiwire ckts to same device **F36**	[210.4B]	{210.4B}
☐ Handle tie for two ckts to receps on same yoke	[210.7C][25]	{210.7B}
☐ Missing twistouts must have fill plates (not tape)	[110.12A]	{110.12A}

OCPDs & Wiring (cont.)	2002	2005
☐ Antioxidant on AL conds PMI	[110.14]	{110.14}
☐ Each neutral cond req's individual terminal	[408.21][26]	{408.41}

Breakers (OCPDs) serve 4 primary functions:

1. Provide a disconnecting means for a circuit
2. Open circuits (stops current flow) if conductor overloaded
3. Open circuits (stops current flow) if short circuit occurs
4. Open circuits (stops current flow) if ground fault occurs

FIG. 28

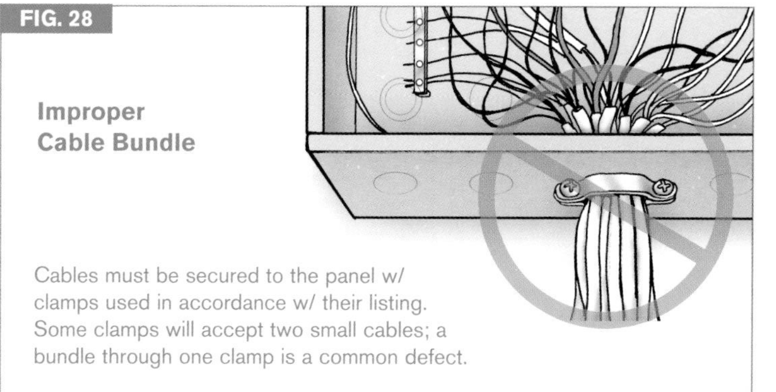

Improper Cable Bundle

Cables must be secured to the panel w/ clamps used in accordance w/ their listing. Some clamps will accept two small cables; a bundle through one clamp is a common defect.

SERVICE & FEEDER CALCULATIONS

The guidelines in the NEC help anticipate the size and number of circuits we need to supply the loads in a typical residence. As a general rule, service conductors, feeders, and branch circuits must be sufficient to serve the connected load. Where loads may vary depending on the particular user, the NEC allows us to consider "demand factors," i.e., to take into account that not everything will be operated at the same time. The procedures below show how these demand factors are calculated.

Steps for Sizing a Service	2002	2005
☐ **1.** Determine the sq.ft area of the residence & multiply by 3W (exclude garage & covered patios) ______	[220.3A]	{220.12}
☐ **2.** Min of 2 1,500W small-appliance ckts ______	[220.16]	{220.52A}
☐ **3.** Each addl kitchen–min 2 small appl ckts at 1,500W per ckt ______	[210.52B3]	{210.52B3}
☐ **4.** Min 1 laundry ckt at 1,500W ______	[220.16B]	{220.52B}
☐ **5.** Total small appl loads & general lighting (enter in middle column) ______	[220.11]	{220.42}
☐ **6.** Subtract 3,000W from line 5 ______	[T220.11]	{T220.42}
☐ **7.** Enter difference in middle column, **T4** multiply middle column by 35%, & enter in right column ______	[T220.11]	{T220.42}
☐ **8.** Range loads are calculated at nameplate rating. If a single range is >8,000W & <12,000W, it still counts as 8,000W (8kW). Above 12,000W, add 5% of each addl 1,000W of nameplate load. The nameplates of a counter-mounted range & up to 2 wall ovens can be added together and computed as if they were 1 range. Enter in column 3 **T4** ______	[220.19]	{220.55}

Steps for Sizing a Service (cont.)	2002	2005
☐ **9.** Enter dryer nameplate rating or 5,000W, whichever is greater, in column 3 ______	[220.18]	{220.54}
☐ **10.** Enter the larger of the fixed space heating or AC load at nameplate rating in column 3 ______	[220.21]	{220.60}
☐ **11–18.** Enter the nameplate ratings of appls that are fixed in place. To determine the load of appls rated in amps, multiply by the voltage. Enter the actual nameplate ratings; the numbers in the first column are typical examples	[220.17]	{220.53}
☐ **19.** Enter the total of fixed appls in the middle column ______	[220.17]	{220.53}
☐ **20.** If there were <4 fixed appls, enter the number from line 19 in the right column ______	[220.17]	{220.53}
☐ **21.** If there were ≥4 fixed appls, multiply line 19 by 75% & enter in the right column ______	[220.17]	{22053}
☐ **22.** Add 25% of the largest motor load. If a nameplate-rated AC is the largest load, this number has already been factored in, & this step is omitted ______	[220.14]	{220.18A}
☐ **23.** Add the numbers in column 3 ______	[220.10]	{220.40}
☐ **24.** Divide line 23 by 240 to find req'd min amperage ______	[220.10]	{220.40}

TABLE 4 **SIZING ELECTRIC SERVICES**

General Lighting & Receptacle Loads (NEC 220-12):			
Sq. ft. × 3W			1
Small Appl. & Laundry Loads (NEC 220-52A,B)			
2 small appliance ckt.	3,000		2
Addl. small appliance			3
Laundry ckt.	1,500		4
Subtotal general light, small appliance & laundry			5
1st 3,000W @100%	3,000	3,000	6
Balance @35%			7
Special Appliance Loads			
Range (NEC 220-55)	8,000 up to 12kW nameplate		8
Dryer (NEC 220-54)	5,000 (or nameplate if >)		9
Heating or AC @100%			10

Appliances Fastened in Place (NEC 220-53):			
Water heater 4,500[A]			11
Microwave 1,300[A]			12
DW 1,500[A]			13
Compactor 900[A]			14
Disposer 800[A]			15
Attic fan 1,600[A]			16
Spa–per manu.			17
Other			18
Subtotal			19
If <4 appliances, enter subtotal @100% *or*			20
If ≥4 appliances, enter subtotal × 75%			21
Largest motor × 25%			22
Total load			23
Total load ÷ 240V = SERVICE AMPS			24

A. Common ratings–use actual nameplate rating of appliances.

Load Calculations

	2002	2005
☐ Continuous load = 3hr or more	[100]	{100}
☐ Service conductors sufficient for load	[230.42]	{230.42}
☐ Min rating 100 for SFD	[230.79C]	{230.79C}
☐ Feeders sufficient for load	[215.2A1]	{215.2A1}
☐ Branch ckts sufficient for connected load	[210.19A1]	{210.19A1}
☐ [Min feeder size 30A to subpanel]	[215.2A2]	{n/a}[27]

For the simplified "optional method" of load calculation, refer to *Code Check, 5th ed.*

3-WIRE EDISON CIRCUITS

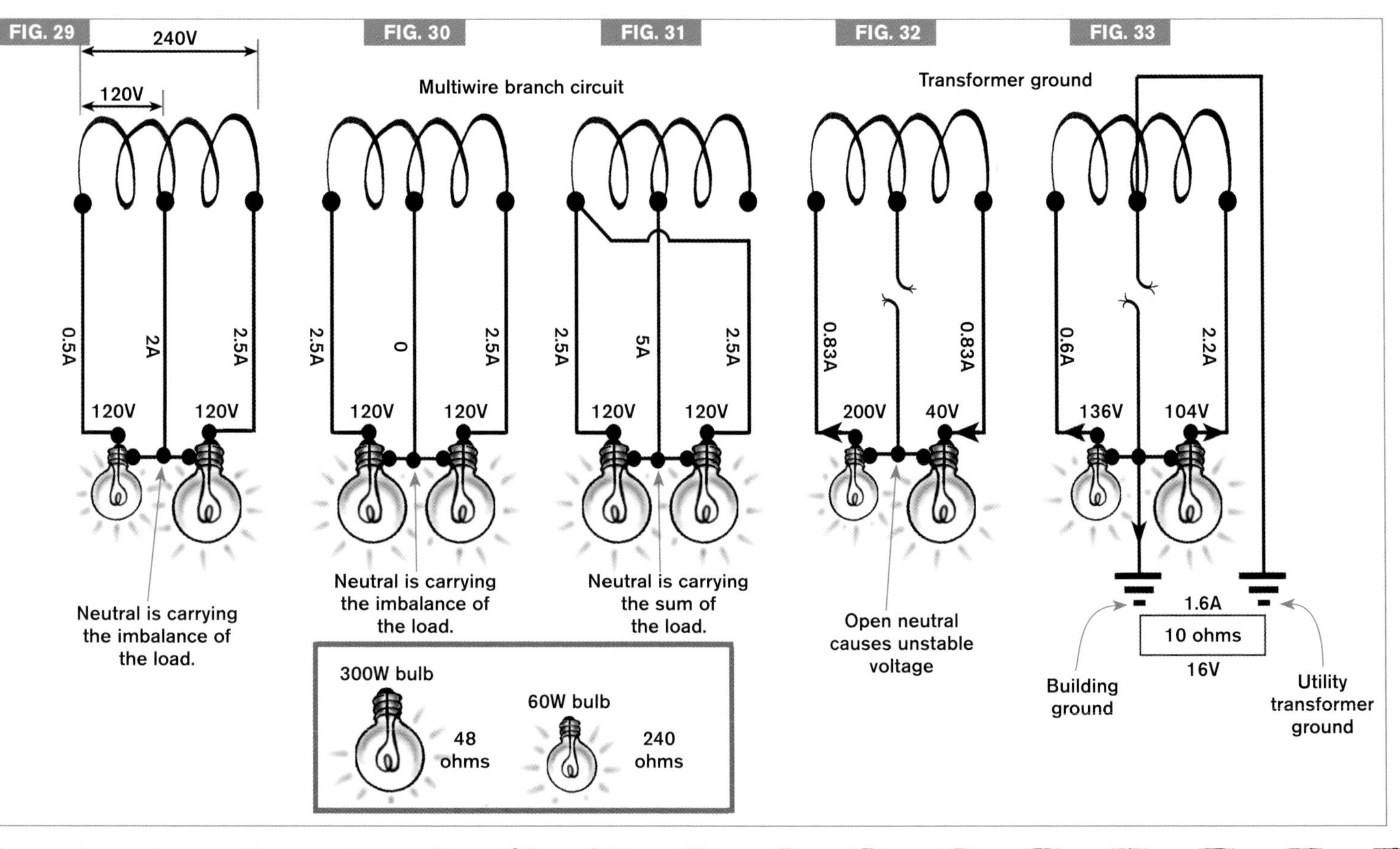
FIG. 29
240V
120V
0.5A
2A
2.5A
120V
120V
Neutral is carrying the imbalance of the load.
FIG. 30
Multiwire branch circuit
2.5A
0
2.5A
120V
120V
Neutral is carrying the imbalance of the load.
FIG. 31
2.5A
5A
2.5A
120V
120V
Neutral is carrying the sum of the load.
300W bulb
48 ohms
60W bulb
240 ohms
FIG. 32
Transformer ground
0.83A
0.83A
200V
40V
Open neutral causes unstable voltage
FIG. 33
0.6A
2.2A
136V
104V
1.6A
10 ohms
16V
Building ground
Utility transformer ground

3-WIRE EDISON CIRCUITS (MULTIWIRE)

Standard electrical services to one- and two-family dwellings originate at a utility transformer with two ungrounded "hot" conductors and a neutral derived from the center of the transformer's secondary coil. The neutral is connected to earth and is referred to as the "grounded" conductor. The neutral limits the voltage on either of the hot conductors to 120V to ground. If the neutral is broken or loose, voltages become erratic as in F32. TV sets, motors, and computers don't do well with fluctuating voltages. The utility company should be notified if there are signs of unstable voltage. Not only is the service to the house a "3-wire" circuit, 120V branch circuits are often installed with shared neutrals and are then known as multiwire circuits.

Multiwire Circuits	**2002**	**2005**
☐ Hot conductors must originate from opposite poles ______	[100]	{100}
☐ All 120V same-amp ckts allowed to be multiwire ______	[210.4A]	{210.4A}
☐ All conductors must originate from same panel ______	[210.4A]	{210.4A}
☐ Multiwire neutrals may not feed through devices such as receps (should be pigtail neutral in box) __________	[300.13B]	{300.13B}

F29 Two unequal loads are fed by a 3-wire ckt. The neutral carries the imbalance between the two loads.

F30 Two equal loads are fed by a 3-wire ckt. There is no imbalance for the neutral to carry.

F31 Without voltage potential between hot conductors, the neutral carries the sum of the loads. In a 3-conductor Romex cable, the black & red wires must originate from different poles or the neutral can be overloaded because it carries the sum of the currents.

F32 Two unequal loads in series across the full voltage of the transformer, 240V. The load with the lowest resistance sees the greater voltage drop.

F33 Transformer is grounded to earth. A second connection to earth is made at the bldg. being fed by the transformer. In this example, the neutral connection between transformer and load has opened. The earth becomes an available path for current to return to the transformer. The earth resistance in this example is 10ohms. This secondary ground helps reduce the voltage imbalance but does not eliminate it.

BRANCH CIRCUITS

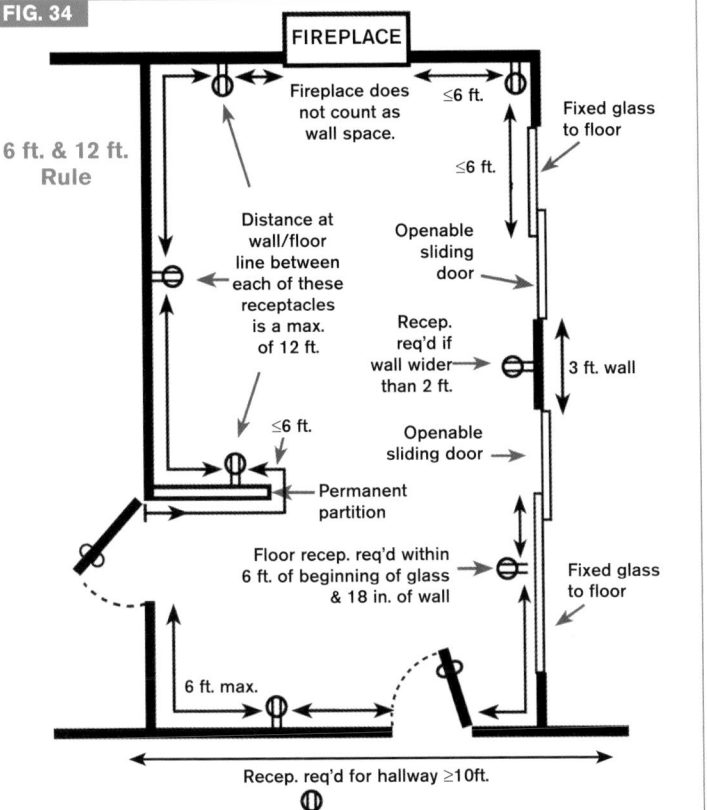

The requirements for branch circuits anticipate the loads we are likely to connect to the system. An insufficient number of outlets might lead to the dangerous substitution of extension cords for permanent wiring. The requirements of the NEC are minimal, and in some cases more circuits and outlets are needed for convenience and safety.

General-Purpose Receptacle Outlets	2002	2005
☐ Min size for branch ckt wiring 14AWG	[210.19A4]	{210.19A4}
☐ Rule of thumb: min 1 general-purpose ckt per 500sq.ft	[210.11A]	{210.11A}
☐ Distance along wall to recep 6ft max F34	[210.52A1]	{210.52A1}
☐ Walls ≥2ft req recep F34	[210.52A2]	{210.52A2}
☐ Hallways ≥10ft req recep F34	[210.52H]	{210.52H}
☐ Permanent partitions & railings count as wall space	[210.52A2]	{210.52A2}
☐ No cords or cables through or stapled to walls	[400.8]	{400.8}
Garage & Unfinished Basement		
☐ Min 1 wall-switched lighting outlet in garage	[210.70A2a]	{210.70A2a}
☐ Min 1 general-purpose (not laundry) recep	[210.52G]	{210.52G}
Bathrooms		
☐ Recep req'd on wall within 3ft of each basin EXC	[210.52D]	{210.52D}
May be in cabinet side or face ≤12in below countertop	[n/a]	{210.52DX}[28]
☐ No face-up outlets on vanity countertop	[406.4E][29]	{406.4E}
☐ Separate 20A ckt for bath receps only OR	[210.11C3]	{210.11C3}
Dedicated 20A ckt to each bathroom	[210.11C3X]	{210.11C3X}
☐ Max rating of space heater on general lighting ckt 15A ckt - 900W, 20A ckt–1,200W	[210.23A2]	{210.23A2}
☐ No panelboards in bathrooms	[240.24E]	{240.24E}

FIG. 34

6 ft. & 12 ft. Rule

FIREPLACE
Fireplace does not count as wall space.
≤6 ft.
≤6 ft.
Fixed glass to floor
Distance at wall/floor line between each of these receptacles is a max. of 12 ft.
Openable sliding door
Recep. req'd if wall wider than 2 ft.
3 ft. wall
≤6 ft.
Openable sliding door
Permanent partition
Floor recep. req'd within 6 ft. of beginning of glass & 18 in. of wall
Fixed glass to floor
6 ft. max.
Recep. req'd for hallway ≥10ft.

Hydromassage (Whirlpool) Tub	2002	2005
☐ GFCI protection req'd for all tub equipment incl motor	[680.71]	{680.71}
☐ Electrical equipment (incl motor) req'd to be accessible	[680.73]	{680.73}
☐ Accessible disconnect req'd in sight of motor	[422.32]	{422.32}
☐ Bond together all grounded metal parts in contact w/ circulation water w/ solid Cu 8AWG	[680.74]	{680.74}[30]
Laundry		
☐ Min 1 20A circuit to laundry recep(s)	[210.11C2]	{210.11C2}
☐ Recep within 6ft of intended appl location	[210.50C]	{210.50C}
☐ Electrical dryer min 30A ckt (10AWG CU, 8AWG AL)	[220.18]	{220.54}
☐ New electrical dryer req's 4-conductor branch ckt	[250.138A]	{250.138A}
☐ Receps within 6ft of laundry sinks req GFCI protection	[n/a]	{210.8A7}7
Outdoors		
☐ Luminaire req'd on ext side of all grade-level personnel doors (not incl garage vehicle door)	[210.70A2B]	{210.70A2B}
☐ Accessible receps req'd front & rear max 62ft above grade	[210.52E]	{210.52E}

FIG. 35

Receptacle & Switch Heights for Handicapped Access

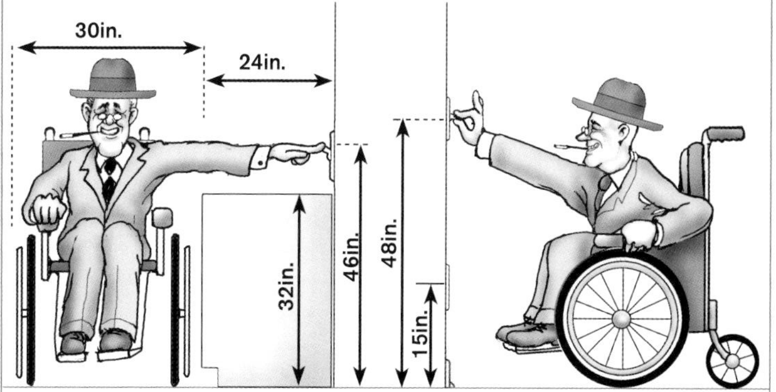

In most cases, the NEC does not have requirements for switch or recep. height. The Americans with Disabilities Act (ADA) provides requirements for accessibility. Local building departments often serve as the ADA compliance officers, & when a bldg. or unit req's accessibility, the receps. must be at least 15 in. above the floor & the switches 48 in. from the center of the switch to the floor.

KITCHENS

Since the early 1990s, appliances have shorter cords, so they are not as likely to be run across cooktops or sinks or to hang down in the reach of children. As a result, more kitchen counter receptacles are needed. At least two 20A circuits are required for countertops and other food preparation areas because of the likelihood of using appliances with large loads.

Branch Circuits	2002	2005
☐ Min two 20A small-appl ckts req'd	[210.11C]	{210.11C}
☐ Small appl ckts must serve refrig & all countertop & exposed wall receps in kitchen, dining room, & pantry	[210.52B1]	{210.52B1}
☐ No other outlets (incl lights) on branch ckts EXC	[210.52B2]	{210.52B2}
Clock or gas range ignition	[210.52B2X]	{210.52B2X}
☐ Refrig may be on individual branch ckt ≥15A	[210.52B1X2]	{210.52B1X2}
☐ Multiwire ckt split recep req's breaker handle tie **F36**	[210.4B]	{210.4B}
☐ DW & disposer usually req separate ckts	[210.23A2]	{210.23A2}

Lighting & Receptacle Outlets	2002	2005
☐ All individual counter spaces ≥12in req receps **F37**	[210.52C1]	{210.52C1}
☐ Individual counter space = undivided by sink or cooktop **F37**	[210.52C4]	{210.52C4}
☐ Spacing so no point >24in from recep **F37**	[210.52C1]	{210.52C1}
☐ Recep not req'd for area behind sink unless deeper than 12in for straight wall or 18in for corner sink	[n/a]	{210.52C1X}[31]
☐ Max 20in above countertop	[210.52C5][32]	{210.52C5}
☐ Island & peninsula counter spaces min 1 recep **F38**	[210.52C2,3]	{210.52C2,3}
☐ Islands & peninsulas w/ no backsplash or overhead cabinet may be mounted no more than 12in below counter if max 6in counter overhang **F38**	[210.52C5X][33]	{210.52C5X}

Lighting & Receptacle Outlets (cont.)	2002	2005
☐ No face-up countertop receps	[406.4E][29]	{406.4E}
☐ GFCI protect on all recep's serving countertops	[210.8A6]	{210.8A6}

FIG. 36

3–Wire Circuit to a Duplex

A common 3-wire ckt. in a residence is run to a duplex receptacle for the dishwasher and the garbage disposal. By breaking the connecting tab on a duplex receptacle, each half can be fed by separate ckts. w/ a shared neutral. When 2 ckts. supply a single device, their breakers req. a common handle tie. It is essential that there be 240V potential between the ckts., otherwise the neutral can be overloaded. See **F31**.

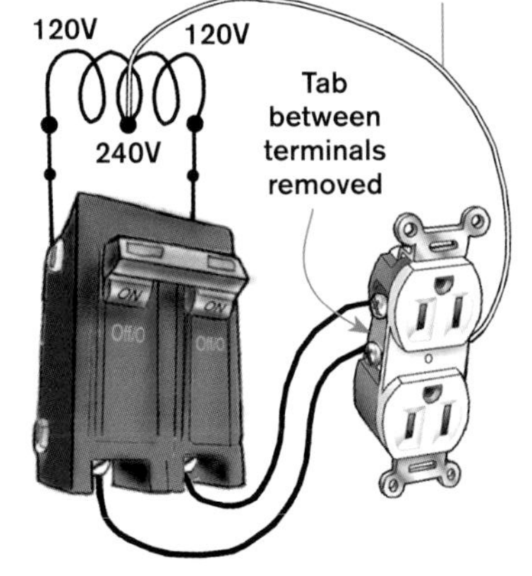

FIG. 37

Receptacle Rules for Kitchen Countertops

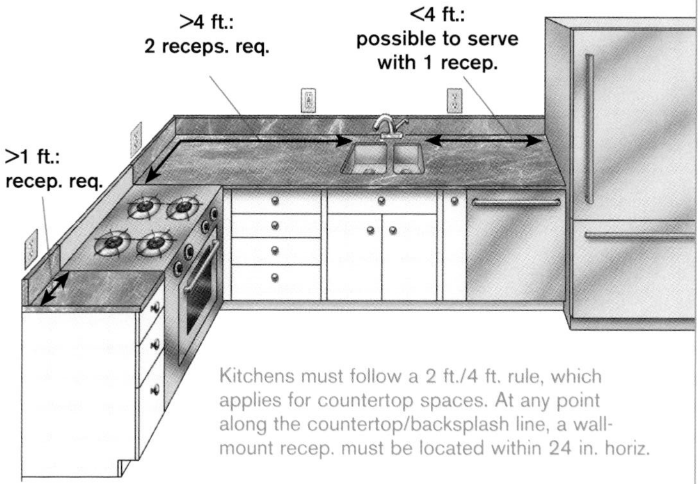

>4 ft.:
2 receps. req.

<4 ft.:
possible to serve
with 1 recep.

>1 ft.:
recep. req.

Kitchens must follow a 2 ft./4 ft. rule, which applies for countertop spaces. At any point along the countertop/backsplash line, a wall-mount recep. must be located within 24 in. horiz.

The "small appl." ckts. supplying these receps must be rated 20A and must serve all receps. in the kitchen, dining area, & pantry. No other outlets, such as lights or permanently affixed appls., are allowed on these ckts.

FIG. 38

Islands & Peninsulas

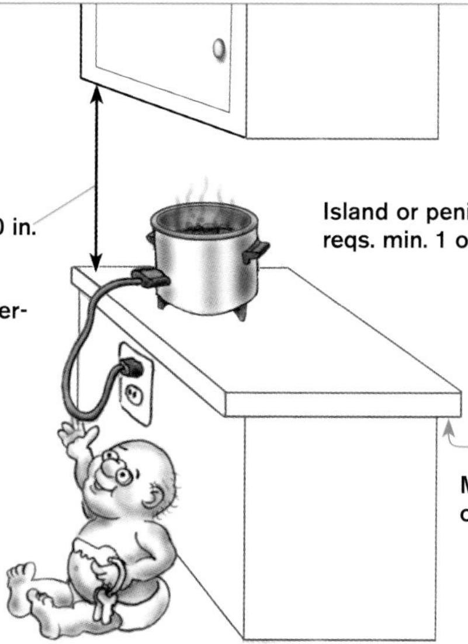

If cabinet is ≤20 in. above counter, receps. can be installed in underside of cabinet.

Island or peninsula reqs. min. 1 outlet

Max. 12 in. below counter

Max. 6 in. overhang

AMPACITY OF WIRE

As current flows through a wire, voltage (pressure) is converted to heat as a squared function of the current. In other words, 4A gives off 16 times as much heat as 1A would through the same wire. Heat can be expressed mathematically as watts = I^2R. Too much current through a wire causes it to get hot. Paper burns at about 451°F. Many electrical fires can be traced back to dangerous I^2R losses. Minimizing these losses is at the heart of the NEC. When wire overheats, its insulation begins to break down, and we say the wire has exceeded its ampacity.

Wire ampacity is based on the temperature rating of its insulation. Conductor insulation has various temperature ratings, 60°C, 75°C, and 90°C being the most common. Typical breakers and equipment terminations also have a temperature rating, typically 60°C and/or 75°C (many are dual rated). One of the mechanisms inside every circuit breaker is a bimetallic element that opens (shuts off) the breaker if it gets too hot (above the overload rating of the breaker).

In determining the overall ampacity of a circuit, ampacity is limited by the lowest rated device or conductor in the circuit. In residential wiring this generally means a 60°C rating. Until recently, most breakers under 100A were only rated 60°C. These 60°C breakers would limit the ampacity of the conductor to a 60°C rating. In addition, Romex is restricted to a 60°C rating, despite containing 90°C-rated conductors.

The ambient temperature (T5) must also be considered in determining the ampacity of a wire, as in the following example:

Ambient Temperature Example: An abandoned 30A NM-fed (60°C wire) dryer circuit runs through a 115°F (46°C) attic (a common summertime occurrence). Could we reuse it to feed a 22A AC unit? Because of the contributed heat of the attic, this 60°C Romex (TW wire) must be derated to 58% of its ampacity or 17.4A, the maximum current flow that can be safely carried by the conductors. To carry a 22A load through the attic would have required that the old NM cable be 8AWG (40 x 0.58 = 23.2).

TABLE 5 CORRECTING FOR HIGH AMBIENT TEMPERATURES

Ambient Temp., °C	For Ambient Temps. >30°C (86°F), Multiply the Allowable Ampacities in T6 by the Following Percentages:			Ambient Temp., °F
	60°C	75°C	90°C	
31–35	0.91	0.94	0.96	87–95
36–40	0.82	0.88	0.91	96–104
41–45	0.71	0.82	0.87	105–113
46–50	0.58	0.75	0.82	114–122
51–55	0.41	0.67	0.76	123–131
56–60	–	0.58	0.71	132–140
61–70	–	0.33	0.58	141–158

This table may have little effect on post-1984 90°C-based NM-B wiring. It can be important in remodels w/ older 60°C wire.

Wire Codes in T6

T = Thermoplastic
W = Wet
H = Heat resistant
U = Underground
N = Nylon jacketed
U = Underground
–2 = 90°C dry or wet
X = Polyethylene

FIG. 39

THNN Wire

THHN is the most common copper wire insulation used in raceways.

TABLE 6 WIRE AMPACITIES (BASED ON NEC T310.16)

Size	60°C 140°F	75°C 167°F	90°C 194°F	60°C 140°F	75°C 167°F	90°C 194°F	Size
TYPES							
AWG kcmil	TW UF	THHW THW THWN USE	THHN THHW THW-2 THWN2 USE-2	TW UF	XHHW USE	USE-2, XHHW-2	AWG kcmil
COPPER				ALUMINUM			
14A	20	20	25	–	–	–	–
12A	25	25	30	20	20	25	12
10A	30	35	40	25	30	35	10
8	40	50	55	30	40	45	8
6	55	65	75	40	50	60	6
4	70	85	95	55	65	75	4
3	85	100	110	65	75	85	3
2	95	115	130	75	90	100	2
1	110	130	150	85	100	115	1
1/0	125	150	170	100	120	135	1/0
2/0	145	175	195	115	135	150	2/0
3/0	165	200	225	130	155	175	3/0
4/0	195	230	260	150	180	205	4/0
250	215	255	290	170	205	230	250

T6 is at the heart of the National Electrical Code. Use the 60°C column for up to 100A or #1 conductors & the 75°C column for larger conductors. The 75°C and 90°C columns are also used for derating.

A. Max 30A OCPD for #10CU, 20A for #12, & 15A for #14. For AL, max 25A for #10 and 20A for #12.

Derating Example: 12 current-carrying conductors through a conduit supply a wood shop. Wire ampacity of a #12 THHN wire would be found by multiplying its 90°C rating, 30A × 0.50 (ET-4) = 15amps. The #12 wires in this conduit must not carry more than 15A & must be protected by 15A breakers.

TABLE 7 CORRECTING FOR WIRES IN CONDUIT OR BUNDLED CABLE–PERCENT DERATING FOR MORE THAN THREE CONDUCTORS

# of Current-Carrying Wires	%
4–6	80
7–9	70
10–20	50

With modern 90°C small conductors this table becomes significant only when there are more than 9 current-carrying conductors in a conduit or Romex bundle. When Romex is bundled more than 2 ft., it is subject to the above derating. Be careful about mixing the newer 90°C wire w/ older 60°C wire, such as TW. Adding the lower-rated wire means that the higher-rated wire must be rated like the lower-rated conductors.

SIZING A MOTOR CIRCUIT

Motor circuits require two forms of overcurrent protection. Short-circuit and ground-fault protection are supplied at the source of the circuit. The conductors and motor are protected against overload by thermal devices located either in the equipment or in a separate motor controller.

For a brief instant, when motors start, they draw a large amount of current, and the OCPD ahead of the motor must be sized to accommodate this "locked rotor" in-rush current. A breaker typically would be sized for 250% of the full-load current (FLC) based on the motor hp, or a time-delay (TD) fuse would be sized to 175% of the FLC. These oversize breakers or fuses do not protect the motor or conductors from burning up due to an overload. Some motors or equipment, such as air-conditioning condensers, have thermal protection in the motor.

Larger motors will have melting alloy "heaters" in the motor controller. This thermal protection protects the motor and wire from sustained overheating. A disconnect must be installed ahead of the motor controller so it can be safely serviced; the disconnect is not intended to be used to turn the motor on and off.

Small motors may require overload protection in the form of an "SSU switch" shown in **F66**. Listed appliances such as air-conditioners have all these calculations taken into account before listing their required conductor and OCPD sizes on the nameplate.

Motor Branch Circuits and OCPDs	**2002**	**2005**
☐ Use nameplate FLA only for overload protection size	[430.32]	{430.32}
☐ Find ckt FLC from hp & voltage per **T8**	[430.6A1]	{430.6A1}
☐ Size ckt conductor to 125% of value in **T8**	[430.22A]	{430.22A}
☐ Breaker size 250% FLC or TD fuse 175%	[430.52C]	{430.52C}
☐ Next higher standard OCPD OK	[430.52X]	{430.52X}

TABLE 8	FULL-LOAD CURRENT FOR SINGLE-PHASE MOTORS [T430.248]			
HP	**115V**	**200V**	**208V**	**230V**
1/6	4.4	2.5	2.4	2.2
1/4	5.8	3.3	3.2	2.9
1/3	7.2	4.1	4.0	3.6
1/2	9.8	5.6	5.4	4.9
3/4	13.8	7.9	7.6	6.9
1	16	9.2	8.8	8
1 1/2	20	11.5	11	10
2	24	13.8	13.2	12
3	34	19.6	18.7	17
5	56	32.2	30.8	28
7 1/2	80	46	44	40
10	100	57.5	55	50

TABLE 9 SIZING CONDUCTORS

Fuse or Breaker	Branch Circuits or Feeders Wire Size[A]		Service Conductors Wire Size[B]	
	Cu	AL	Cu	AL
15	14	12	n/a	n/a
20	12	10	n/a	n/a
25	10	10	n/a	n/a
30	10	8	n/a	n/a
35	8	6	n/a	n/a
40	8	6	n/a	n/a
45	6	4	n/a	n/a
50	6	4	n/a	n/a
60	6	3	n/a	n/a
70	4	2	n/a	n/a
80	3	1	n/a	n/a
90	2	1/0	n/a	n/a

Fuse or Breaker	Branch Circuits or Feeders Wire Size[A]		Service Conductors Wire Size[B]	
	Cu	AL	Cu	AL
100	2	1/0	4	2
110	1	1/0	3	1
125	1/0	1/0	2	1/0
150	1/0	2/0	1	2/0
175	2/0	3/0	1/0	3/0
200	3/0	4/0	2/0	4/0
225	4/0	250kcmil	3/0	250kcmil
250	4/0	300kcmil	4/0	300kcmil
300	300kcmil	400kcmil	250kcmil	350kcmil
350	400kcmil	600kcmil	350kcmil	500kcmil
400	500kcmil	700kcmil	400kcmil	600kcmil

A. Branch ckt. & feeder wire sizes are based on table 310.16 of the NEC. The 60°C column is used for sizes #1 or smaller, & the 75°C column is used for larger sizes.
B. Service conductor sizes are based on the wire types in NEC table 310.15(B)(6).

Motor circuit size example:

Motor nameplate = 3hp, 1Ø, FLA = 15A, 230V, SF = 1.15

Overload protection = 15 × 125% = 18.75

Full load current (FLC) = 17A (T8)

Conductor ampacity ≥17A × 125% = 21.25A

Breaker size = 17A × 250% = 42.5A

TD fuse size = 17A× 175% = 38.75A

A 12AWG THW wire is OK (75°C column).

A 45A breaker is OK (next higher standard size).

A 40A TD fuse is OK (next higher standard size).

CABLE SYSTEMS

Cable systems are the most common residential wiring methods. Cables contain all conductors of the circuit inside a protective outer sheath of metal or plastic.

Cable Protection Indoors (NM, AC, MC, UF, SE) 2002 2005

- ☐ Protect cables w/ 1⁄16th steel plate {or L&L plate} if closer than 14in to framing surfaces F50 ____ [300.4A,D] {300.4A,D}[34]
- ☐ Guard strips req within 6ft of attic scuttle F48 __ [320,30,34.23]{320,30,34.23}
- ☐ Provide guard strips up to 7ft high in attic w/ ladder or permanent stairs ____ [320,30,34.23]{320,30,34.23}

FIG. 40

NM (Nonmetallic-Sheathed Cable) "Romex"

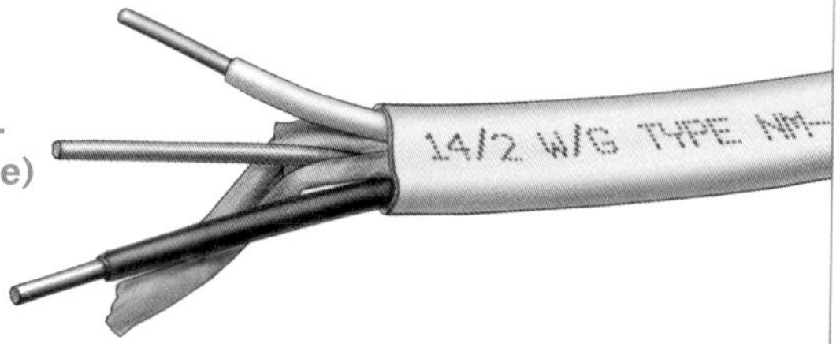

14/2 W/G TYPE NM-

NM–Nonmetallic-Sheathed Cable "Romex" F40

- ☐ OK in normally dry locations only ____ [334.10A] {334.10A}
- ☐ Protect exposed cable from damage ____ [334.15B] {334.15B}
- ☐ Listed grommets for holes through metal framing__ [300.4B1][35] {300.4B1}
- ☐ OCPD selection based on 60° column ____ [334.80] {334.80}
- ☐ OCPD derating & temp correction based on 90° rating for cables marked w/ "-B" ____ [334.80] {334.80}
- ☐ Derate cable bundles through insulation or firestop caulking [n/a] {334.80}[36]

NM–Cable "Romex" (cont.) 2002 2005

- ☐ Secure to box w/ approved NM clamp EXC F47 __ [314.17B,C] {314.17B,C}
  Single gang (24 by 4in) plastic box if stapled within 8in [314.17CX] {314.17CX}
- ☐ Min 4in sheathing into plastic boxes ____ [314.17C] {314.17C}
- ☐ Secure within 12in of box & max 42ft intervals ____ [334.30] {334.30}
- ☐ Do not overdrive staples or place flat cable on its edge[334.30] {334.30}
- ☐ Bends gradual (min 5× cable diameter) ____ [334.24] {334.24}
- ☐ Running board for small cable under joists F49 ____ [334.15C]{334.15C}41

FIG. 41

BX Armored Cable

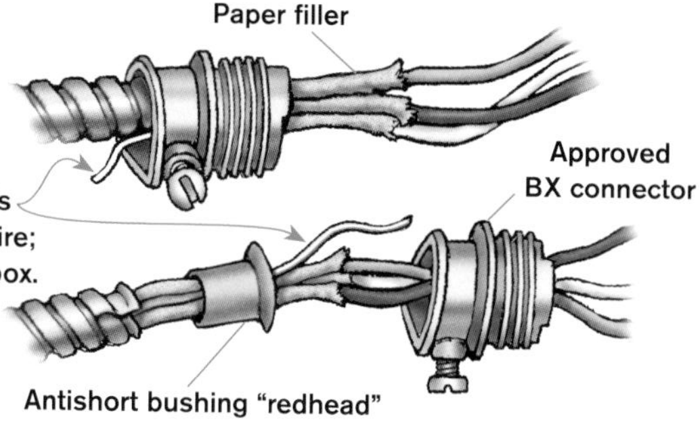

Paper filler

Approved BX connector

Bonding wire is not grounding wire; don't bring into box.

Antishort bushing "redhead"

AC–Armored Cable (BX) F41

- ☐ Dry location only ____ [320.10(3)] {320.10(3)}
- ☐ Secure within 12in of box & max 42ft intervals EXC [320.30A] {320.30B}
  2ft where flexibility needed (motors) ____ [320.30B] {320.30D}
- ☐ Insulated bushing at terminations F41 ____ [320.40] {320.40}
- ☐ Armor is EGC–don't bring bond wire into box F41 __ [320.108] {320.108}
- ☐ Underside of joists–secure at each joist F49 ____ [320.15] {320.15}

FIG. 42

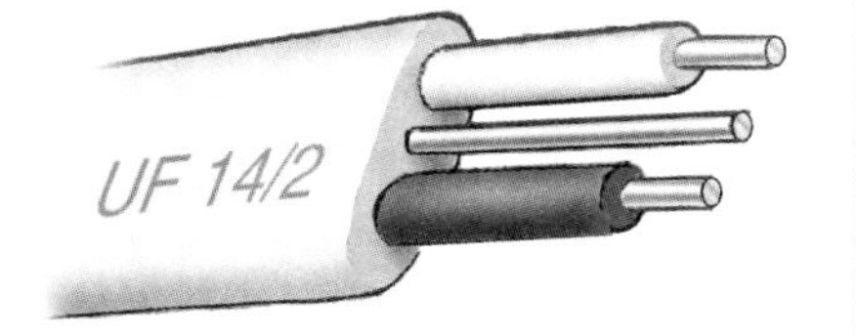

UF (Underground Feeder) Cable

UF–Underground Feeder F42

	2002	2005
☐ Interior installation same rules as NM	[340.10]	{340.10}
☐ May be buried in earth with cover per **T3, F46**	[340.10]	{340.10}
☐ Protect where emerging from earth from 18in below grade to 8ft above **F46**	[300.5D1]	{300.5D1}
☐ Single conductors in trench must be grouped	[340.10]	{340.10}
☐ UV-resistant type OK exposed to sunlight	[340.12]	{340.12}
☐ May not be strung through air w/o support messenger	[340.12]	{340.12}

FIG. 43

Cable with Bare Wire and 4-Wire SE Cable

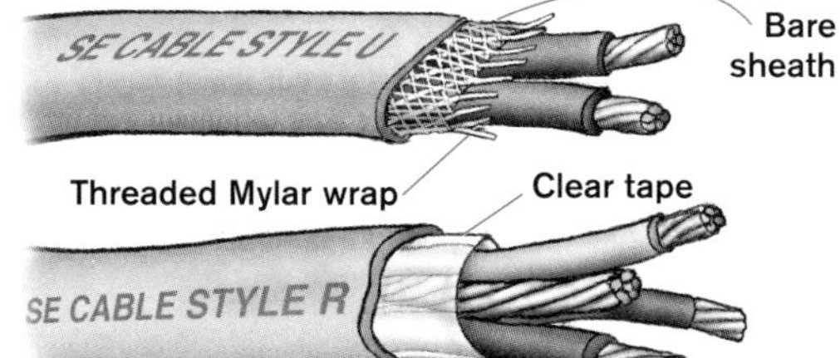

SE–Service Entrance Cable F43

	2002	2005
☐ OK as service entrance conductor (see **p.198**)	[338.10A]	{338.10A}
☐ Interior installation same rules as NM EXC 60° limit does not apply if terminals rated >60°	[338.10B4a]	{338.10B4a}
☐ Direct buried cable same rules as UF cable	[338.10A,B4]	[338.10A,B4]
☐ Insulated neutral (type SER) req'd after service EXC	[338.10B1]	{338.10B1}
Bare conductor in type SEU OK as EGC or in feeder to separate bldg if no continuous metal path between bldgs (see **p. 198**)	[338.10B2]	{338.10B2}
☐ Bends gradual (min 5× cable diameter)	[338.24]	{338.24}

FIG. 44 FIG. 45

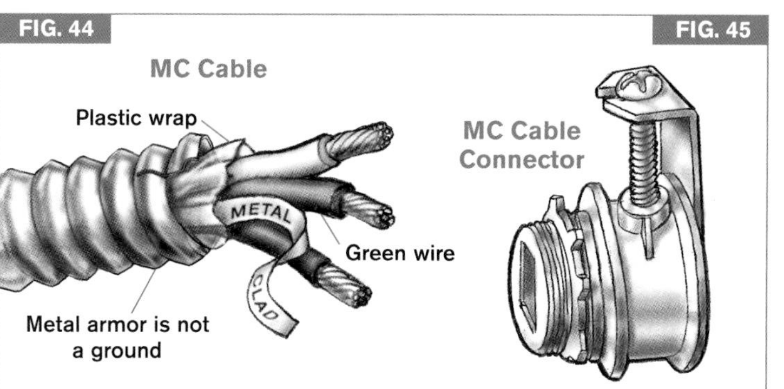

MC–Metal-Clad Cable F44

	2002	2005
☐ Secure within 12in of box & max 6ft intervals EXC >10AWG or where fished or to lum above ceiling	[330.30A,C] [330.30B]	{330.30B,C} {330.30D}
☐ Bends gradual (min 7× cable diameter)	[330.24]	{330.24}

Voltage Drop

When laying out wiring, consider the voltage drop caused by long runs of wire. The NEC recommends (though it does not require) a maximum voltage drop of 3% on branch circuits and a 5% overall voltage drop including the feeders. One way to overcome a voltage drop problem is to use larger wire than the minimum size **T10** and to make sure that all connections are tight. Using fewer than the maximum number of outlets on each circuit will help prevent overloading. Compliance with modern codes now requires separate runs of wire for bedroom outlets (so the entire circuit can be AFCI protected) and for the 20A bathroom receptacle circuits.

TABLE 10 CABLE LENGTH TO LIMIT VOLTAGE DROP TO 3%

Wire Size	CU Distance	AL Distance
14AWG	50 ft.	N/A
12AWG	60 ft.	36 ft.
10AWG	64 ft.	38 ft.
8AWG	76 ft.	46 ft.
6AWG	94 ft.	57 ft.

Based on 120V and 80% ckt. loading for normal OCPD. Distance doubles for 240V.

FIG. 46

Protecting Underground Cable

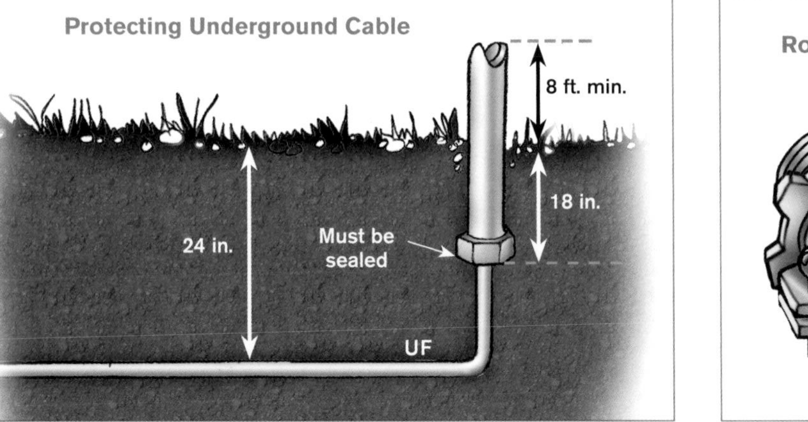

FIG. 47

Romex Clamps

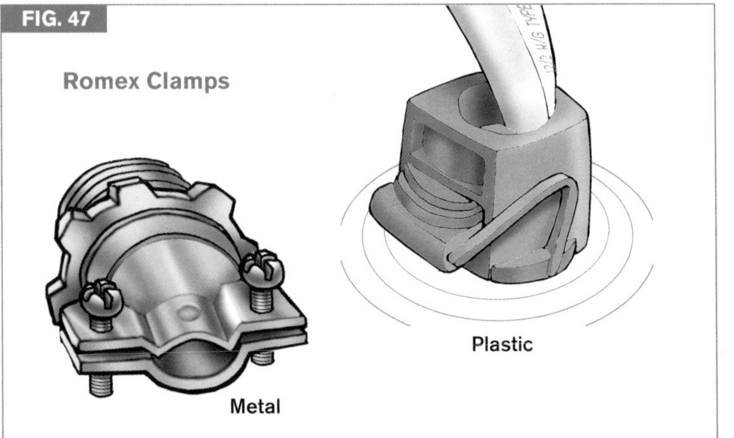

FIG. 48 Cable in an Attic

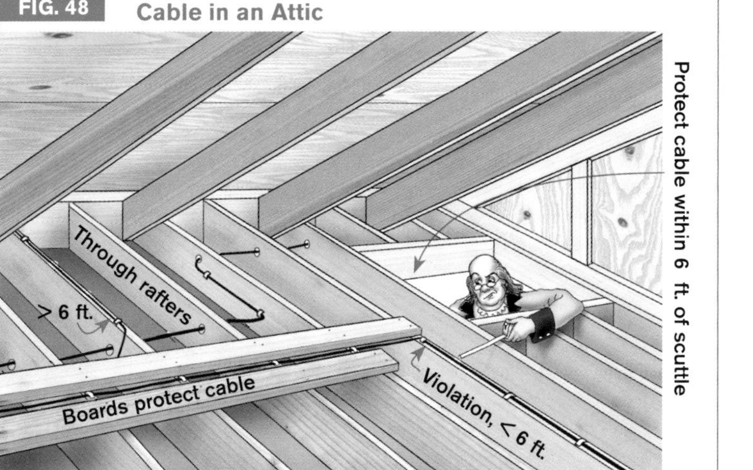

FIG. 49 Cable under Floor

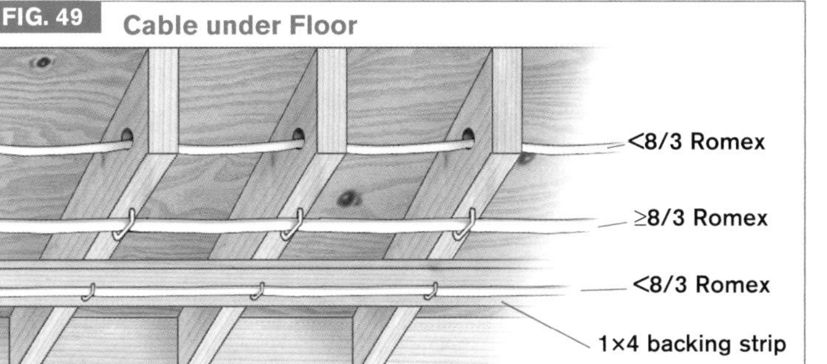

FIG. 50 Nail-Plate Protection

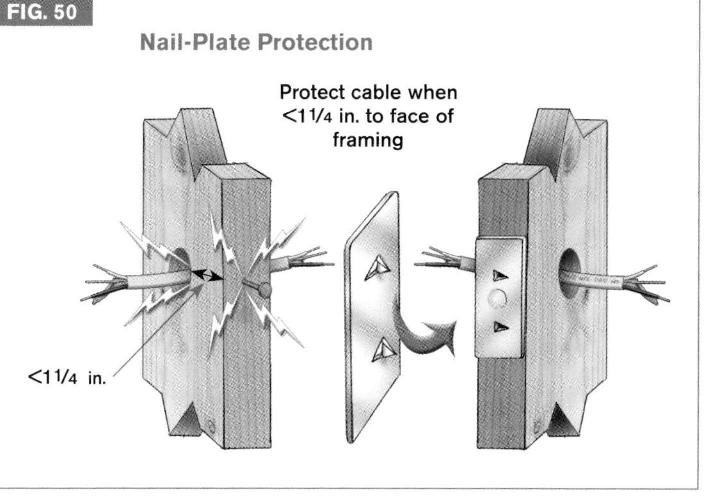

FIG. 51 Stand-off Clamp

Used to maintain clearances to stud or joist edge

RACEWAYS

Raceways are complete systems of conduit or tubing through which conductors are installed. In the 2002 & 2005 NEC numbering system, all articles pertaining to raceways have a parallel numbering system so the portion after the article number is the same for all types. Article numbers are the first 3 digits before the period inside each section number. In the first four three code citations below, the designation "***" is used for articles that share a common ending. For example, the rule for maximum bends is found in 344.26, 348.26, 350.26, etc., all ending with "26" after the period. See **T20**

Installation Requirements (All Conduits & Tubing)	2002	2005
☐ 360° max bends between pull points **F58**	[***.26]	{***.26}
☐ Raceway must be reamed after cutting	[***.28]	{***.28}
☐ Bends req'd to have even radius–no kinks **F58**	[***.24]	{***.24}
☐ Box & conduit body covers must remain accessible	[314.29]	{314.29}
☐ No plastic boxes w/ metal cables or raceways unless bonded through box	[314.3X]	{314.3X}
☐ No splicing in conduit bodies exc conduit bodies w/sufficient volume per marking	[314.16C2]	{314.16C2}
☐ Fittings req plastic bushing if conductors ≥4AWG	[300.4F]	{300.4F}
☐ Max 40% fill if >2 conductors (incl EGC) **T11–16**	[9-T1]	{9-T1}
☐ Derate conductors per bundling & ambient temp **T6**	[310.15B2a]	{310.15B2a}

FIG. 52 EMT (Electrical Metallic Tubing)

Raintight wet location

Dry location

EMT–Electrical Metallic Tubing F52	2002	2005
☐ EMT not OK for burial in soil or concrete	[358.10B]	{358.10B}
☐ Raintight couplings & connectors req'd in wet locs **F52**	[358.42]	{358.42}
☐ Secure in place max 10ft intervals & 3ft from terminations (box, conduit body, cabinet)	[358.30A]	{358.30A}
☐ Horiz runs OK to support in framing holes if securely fastened within 3ft of terminations	[358.30B]	{358.30B}
☐ OK to support conduit bodies, not boxes	[358.12]	{358.12}

RMC–Rigid Metal Conduit F53		
☐ OK for burial in soils	[344.10B]	{344.10B}
☐ Coat buried field cut threads w/ L&L compound	[300.6A]	{300.6A}
☐ Provide bushing or fitting at box connection **F53**	[344.46]	{344.46}
☐ Plastic bushings if any conductor ≥4AWG	[300.4F]	{300.4F}
☐ No threadless connectors on threaded conduit ends	[344.42]	{344.42}
☐ Secure in place within 3ft of termination	[344.30A]	{344.30A}
☐ Horiz support spacing max 10ft	[344.30B]	{344.30B}

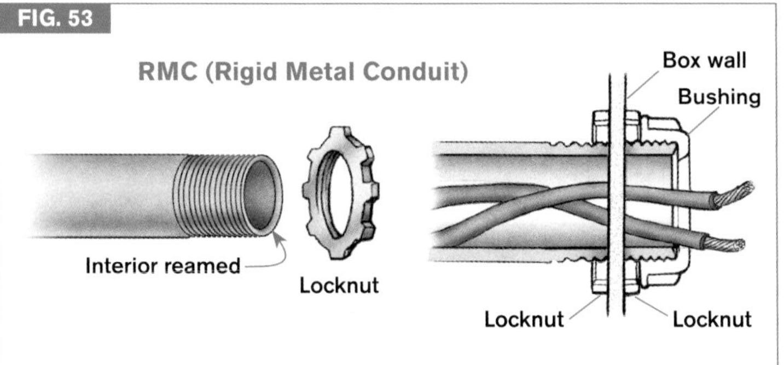

FIG. 53 RMC (Rigid Metal Conduit)

FMC–Flexible Metal Conduit F54

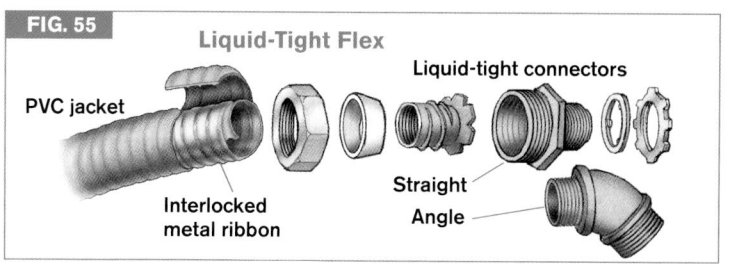

	2002	2005
☐ Dry locations only EXC	[350-5]	{348.12}
OK wet w/ drip loop and "W"-rated conductors F68	[348.12]	{348.12}
☐ Supports: General–42ft & within 12in of boxes EXC	[348.30A]	{348.30A}
• When fished–not req'd	[348.30AX1]	{348.30AX1}
• Lighting thermal whip conductor OK to 6ft F72	[348.30AX3]	{348.30AX3}
• To lums or other equipment in drop ceiling OK to 6ft	[n/a]	{348.30AX4}[37]
• 2 to 14in sizes OK to 3ft where flexibility req'd	[348.30AX2]	{348.30AX2}[38]
• 12 to 2in sizes OK to 4 ft where flexibility req'd	[348.30AX2]	{348.30AX2}[38]
☐ OK as EGC if fittings L&L, ckt ≤20A, no motors, & ≤6ft long	[250.118]	{250.118}
☐ Angle connections may not be concealed F54	[348.42]	{348.42}

FIG. 54 Flex with Fittings

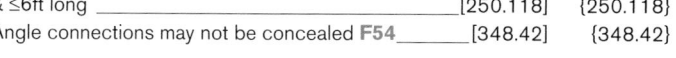

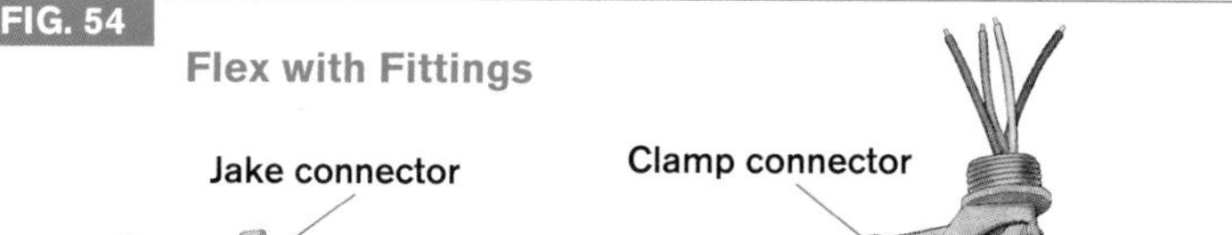

ENT–Electrical Nonmetallic Tubing (Smurf) F57

	2002	2005
☐ OK embedded in concrete w/ approved fittings	[362.10]	{362.10}
☐ Not permitted in environments >50°C (122°F)	[362.12]	{362.12}
☐ Not permitted for direct burial	[362.12]	{362.12}
☐ Must be identified as sunlight resistant if outdoors	[362.12]	{362.12}
☐ Secure or support every 3ft EXC	[362.30A,B]	{362.30A,B}
To lums or other equipment in drop ceiling OK to 6ft	[n/a]	{362.30AX2}[39]

FIG. 55 Liquid-Tight Flex

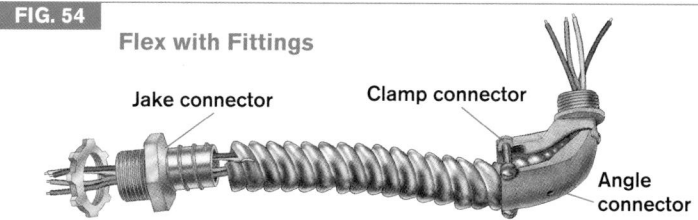

PVC jacket
Interlocked metal ribbon
Liquid-tight connectors
Straight
Angle

LFMC–Liquid-tight Flexible Metal Conduit F55

	2002	2005
☐ Same as flex exc OK for wet locations	[350.10]	{350.10}
☐ OK as EGC if fittings L&L, no motors, ≤6ft long, & ckt ≤20A up to 2in size or <60A for 3/4–1 1/4in size	[250.118]	{250.118}

FIG. 56 PVC 80 Conduit & Connector

FIG. 57 ENT & Connector

RNC–Rigid Nonmetallic Conduit (PVC) F56

	2002	2005
☐ Burial depth	[300.5A]	{300.5A}
☐ Support to prevent sags & within 3ft of box	[352.30B]	{352.30B}
☐ Expansion joints req'd if subject to ≥4in movement OR	[352.44]	{352.44}
If lateral riser is in soil subject to frost or upheaval	[300.5J]	{300.5J}
☐ Not permitted in environments >50°C (122°F)	[352.12D]	{352.12D}

These tables apply if all the conductors in the pipe are the same size. Based on NEC Chapter 9 & Annex C

TABLE 11
EMT FILL

Size AWG kcmil	Number of Conductors in THHN, THWN					
•	½	¾	1	1¼	1½	2
14	12	22	35	61	84	138
12	9	16	26	45	61	101
10	5	10	16	28	38	63
8	3	6	9	16	22	36
6	2	4	7	12	16	26
4	1	2	4	7	10	16
3	1	1	3	6	8	13
2	1	1	3	5	7	11
1	1	1	1	4	5	8
1/0	1	1	1	3	4	7
2/0	0	1	1	2	3	6
3/0	0	1	1	1	3	5
4/0	0	1	1	1	2	4
250	0	0	1	1	1	3

TABLE 12
EMT FILL

Size AWG kcmil	Number of Conductors in XHHW (Compact Stranding)					
•	½	¾	1	1¼	1½	2
14	–	–	–	–	–	–
12	–	–	–	–	–	–
10	–	–	–	–	–	–
8	3	5	8	15	20	34
6	1	4	6	11	15	25
4	1	3	4	8	11	18
3	–	–	–	–	–	–
2	1	1	3	6	8	13
1	1	1	2	4	6	10
1/0	1	1	1	3	5	8
2/0	1	1	1	3	4	7
3/0	0	1	1	2	3	6
4/0	0	1	1	1	3	5
250	0	1	1	1	2	4

TABLE 13
RIGID NONMETALLIC CONDUIT SCHEDULE 80 PVC

Size AWG kcmil	Number of Conductors in THHN, THWN					
•	½	¾	1	1¼	1½	2
14	9	17	28	51	70	118
12	6	12	20	37	51	86
10	4	7	13	23	32	54
8	2	4	7	13	18	31
6	1	3	5	9	13	22
4	1	1	3	6	8	14
3	1	1	3	5	7	12
2	1	1	2	4	6	10
1	0	1	1	3	4	7
1/0	0	1	1	2	3	6
2/0	0	1	1	1	3	5
3/0	0	1	1	1	2	4
4/0	0	0	1	1	1	3
250	0	0	1	1	1	3

TABLE 14
RIGID NONMETALLIC CONDUIT PV80

Size AWG kcmil	Number of Conductors in XHHW (Compact Stranding)					
•	½	¾	1	1¼	1½	2
14	–	–	–	–	–	–
12	–	–	–	–	–	–
10	–	–	–	–	–	–
8	3	5	8	14	20	33
6	1	4	6	11	15	25
4	1	2	4	8	11	18
3	–	–	–	–	–	–
2	1	1	3	5	8	13
1	1	1	2	4	6	9
1/0	1	1	1	3	5	8
2/0	1	1	1	3	4	7
3/0	0	1	1	2	3	5
4/0	0	1	1	1	3	5
250	0	0	1	1	1	4

Conduit Fill Calculations

When different sized conductors are used, use **T15** to find the wire areas, add them up, & use **T16** to find the minimum size conduit. Example: 3 2AWG THHN + 3 8AWG XHHW in FMC. (3 × 0.1158) + (3 × 0.0347) = 0.4515, next > size in 40% column is 0.511. Therefore a min. 1¼ in. size is needed.

TABLE 15 — SQ. IN. AREA OF CONDUCTORS (BASED ON NEC T5 CHAPTER 9)

	14	12	10	8	6	4	2	1	1/0	2/0	3/0	4/0	250
TW	.0139	.0181	.0243	.0437	.0726	.0973	.1333	.1901	.2223	.2624	.3117	.3718	.4596
THHN	.0097	.0133	.0211	.0366	.0507	.0824	.1158	.1562	.1855	.2223	.2679	.3237	.3970
XHHW	.0139	.0181	.0243	.0437	.0590	.0814	.0962	.1146	.1534	.1825	.2190	.2642	.3197

TABLE 16 — CONDUIT AND TUBING FILL (BASED ON NEC T4 CHAPTER 9)

Trade Size	Internal Diameter								2 wire sq. in. fill 31%								>2 wire sq. in. fill 40%							
	EMT	EN	FMC	LTC	IMC	RMC	P80	P40	EMT	EN	FMC	LTC	IMC	RMC	P80	P40	EMT	EN	FMC	LTC	IMC	RMC	P80	P40
3/8	–	–	.384	.494	–	–	–	–	–	–	.036	.059	–	–	–	–	–	–	.046	.077	–	–	–	–
1/2	622	.560	.635	.632	.660	.632	.526	.602	.094	.076	.098	.097	.106	.097	.067	.088	.122	.099	.127	.125	.137	.125	.087	.114
3/4	.824	.760	.824	.830	.864	.836	.722	.804	.165	.141	.165	.168	.182	.170	.127	.157	.213	.181	.213	.216	.235	.220	.164	.203
1	1.049	1.000	1.020	1.054	1.105	1.063	.936	1.029	.268	.243	.253	.270	.297	.275	.213	.258	.346	.314	.327	.349	.384	.355	.275	.333
1¼	1.380	1.340	1.275	1.395	1.448	1.394	1.255	1.360	.464	.437	.396	.474	.510	.473	.383	.450	598	.564	.511	.611	.658	.610	.495	.581
1½	1.610	1.570	1.538	1.588	1.683	1.624	1.476	1.590	.631	.600	.576	.614	.689	1.056	.530	.616	.814	.774	.743	.792	.889	.829	.684	.794
2	2.067	2.020	2.040	2.033	2.150	2.083	1.913	2.047	1.040	.994	1.013	1.006	1.125	1.508	.891	1.020	1.342	1.282	1.307	1.298	1.452	1.363	1.150	1.316

FIG. 58 Too Many Bends

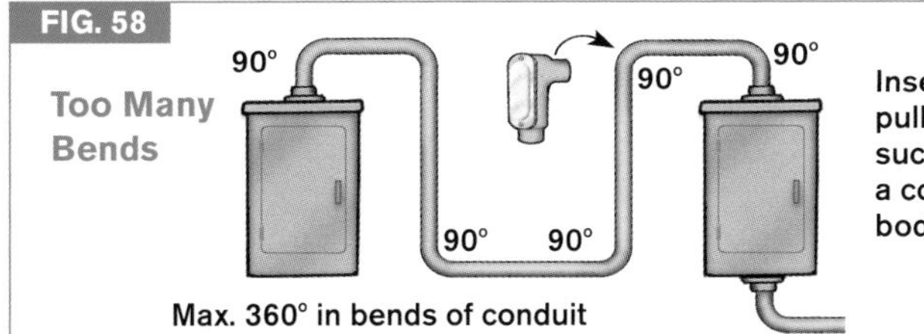

Max. 360° in bends of conduit

FIG. 59 Conduit Bodies

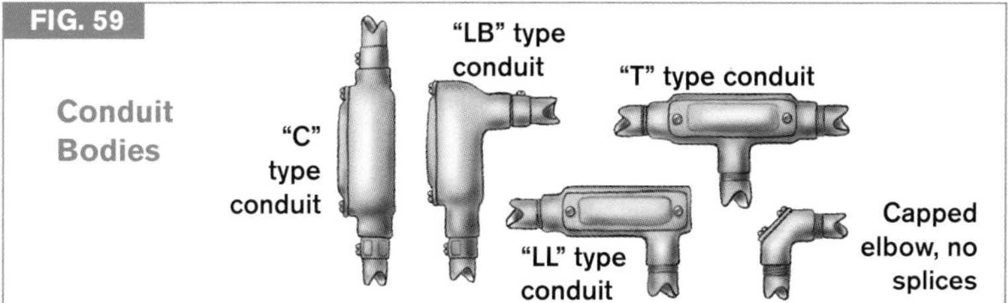

BOXES

Boxes are necessary to safely enclose and protect wiring splices and to support devices and luminaires (fixtures).

General

	2002	2005
☐ Metal boxes must be grounded	[314.4]	{314.4}
☐ Box & conduit body covers must remain accessible	[314.29]	{314.29}
☐ Max 1/4in setback from noncombustible surface **F61**	[314.20][40]	{314.20}
☐ Box extenders OK to correct excess setback	[local]	{local}
☐ Boxes flush w/ combustible surface **F61**	[314.20]	{314.20}
☐ No wallboard gaps >8in {for boxes w/ flush-type covers}	[314.21]	{314.21}[41]
☐ Min 6in free conductor in box & min 3in past box face	[300.14]	{300.14}
☐ Boxes must be supported	[314.23]	{314.23}
☐ EMT OK for conduit body support, not for box support	[358.12]	{358.12}
☐ PVC OK for conduit body support, not for box support	[352.12B]	{352.12B}
☐ Outdoor boxes must prevent water entry **F65**	[314.15A]	{314.15A}
☐ Outdoor wet location recep boxes req in-use covers **F65**	[406.8B1][42]	{406.8B1,2}

Box Fill

	2002	2005
☐ Size must be OK to provide free space for conductor	[314.16]	{314.16}
☐ 4in (6cu.in) pancake box only OK for end of 14/2 run **F60**	[314.16B]	{314.16B}
☐ 3in (4cu.in) pancake box too small for any splices **T17,18**	[314.16B]	{314.16B}
☐ 18cu.in box too small for 3 12/2 Romex **T18,F62**	[314.16B]	{314.16B}

Factors to Consider for Box Fill

	2002	2005
☐ Count number & size of conductors exiting box **T17**	[314.16B1]	{314.16B1}
☐ Pigtail conductors to devices don't count	[314.16B1]	{314.16B1}
☐ Support fittings count as 1 conductor for each fitting type based on largest conductor in box	[314.16B3]	{314.16B3}
☐ Internal clamps–count only 1 based on largest conductor in box	[314.16B2]	{314.16B2}
☐ Devices = 2 conductors per connected wire size	[314.16B4]	{314.16B4}
☐ All EGCs count as only 1 based on largest	[314.16B5]	{314.16B5}

TABLE 17 BOX FILL WORKSHEET

Item	Size	#	Total
#14 conductors exiting box	2.00		
#12 conductors exiting box	2.25		
#10 conductors exiting box	2.50		
#8 conductors exiting box	3.00		
#6 conductors exiting box	5.00		
Largest grounding conductor–count only one		1	
Devices–2× times connected conductor size			
Internal clamps–one based on largest wire present		1	
Fixt. fittings–one for each type based on largest wire			
TOTAL			

Based on NEC 314.16(B)

FIG. 60

Pancake Boxes

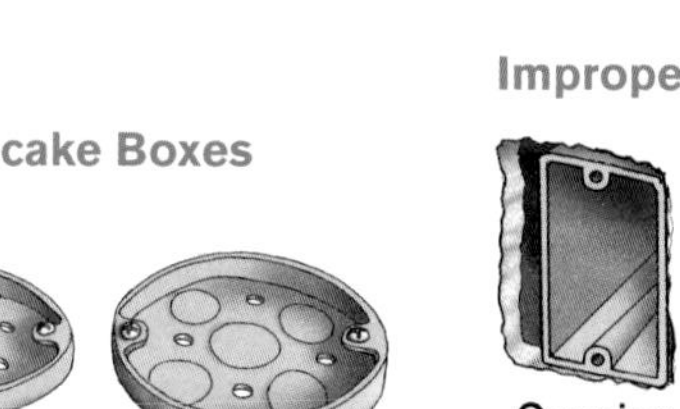

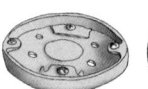

FIG. 61

Improper Box Installation

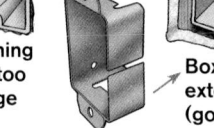

Opening cut too large

Box set back too deeply

Box extender (goof ring)

TABLE 18	BOX FILL EXAMPLE		
Item	Size	#	Total
#14 conductors exiting box	2.00		
#12 conductors exiting box	2.25	6	13.50
#10 conductors exiting box	2.50		
#8 conductors exiting box	3.00		
#6 conductors exiting box	5.00		
Largest grounding conductor–count only one	2.25	1	2.25
Devices–2× times connected conductor size	4.50	1	4.50
Internal clamps–one based on largest wire present	2.25	1	2.25
Fixture fittings–one for each type based on largest wire			
TOTAL			22.5

3 12/2 + G Romex + device overfills 18 cu. in. box.

FIG. 62

Overfilled 18 cu. in. Box

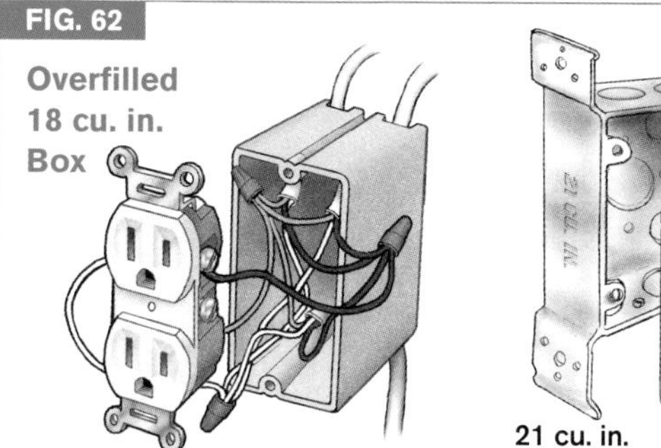

FIG. 63

4-Square Steel/Side Bracketa

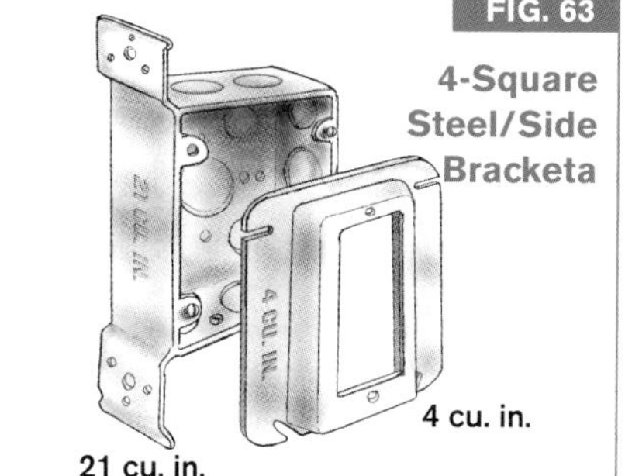

FIG. 64

Weatherproof Boxes Are Threaded, Switch Covers Raintight

FIG. 65

In-Use Covers

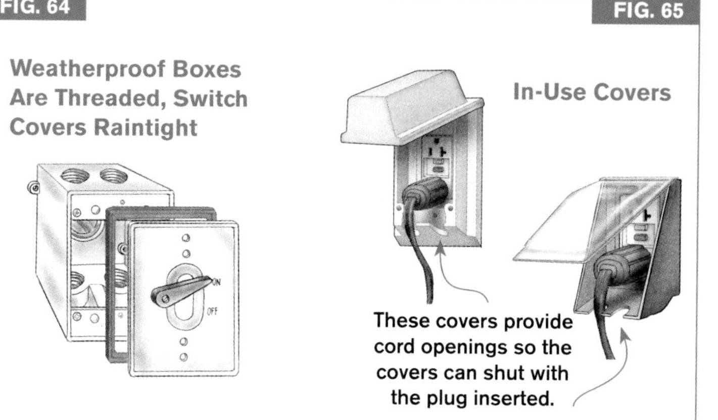

ON
OFF
These covers provide cord openings so the covers can shut with the plug inserted.

FIG. 66

SSU-Fused Disconnect

FIG. 67

Standard Outdoor Cover

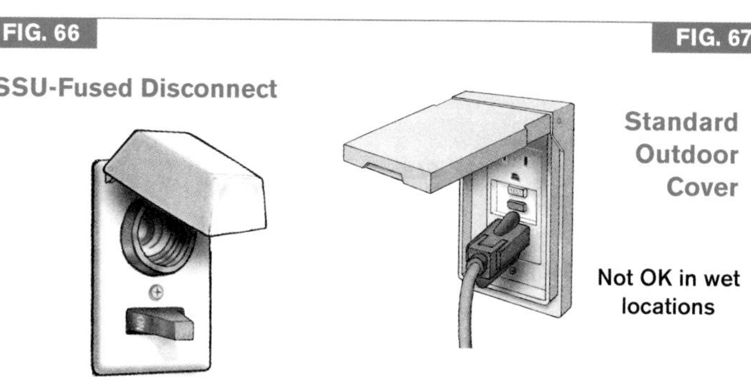

APPLIANCES

The term *appliances* refers to equipment (other than lighting) that uses electricity. All appliances require some means of disconnecting the hot conductors so the appliance can be safely serviced or replaced.

Acceptable Disconnecting Means

	2002	2005
☐ C&P if accessible (behind DW OK)	[422.33A]	{422.33A}
☐ Breaker alone OK for appliances <300VA or 8hp	[422.31A]	{422.31A}
☐ In-sight switch or breaker req'd if ≥300VA or 8hp OR Lockable out-of-sight breaker (exc AC) **F70**	[422.31B]	{422.31B}
☐ 2005 req's permanent hasp breaker lockouts–no temporary handle locks **F70**	[n/a]	{422.31B}[43]
☐ Unit switch that opens all ungrounded conductors	[422.34]	{422.34}

Kitchens

	2002	2005
☐ Garbage disposer cord min 18in max 36in	[422.16B1]	{422.16B1}
☐ DW cord min 3ft max 4ft measured from back	[422.16B2]	{422.16B2}
☐ Cords must be flexible (no NM cable)	[422.16B1,2]	{422.16B1,2}
☐ Electric range ckt min 40A	[210.19A3]	{210.19A3}
☐ Range hoods (can incl microwave) OK for C&P if cord 18in–36in & accessible recep not subject to damage on individual ckt	[n/a]	{422.16B4}[44]

Air-Conditioning

	2002	2005
☐ Multimotor equipment wiring per nameplate, (ex: min circuit ampacity & max fuse or breaker)	[440.4B]	{440.4B}
☐ Disconnect on or in sight of condenser **F68**	[440.14]	{440.14}
☐ Disconnect not OK on compressor access panel	[440.14][45]	{440.14}
☐ Working space req'd in front of disconnect	[110.26A]	{110.26A}
☐ Room AC plug disconnect OK if controls ≤6ft of floor	[440.63]	{440.63}
☐ Max cord length 120V=10ft, 240V=6ft	[440.64]	{440.64}
☐ AFCI or LCDI protection req'd for C&P room AC units	[440.65]	{440.65}

Central Fuel-Burning Furnace

	2002	2005
☐ In-sight disconnect req'd (w/ fuses if req'd by manu)	[422.32]	{422.32}
☐ Lighting outlet switched at entry to eqpmt space	[210.70A3]	{210.70A3}
☐ Central furnace must be on individual ckt	[422.12]	{422.12}
☐ 120V recep req'd within 25ft on same elevation	[210.63][46]	{210.63}

Electric Furnaces & Space Heaters

	2002	2005
☐ Branch ckt 125% load (heat watts + motor FLC)	[424.3B]	{424.3B}
☐ Disconnect in sight or lockable breaker **F70**	[424.19]	{424.19}
☐ Unit switch that opens all ungrounded conductors OK as disconnect for space heater w/ no motor >8hp	[424.19C]	{424.19C}

Water Heater

	2002	2005
☐ In-sight or lockable breaker or switch OK **F70**	[422.31B]	{422.31B}
☐ Breaker lockout hasp req'd to remain when not in use	[n/a]	{422.31B}[43]
☐ Bond hot, cold, & gas pipes	[250.104A,B]	{250.104A,B}

Paddle Fans

	2002	2005
☐ Listed box for fan support (no standard boxes)	[314.27D]	{314.27D}
☐ Listed boxes or box systems OK to 70lb **F69**	[422.18BX]	{314.27D}[47]
☐ Independent support for fans >70lb **F69**	[422.18B]	{314.27D}[47]
☐ Max weight marked on boxes rated >35lb	[422.18BX]	{314.27D}[48]

Hydromassage Tub (Circulating Bathtub)

	2002	2005
☐ GFCI protection req'd	[680.71]	{680.71}
☐ GFCI protect 125V receps within 5ft horiz of tub	[680.71]	{680.71}
☐ GFCI protect 125V receps in same room as tub	[680.72]	{680.72}
☐ Electrical equipment (pump motor) must be accessible	[680.73]	{680.73}
☐ Disconnecting means (C&P OK) req'd in sight of motor	[430.102B]	{430.102B}
☐ Bond metal circulating piping to motor bond lug w/ solid Cu 8AWG EXC Bonding prohibited if listed double-insulated motor	[680.74]	{680.74}[49]

Outdoor De-Icing & Snow Melting Equipment	2002	2005
☐ GFPE protection req'd for de-icing equipment	[426.28]	{426.28}
☐ AFCIs may be used as GFPE for de-icing equipment	[426.28]	{426.28}

FIG. 68

Air-Conditioning Condenser

All ACs req. an in-sight disconnect.

Switch not to be installed directly behind condenser.

FIG. 69

Paddle Fan Support

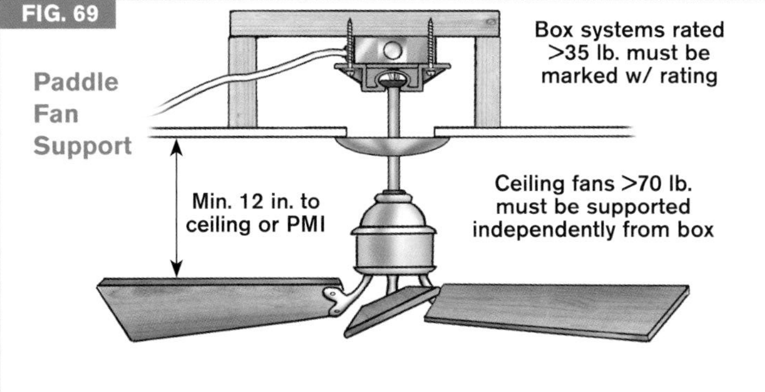

FIG. 70

Breaker Lockout

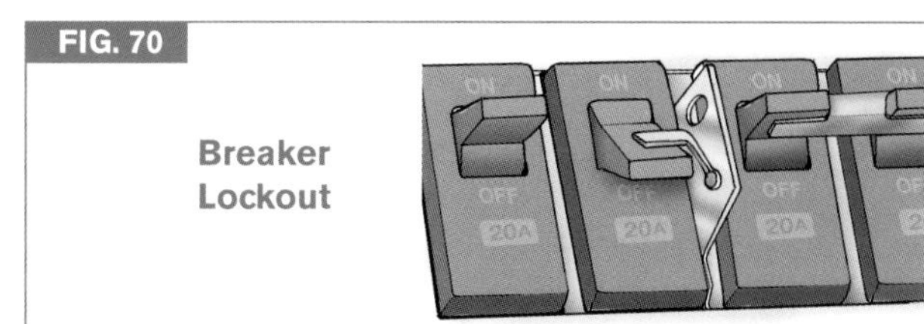

FIG. 71

Smoke Alarms Required in Each Bedroom

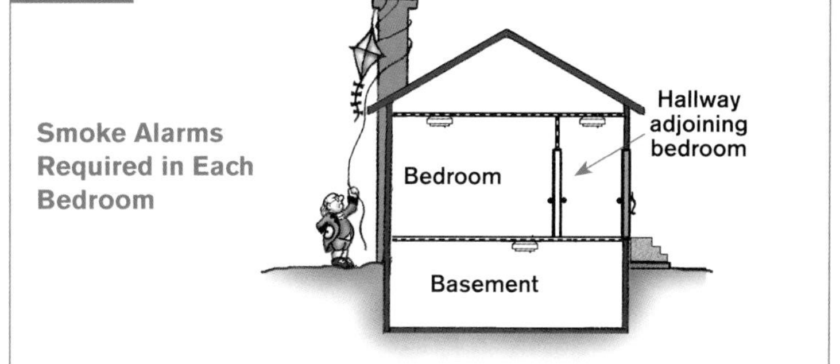

Smoke Alarms	06 IRC
☐ New construction hard wired w/ battery backup	[313.3]
☐ Req'd in each bedroom & adjoining hall **F71**	[313.2]
☐ At least one req'd each story & basement **F71**	[313.2]
☐ Must be interconnected & audible from sleeping rooms	[313.2]
☐ Compliance req'd for interior remodeling req permit	[313.2.1]
☐ Interconnection & hard wire not req'd on remodel if finishes would have to be removed	[313.2.1X]

LIGHTING

Luminaires (lighting fixtures) produce heat, so precautions must be taken in their location and installation to protect combustible materials.

Required Locations	**2002**	**2005**
☐ Wall-switched lighting outlets req'd in all habitable rooms & bathrooms	[210.70A1]	{210.70A1}
☐ Lighting outlet may be switched recep exc in kitchens & bathrooms	[210.70A1X1]	{210.70A1X1}
☐ Switched lighting outlet on ext side of all grade level doors exc garage vehicle door	[210.70A2b]	{210.70A2b}
☐ Switched light in garage, hall & stairs	[210.70A2b]	{210.70A2b}
☐ Switched outlet at entrance for utility rooms, basements, crawlspaces, or attics containing equipment req servicing	[210.70A3]	{210.70A3}

Recessed Light Clearances		
☐ Non-Type IC rated min 3in from insulation	[410.66B]	{410.66B}
☐ Non-Type IC rated min 2in from combustibles	[410.66A1]	{410.66A1}
☐ IC rated OK in contact w/ combustible material	[410.66A2]	{410.66A2}
☐ IC rated OK in contact w/ insulation	[410.66B]	{410.66B}
☐ Max bulb wattage must be visible when relamping	[410.70]	{410.70}
☐ Isolate old low-temp rated wiring from fixt **F72**	[410.67B,C]	{410.67B,C}

FIG. 72

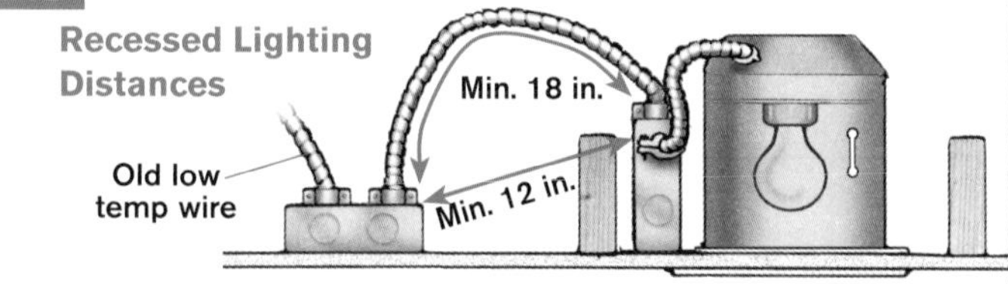

Closet Light Clearances F73	**2002**	**2005**
☐ Incandescent fixts bulbs req'd to be fully enclosed	[410.8C]	{410.8C}
☐ Surface incandescent fixts min 12in from storage	[410.8D1]	{410.8D1}
☐ Surface fluorescent min 6in from storage	[410.8D2]	{410.8D2}
☐ Surface fixts on wall only OK over door	[410.8D1,2]	{410.8D1,2}
☐ Recessed fixts (wall or ceiling) min 6in from storage	[410.8D3,4]	{410.8D3,4}

FIG. 73

Closet Lights

Surface fluorescent or recessed incandescent

Surface incandescent

Surface wall lights OK only over door

6 in.

12 in.

12 in.

Storage area

72in.

12 in.

Storage area

Shaded areas are designated as storage. The storage area above the shelf is the shelf width or 12 in., whichever is greater.

Switches		
☐ All switching in ungrounded conductors **F74,75**	[404.2A,B]	{404.2A,B}
☐ Switches req grounding EXC		
Replacement switches w/ plastic faceplates OR	[404.9B]	{404.9B}
Provide GFCI protection	[n/a]	{404.9BX}[50]
☐ 3-way switch req'd at stairs w/ 6 or more risers	[210.70A2c]	{210.70A2c}
☐ Faceplate must completely cover wall opening	[404.9A]	{404.9A}
☐ Dimmers only for incandescent fixts (not receps)	[404.14E][51]	{404.14E}
☐ Current-carrying conductors of ckt grouped **F74,75**	[300.20A]	{300.20A}
☐ Reidentify ungrounded white or gray wires at each box	[200.7C][52]	{200.7C}

3-Way & 4-Way switches

FIG. 74

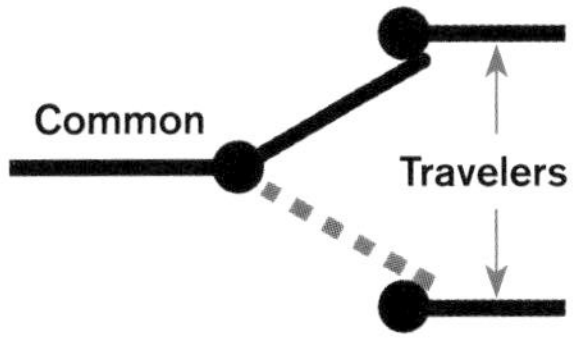

Common
Travelers

3-Way Switch

A "3-way" switch is a somewhat misleading name for a single-pole double-throw switch. There are actually only two positions. Switching takes place from a common terminal to one or the other "traveler" terminals.

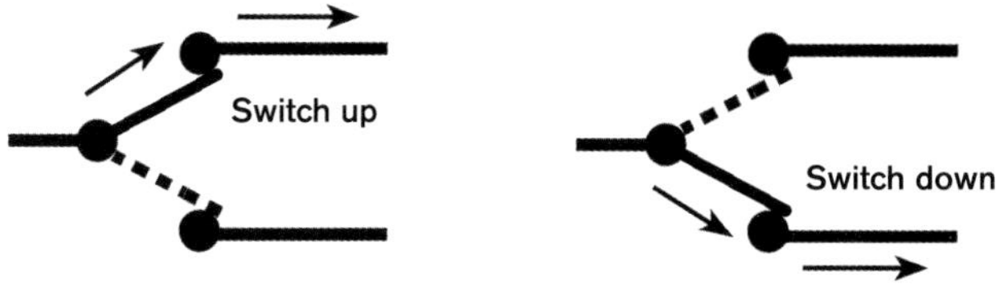

Switch up
Switch down

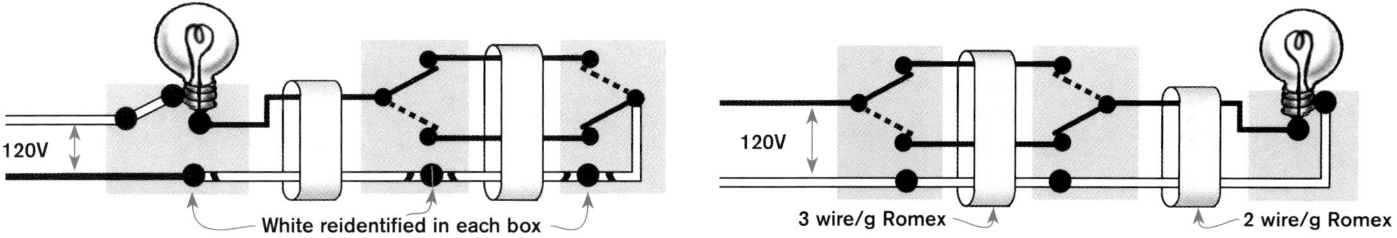

120V
White reidentified in each box
120V
3 wire/g Romex
2 wire/g Romex

FIG. 75

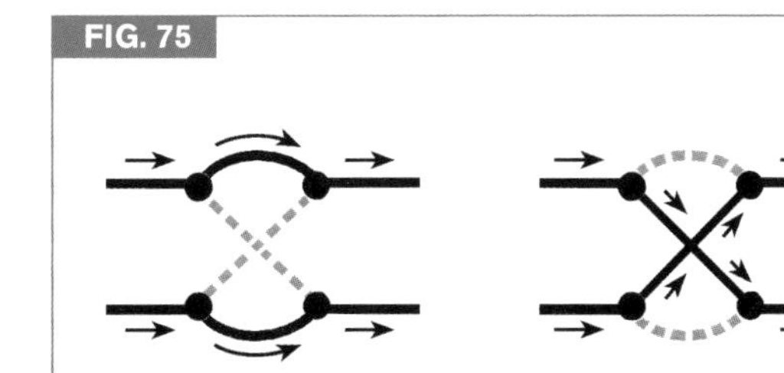

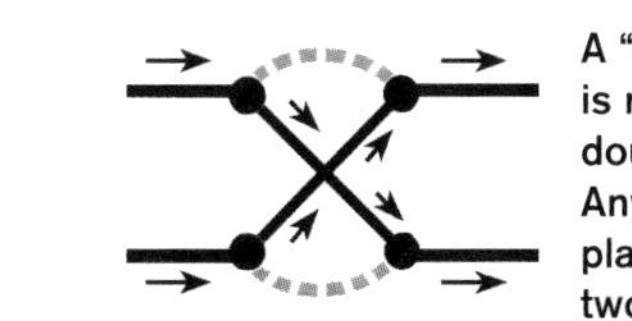

4-Way Switch

A "4-way" switch is really a double-pole double-throw switch. Any number can be placed between the two 3-ways.

4-way interrupts travelers.
120V

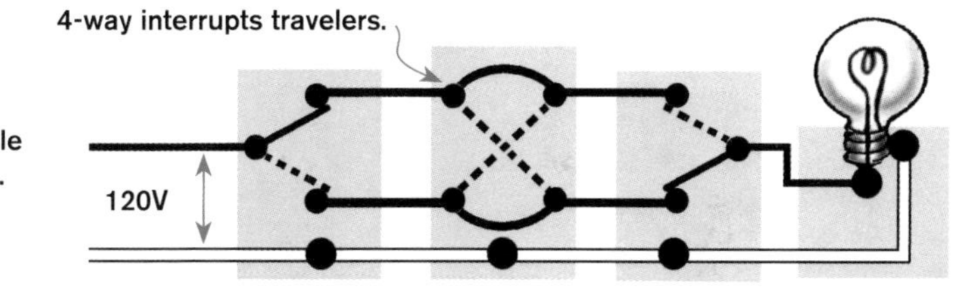

SWIMMING POOLS

Electricity and water don't mix. Equipment grounding conductors must be insulated to protect them from corrosive chemicals and must land on terminal bars, not on wire nuts. In addition to the hazards created by wiring and equipment, pools also require bonding to eliminate voltage gradients even when there is no electrical equipment in the pool area. For GFCI requirements, see **p. 196**.

Overhead Clearances	**2002**	**2005**
☐ Triplex service drop above or within 10ft of pool req's 22½ft clearance in any direction from water	[680.8A][53]	{680.8A}
☐ Clearance from diving platform 14½ft	[680.8A][54]	{680.8A}

Wiring		
☐ Nonpool underground wiring min 5ft from pool EXC	[680.10]	{680.10}
☐ RMC or IMC w/6in cover or RNMC w/ 18in cover OK when space limitations leave no choice	[680.10]	{680.10}
☐ Feeder in RMC, IMC, LFNMC, or PVC only **F76**	[680.25A]	{680.25A}
☐ EMT only OK for feeder on or within bldg	[680.25A]	{680.25A}
☐ Motor connection OK in LFMC or LFNMC	[680.21A3]	{680.21A3}
☐ Motors inside SFD any approved wiring method OK	[680.21A4]	{680.21A4}

Equipment Grounded Conductors (EGCs)		
☐ Min size ckt EGC 12AWG	[680.23F2]	{680.23F2}
☐ No splices (must land on terminals) **F76**	[680.23F2]	{680.23F2}
☐ New feeders must be insulated EGC **F76**	[680.25B]	{680.25B}

Equipotential Bonding		
☐ Bond all parts of pool structure & equip EXC **F77**	[680.26B]	{680.26B}
Small isolated parts <4in & <1in into plaster	[680.26B3]	{680.26B3}
☐ Bond motors exc listed & double insulated type	[680.26B4]	{680.26B4}
☐ Bonding conductor min #8 solid CU	[680.26C]	{680.26C}

FIG. 76

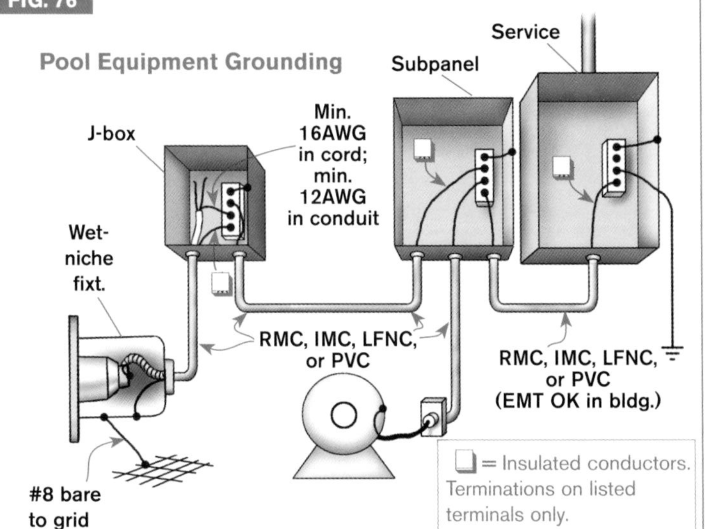

FIG. 77

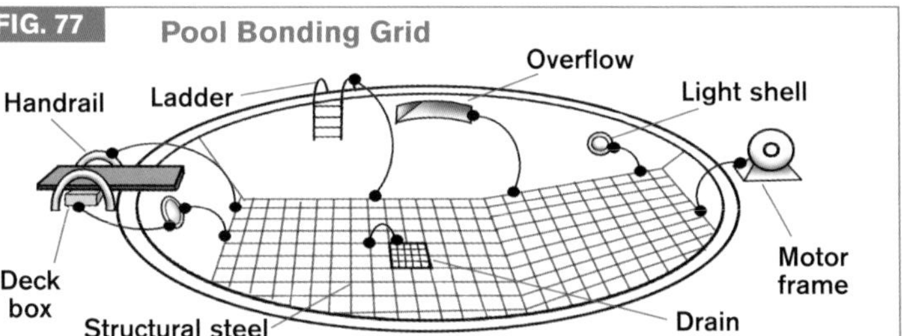

Underwater Wet-Niche Lighting

Item	2002	2005
☐ Min 18in below water level **F78**	[680.23A5]	{680.23A5}
☐ Fixt bonded & secured to shell with locking device req a tool for removal	[680.23B5]	{680.23B5}
☐ Low-voltage transformer L&L for pool	[680.23A2]	{680.23A2}
☐ Low-voltage & GFCI wires not in same raceway or box as non-GFCI wires	[680.23F3]	{680.23F3}
☐ 8AWG bonding conductor req'd in LFNMC or RNMC to wet niche	[680.23B2]	{680.23B2}
☐ Bonding connection in wet niche must be potted	[680.23B2]	{680.23B2}
☐ Min 16AWG EGC in cord to wet-niche fixt **F76**	[680.23B3]	{680.23B3}

FIG. 78 Underwater Pool Lighting

Receptacles (See p. 196 for GFCI Requirements)

Item	2002	2005
☐ Min 1 recep <20ft from pool walls	[680.22A3]	{680.22A3}
☐ Min distance from pool wall 10ft EXC	[680.22A3]	{680.22A3}
Reduction to not <5ft horiz OK if space restricted	[680.22A4][55]	{680.22A4}
☐ Pump motor recep not <10ft from pool wall EXC 5ft OK if twist-lock single recep GFCI protected	[680.22A1]	{680.22A1}
☐ Dimensions include distance around barriers	[680.22A6]	{680.22A6}

Lighting Outlets (See p. 196 for GFCI Requirements)

Item	2002	2005
☐ Outdoors min 5ft from pool unless 12ft above	[680.22B1]	{680.22B1}
☐ Indoors 7ft 6in above water OK if enclosed & GFCI	[680.22B2]	{680.22B2}

HOT TUB/SPA

Outdoor hot tubs or spas follow the same rules as swimming pools. There are also additional specific rules as shown below for all hot tubs and for indoor hot tubs. A hydromassage tub (**p. 224**) is not a spa, because it is emptied after each use.

General

Item	2002	2005
☐ GFCI-protected package unit OK for cord up to 15ft	[680.42A2]	{680.42A2}
☐ Bands to secure hot tub staves exempt from bonding	[680.42B]	{680.42B}

Indoor Spas

Item	2002	2005
☐ Min one recep 5 to 10ft from inside wall of spa	[680.43A1]	{680.43A1}
☐ No wall switches <5ft from inside wall of spa	[680.43C]	{680.43C}

FIG. 79

Utility Interactive Photovoltaic System

In modern listed systems, the components on this wall are often contained inside a single enclosure. Older "off the grid" systems typically have numerous individual components.

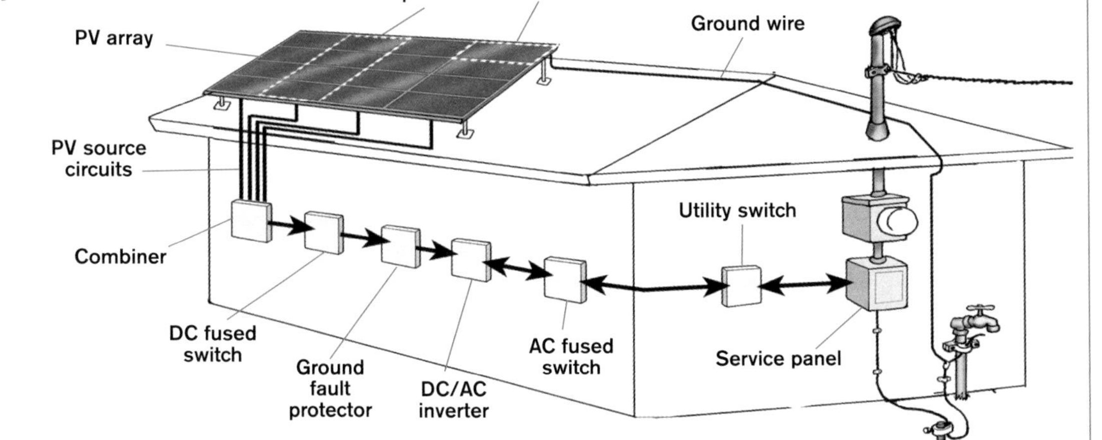

PHOTOVOLTAICS

As energy costs rise, photovoltaic (PV) systems are more popular. They are no longer exclusively used by back-to-the-land homeowners living off the utility grid. Modern utility interactive systems provide a way for consumers to reduce the costs of power consumption while helping the environment. In some states, the utility will rebate a portion of the cost of a PV system. The quality of PV equipment has improved greatly in the last few years, and arrays today have a 30-year expected life. Package systems are now available from major manufacturers. To qualify for utility rebates or be used in an interactive system, equipment should be listed. The standard for modules is UL 1703, and for inverters it is UL 1741. Contact your utility and building department before beginning any project involving renewable energy sources.

Definitions

AC Module: A complete unit, including solar cells, inverter, and other components, that produces AC power.

Array: An assembly of panels that forms the power-producing unit.

Charge Controller: A device to prevent a battery from being overcharged.

Combiner: The location where parallel PV source circuits are connected to create a PV output circuit.

DC Ground-Fault Protection: DC GFP is not intended to prevent shock hazards. It is for fire protection when panels are located on dwelling roofs.

Hybrid System: A system with multiple power sources (not including the utility or batteries). An example would be a system with a generator and a PV source.

Interactive System: A solar PV system that operates in parallel to the utility.

Inverter: Equipment that converts the DC current of a PV output circuit to an AC waveform.

Inverter Output Circuit: change from "load center" to "panelboard" in 2005.

Junction Box: An enclosed terminal block on the back of modules to allow module connection to the electrical system.

Module: A group of PV cells connected together and encapsulated in an environmentally protective laminate–usually tempered glass. Unless otherwise specified, modules have direct current output.

Panel: While this term is more appropriately used for equipment like solar water-heating panels, it is sometimes used to refer to a group of modules that have been preassembled onto a common frame and are designed to be field installed.

Photovoltaic Output Circuit: Conductors between the photovoltaic source circuits and the inverter or DC utilization equipment.

Photovoltaic Source Circuit: Circuits between modules and from modules to the common connection points of the DC system.

Stand-Alone System: A solar PV system that supplies power independent of the utility.

ARRAYS & COMPONENTS

The mounting for a photovoltaic array must consider building codes and structural issues, including the weight of the panels and potential wind uplift. The NEC requirements for lightning protection are minimal, and lightning can severely damage PV equipment. Surge suppressors can be permanently installed for component protection. The module frames must also be connected to the earth. In a roof-mounted system, they must be connected to the house GES.

Modules	2002	2005
☐ Modules req marking of polarity, max OCPD rating for module protection, open ckt voltage, operating voltage, max system voltage, operating current, short ckt current, & max power	[690.51]	{690.51}
☐ AC modules req marking of nominal voltage, frequency, max power, max current, & max OCPD rating for module protection	[690.52]	{690.52}

Photovoltaic Arrays & Inverters

☐ PV ckts must be isolated from other systems	[690.4B]	{690.4B}
☐ Rooftop arrays must be DC ground-fault protected	[690.5]	{690.5}
☐ Inverter must be listed if used in interactive system	[690.60]	{690.60}
☐ Roof-mounted inverters OK in not readily accessible locations (AC output modules) if warning plaque at service	[Ø]	{690.14D}[56]
☐ Interactive systems must automatically disconnect from service in grid outage EXC OK to feed subpanel through transfer switch during outage	[690.61]	{690.61}

Grounding

☐ Module frames & all metal parts grounded	[690.43]	{690.43}
☐ DC 2-wire system >50V req's grounded conductor	[690.41]	{690.41}
☐ DC GEC to same electrode as AC system OR	[n/a]	{690.47C2}[57]
to separate electrodes bonded together	[n/a]	{690.47C1}[57]
☐ Size EGCs of PV output ckt same as ckt conductors OR Size EGCs to **T2** if GFP protected **T2**	[690.45][58]	{690.45}

Overcurrent Protection & Wiring

	2002	2005
☐ Max voltage = sum of rated open-ckt voltage of series connected modules times correction factors for cold temperatures **T19**	[690.7A]	{690.7A}
☐ DC ckt overcurrent protection must be DC rated	[690.9D]	{690.9D}
☐ Single OCPD OK for series-connected string	[690.9E][59]	{690.9E}
☐ OCPD at each source for components with >1 source	[690.9A]	{690.9A}
☐ OCPDs <15A sized in 1A increments	[690.9C][60]	{690.9C}
☐ Rating of PV breakers +utility main breaker ≤120% of panel rating	[690.64B2X]	{690.64B2X}
☐ All currents considered continuous	[690-8b]	{690.8}
☐ PV source ckt currents = sum of parallel module short ckt currents time 125%	[690.8A1]	{690.8A1}
☐ Size conductors for 125% of max PV source short ckt currents	[690.8B1]	{690.8B1}
☐ Consider high ambient temps (use 90° degree wire)	[310.15B]	{310.15B}
☐ No multiwire ckts on 120 Volt supply	[690.10C]	{690.10C}
☐ Single conductor cables type SE, UF, USE, & USE-2 OK in PV source ckts, exposed cables sunlight resistant	[690.31B]	{690.31B}
☐ PV source ckts may pass through bldg interior if in metal conduit	[Ø]	{690.31E}[61]

Disconnects

	2002	2005
☐ Req'd for inverters, batteries, charge controllers	[690.15]	{690.15}
☐ Fuses in PV source ckts req individual disconnects	[690.16]	{690.16}
☐ Ungrounded conductor disconnect must be switch or breaker	[690.17]	{690.17}
☐ Disconnects energized from 2 directions req warning label	[690.17]	{690.17}
☐ Array must be capable of being disabled	[690.18]	{690.18}
☐ Backfed connection to load center–breaker not req'd to be secured in place unless acting as main breaker	[Ø]	{690.64B5}[62]
☐ Don't disconnect grounded conductor	[n/a]	{690.13}[63]

TABLE 19 — VOLTAGE CORRECTION FACTORS (BASED ON NEC T690.7)

Multiply Rated Open Circuit Voltage by the Correction Factor Shown Below	Ambient Temperature (in °F)	Ambient Temperature (in °C)
1.06	77 to 50	25 to 10
1.10	49 to 32	9 to 0
1.13	31 to 14	−1 to −10
1.17	13 to −4	−11 to −20
1.25	−5 to −40	−21 to −40

FIG. 80

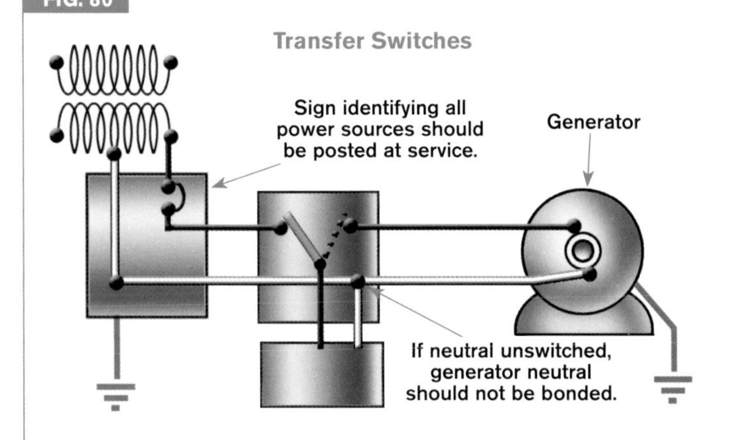
Transfer Switches
Sign identifying all power sources should be posted at service.
Generator
If neutral unswitched, generator neutral should not be bonded.

FIG. 81

Utility Interactive System with Batteries

Component	Connection
Solar array	→ Combiner
Combiner	→ GFP
GFP	→ Switch
Switch	→ Charge controller
Charge controller	→ Battery-powered DC/AC Inverter & charge controller
Battery system	↔ Battery controller
Battery controller	↔ Battery-powered DC/AC Inverter & charge controller
Battery-powered DC/AC Inverter & charge controller	↔ AC fused switch
AC fused switch	↔ Utility switch
Utility switch	↔ Utility service

GENERATORS

Generators provide a source of emergency power during a utility outage. Care must be taken to ensure that the two sources of power–utility and generator–cannot be connected simultaneously. This dangerous condition results from failure to install proper transfer switches and improper use of portable generators.

Generators	2002	2005
☐ Must be suitable for environment, i.e., type 3 if outdoors	[445.10]	{445.10}
☐ Rainproof generators NOT to be enclosed indoors	[110.3B]	{110.3B}
☐ Conductors sized 115% of nameplate current rating	[445.13]	{445.13}
☐ GEC req'd for permanently installed generators	[250.30A2]	{250.30A3}
☐ Must have sufficient capacity for all connected loads	[702.5]	{702.5}
☐ Remove bonding jumper if transfer switch does not switch neutral **F80**	[250.24A5]	{250.24A5}

Transfer Switches	2002	2005
☐ Sign req'd at service indicating generator location **F80**	[702.8A]	{702.8A}
☐ Transfer equip must prevent simultaneous connection of generator & utility service	[702.6]	{702.6}
☐ Electric vehicle OK as standby power source through listed utility interactive connection	[n/a]	{625.26}[64]

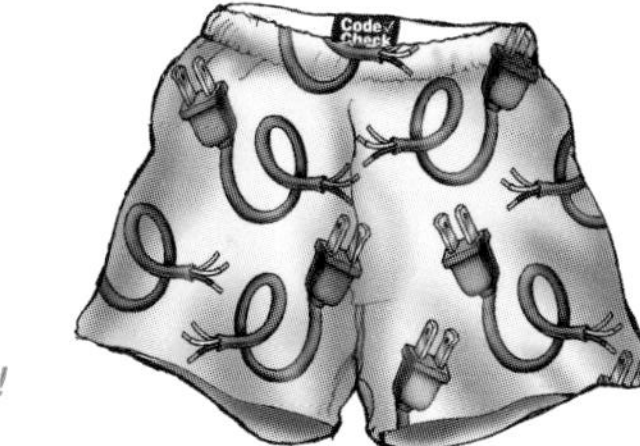

Beware of electrical shorts!

BATTERIES

Batteries are a dangerous portion of on-site power systems due to exposed contacts, off-gassing, chemical spills, and large amounts of stored energy. Most interactive utility systems do not include batteries. Off-grid systems and most hybrid systems do use batteries. It is important that they be kept at a proper temperature (70°F–80°F). At too low a temperature, they lose some of their capacity, and if too warm they will self discharge and eventually be damaged.

Location Requirements	**2002**	**2005**
☐ Racks to be rigid & substantial	[480.8A]	{480.8A}
☐ Trays to be nonconductive & resist deterioration	[480.8B]]	{480.8B}
☐ Working space (min 3ft deep 30in wide) for batteries	[480.9C]	{480.9C}
☐ Working space measured from edge of battery rack	[480.9C]	{480.9C}
☐ No equipment above battery rack	[110.26A3]	{110.26A3}
☐ Segregate other equipment from battery off-gas area	[110.11]	{110.11}
☐ Ventilation to prevent gas buildup	[480.9A]	{480.9A}
☐ Live parts >50V req guarding	[480.9B]	{480.9B}
☐ PV battery systems to operate <50V nominal	[690.71B1]	{690.71B1}
☐ Guard all live parts of PV battery systems	[690.71B2]	{690.71B2}

Wiring

☐ Flex cables ≥2/0AWG OK from battery to junction box	[690.74]	{690.74}
☐ Other wiring from batteries in approved wiring method	[480.3]	{480.3}

Charge Controllers

☐ Should be listed (some are not)	[110.3B]	{110.3B}
☐ Current limiting OCPDs req'd IF output exceeds short circuit rating of other equipment in ckt	[690-71c]	{690.71C}
☐ Series type fused on both input & output sides	[manu]	[manu]

A charge controller prevents a battery from being overcharged and damaged. In a system such as the one in **F81**, the batteries could be charged from either the PV source or the utility source, so a charge controller is needed at each source. Recombinant-type battery caps will greatly reduce the free hydrogen in the air but may need to be removed during equalization. Placing the batteries in plastic trays helps for this procedure. This problem usually occurs only during a very heavy charge or discharge. When doing an equalization charge (controlled overcharge to prevent sulfation buildup), open a window or door.

REPLACEMENT RECEPTACLES

Houses built before adoption of the 1962 NEC will not have 3-hole receptacles in all locations. Appliances with 3-prong cords are designed to be used only with grounded 3-hole receptacles. Two-prong receptacles may be replaced with a GFCI receptacle whether or not an EGC is present and will provide shock protection.

	2002	**2005**
☐ 2-hole OK if no ground present & no GFCI req **F83**	[406.3D3a]	{406.3D3a}
☐ OK to install GFCI where no ground present **F84**	[406.3D3c]	{406.3D3c}
☐ Separate EGC OK from box to service or GEC or ground bar of panel where ckt originates **F85**	[250.130C]	{250.130C}
☐ Not OK to jumper neutral & EGC **F86**	[250.142B]	{250.142B
☐ If location where GFCI req'd, must use GFCI **F84**	[406.3D2]	{406.3D2}
☐ Replacements must be 3-hole if EGC present **F89**	[406.3D1]	{406.3D1}
☐ Bond 3-hole recep to grounded box w/ wire OR **F89,90**	[250.146]	{250.146}
Use grounding-type recep **F88**	[250.146B]	{250.146B}

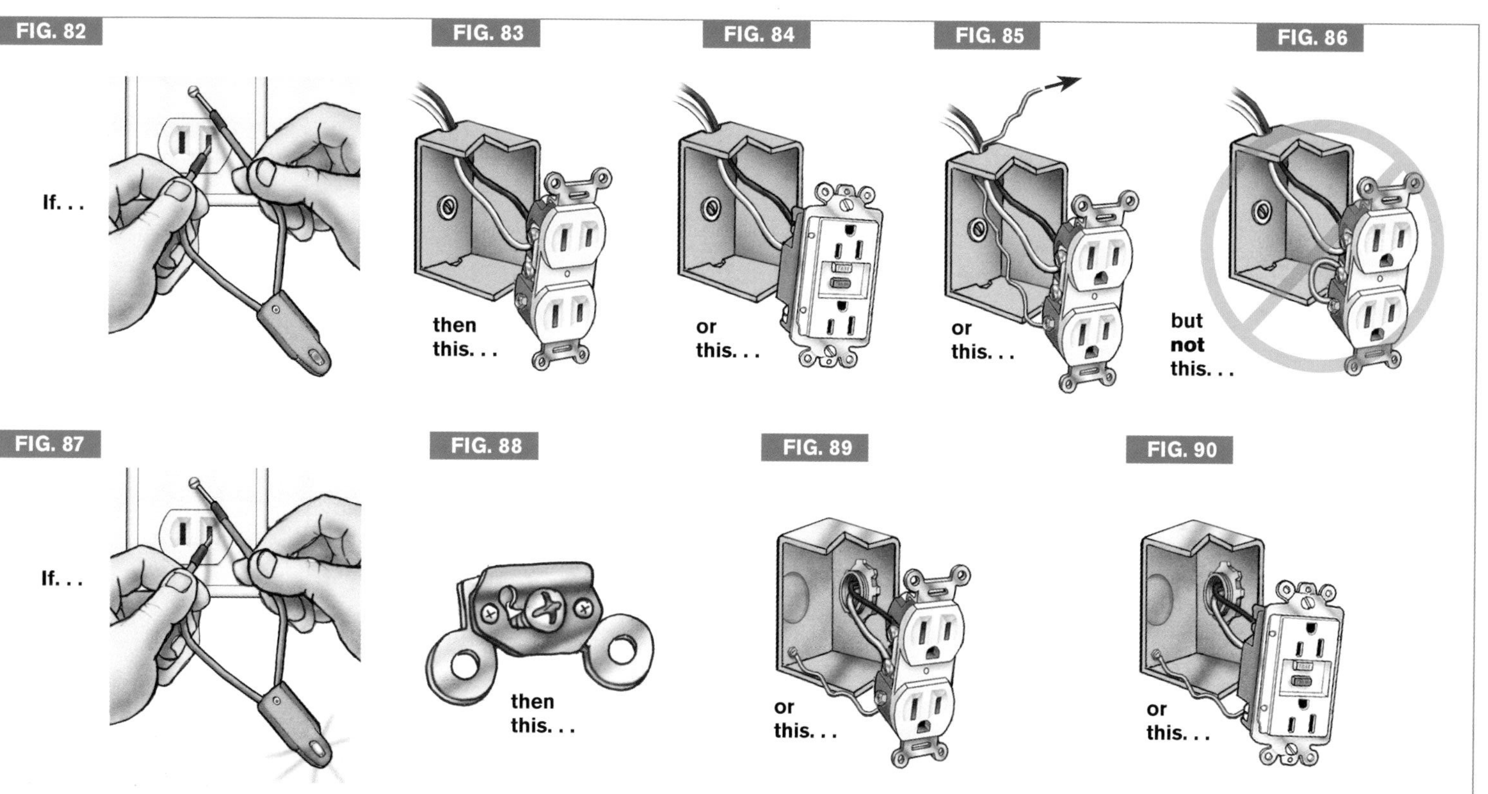
FIG. 82
If. . .
FIG. 83
then
this. . .
FIG. 84
or
this. . .
FIG. 85
or
this. . .
FIG. 86
but
not
this. . .
FIG. 87
If. . .
FIG. 88
then
this. . .
FIG. 89
or
this. . .
FIG. 90
or
this. . .

FUSES

Fuses provide excellent protection if the right size fuse is in place. Unfortunately, they are often sized incorrectly, allowing the wiring to be overloaded or shunted (a penny behind the fuse). Older ceramic fuse panels, and panels with cartridge fuses, also pose a risk of electrocution because of exposed electrical contacts.

	2002	2005
☐ No exposed contacts (must be dead front)	[240.50D]	{240.50D}
☐ Type S req'd if tampering or overfusing exists **F91**	[240.51B]	{240.51B}
☐ S type adapter sized to wire per **T6, F91**	[240.4D]	{240.4D}
☐ No fuses in grounded conductor (neutral) **F91**	[240.22]	{240.22}
☐ No plug fuses for 240V ckts	[240.51A]	{240.51A}

KNOB & TUBE (K&T)

K&T wiring is the oldest wiring method found in American homes. When left in its original state it has proven to be reliable and safe. Electrical safety was inherent in its design. As a wiring method in uninsulated joist and stud cavities it is protected from damage and provided with air circulation. The knobs of K&T maintain 1 in. clearance to wood framing and tubes isolate conductors when passing through wood. Unfortunately, when modified by unqualified persons, the inherent safety of K&T is dangerously compromised.

	2002	2005
☐ No new knob & tube	[394.10]	{394.10}
☐ OK to extend to other wire method w/ proper splices	[394.10]	{394.10}
☐ Splices to other methods must be in box EXC	[300.16A]	{300.16A}
Bushing OK from raceway to open equipment	[300.16B]	{300.16B}
☐ Must be protected w/ loom where entering box	[314.17B,C]	{314.17B,C}
☐ Do not envelop w/ thermal insulation	[394.12]	{394.12}
☐ 3in min between wires, 1in to surfaces **F93**	[394.19A1]	{394.19A1}
☐ Provide protection where exposed <7ft above floor	[398.15C]	{398.15C}
☐ Protect w/ running boards up to 7ft high in attics w/ stairs or permanent ladder **F48**	[394.23A]	{394.23A}

FIG. 91

Old Knob & Tube with Fuses

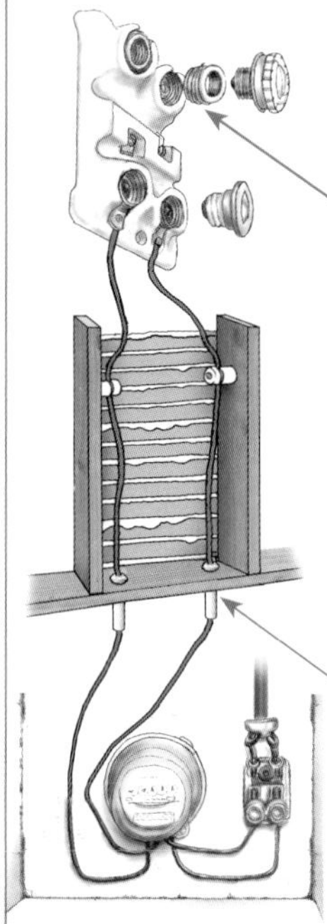

A properly sized type S adapter is req'd when a fuse has been tampered with or improperly sized. Open ceramic fuse panels such as these are no longer allowed because they have exposed contacts.

FIG. 92

Porcelain Knob

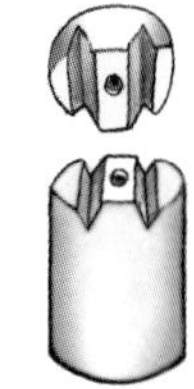

FIG. 93

Splice Wrap

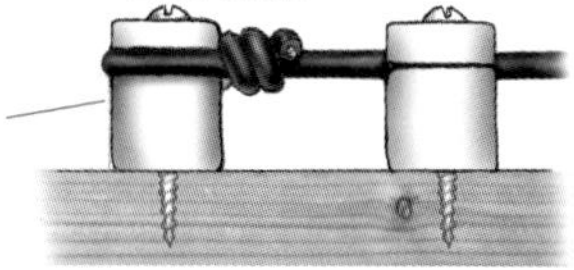

End turns

Solid knob

FIG. 94

Porcelain Tube

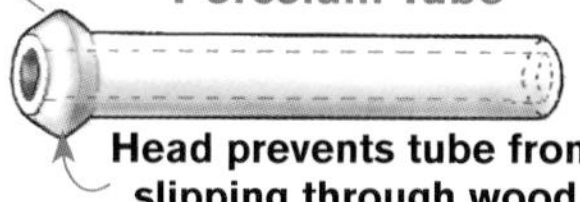

Head prevents tube from slipping through wood

CODE CHANGE SUMMARY

The National Electrical Code is revised on a 3-year cycle. Thousands of public proposals are distilled into the changes in each new edition. We have included the following summary of major residential changes to aid our readers in the transition to the newer code editions. The 2002 edition adopted a decimal numbering system. Measurements were shown with the metric designation first, followed by the "hard conversion" numbers to feet and inches. A *hard conversion* is an approximation to the nearest equivalent feet and inch numbers; a *soft conversion* would be an exact conversion necessitated by safety considerations.

In 2002, the articles for wire, cable, and raceways in chapter 3 were reorganized. The articles for cable, conduit, and tubing were grouped by type and renumbered to follow a parallel system. The suffix numbers (those that follow the article number and the period, e.g., the "10" in 334.10) always cover the same topic regardless of which article they are found in. Numbers ending in ".10" are always the sections dealing with "Uses Permitted" and so on.

A summary of these parallel suffix numbers is listed in **T21** on p. 238.

1 2005 clarified when Ufer was considered present & req'd to be used.
2 2005 added unbonded metal well casings to GEC.
3 2005 specified when bldg steel was considered effectively grounded.
4 2002 clarified that GEC(s) could go to any part(s) of GES.
5 2005 forbids use of sheet-metal screws to attach grounding lugs.
6 2002 did not specify intersystem bonding was for GEC of other system .
7 2005 added receps within 6ft of utility & laundry sinks & expanded wet bar sink reqm't to include receps that are not serving countertop area.
8 02 did not req GFCIs in outdoor areas accessible to public exc residential.
9 2002 was first to req GFCI protection in non-dwelling kitchens.
10 2005 clarified that above rule applies to institutional & commercial kitchens & defined kitchen as an area w/ permanent facilities for food preparation & cooking.
11 2005 added rule for GFCI protection of the req'd receps near commercial HVAC equipment.
12 2002 added rule for GFCI protection of all pool pump motor receps.
13 1999 req'd protection only for bedroom recep outlet ckts–02 entire ckt.
14 02 did not specify combination type–only "branch-feeder" types were available.
15 2002 did not allow anything other than ckt breaker type AFCIs.
16 02 raised clearance from 22ft to 22.5ft when metric numbers took precedence.
17 2005 specified that measurement is taken from max water level.
18 2002 first to prohibit trees from supporting service conductors.
19 2005 coordinated w/ definition of number of supply sources in 225.30.
20 2002 allowed bldg management to have access only to guest suite OCPDs.
21 2002 spec'd only max height of breakers used as switches.
22 05 limits height of all breakers, 02 only limited height when used as switch.
23 05 first to specify plaster gap for panelboard cabinets is same as for boxes.
24 2002 allowed labeling that was generic, 05 req's more specificity
25 2002 first to specify handle tie for two ckts to single device.
26 2002 first to specify each neutral to have individual terminal – synchronizes w/ UL.
27 2005 dropped rule for min 30A feeder.
28 2005 first to allow bath receps on side of vanity rather than wall.
29 2002 extended rule against face-up countertop receps to include all countertops & work surfaces, not just baths, wet bars, & kitchens.
30 05 clarified that bonding req'd only for metal in contact w/ circulating piping.
31 2005 first to specify when receps are req'd behind sinks.
32 2002 raised max height of receps above counters from 18in to 20in.

CODE CHANGE SUMMARY

33 2002 clarified that receps below counter not allowed if overhead cabinet available for recep mounting.
34 2005 allows protective plates < 1/16in thick if L&L.
35 2002 first to req steel framing grommets & bushings to be listed.
36 05 req's derating bundles <24in when passing through firestop or insulation.
37 2002 did not permit end of run of FMC to have 6ft unsupported to lum.
38 02 limited unsupported FMC length to 3ft for flexibility for all sizes of FMC.
39 2002 did not permit end of run of ENT to have 6ft unsupported to lum.
40 2002 clarified that gypsum is considered a noncombustible surface.
41 05 specified that side gap rule on boxes only for flush-type covers/faceplates.
42 2002 req'd in-use covers for all outdoor wet location boxes, not just unattended cords.
43 2002 did not address cord plug range hoods or combo microwave.
44 1999 did not specifically prohibit installing switch on AC access panel.
45 2002 did not specifically prohibit temporary (handle only) lockout devices.
46 2002 req'd service recep for HVAC to be on same elevation.
47 2005 includes editorial clarification for listed box systems over 35lb.
48 2005 clarified that marking is req'd for boxes rated >35lbs.
49 2005 clarified that tubs w/ plastic circulating pipe do not req bonding.
50 2005 introduced alternative of GFCI protection for ungrounded switches.
51 1999 did not specifically prohibit dimmers on receps except PMI.
52 1999 did not recognize the color gray for insulation on grounded conductors
53 2002 raised min service height from 22ft to 22.5ft (metrification).
54 02 raised min height over diving platform from 14ft to 14.5ft (metrification).
55 2002 allowed receps 5ft from pool when space restricted.
56 2002 did not acknowledge rooftop modules w/ AC output.
57 2005 clarified need to either bond separate AC & DC electrodes or to share a common electrode.
58 02 first req'd sizing EGCs to 125% of short ckt rating (same as ckt wiring).
59 2002 first to accept single OCPD for series connected string.
60 2002 first to req 1A increment ratings on supplementary OCPDs.
61 2002 did not allow conductors to pass through building before disconnect.
62 2002 req'd all backfed breakers to be secured in place.
63 2002 did not specifically prohibit disconnects in grounded conductor.
64 2002 prohibited using an electric vehicle as a backup power source.

TABLE 20 COMMON NUMBERING SYSTEM FOR WIRE, CABLE, & RACEWAY ARTICLES (BASED ON NEC CHAPTER 3)

I. GENERAL	II. INSTALLATION			III. CONSTRUCTION SPECIFICATIONS
xxx.1 Scope	xxx.10 Uses Permitted	xxx.26 Bends: # in 1 Run	xxx.44 Expansion Fittings	xxx.100 Expansion Fittings
xxx.2 Definitions	xxx.12 Uses Not Permitted	xxx.28 Reaming & Threading	xxx.46 Bushings	xxx.110 Bushings
xxx.3 Other Articles	xxx.14 Dissimilar Metals	xxx.28 Trimming	xxx.48 Joints	xxx.120 Joints
xxx.4 Listing Requirements	xxx.16 Temperature Limits	xxx.30 Securing & Supporting	xxx.50 Conductor Terminations	xxx.130 Conductor Terminations
	xxx.20 Size	xxx.40 Boxes & Fittings	xxx.56 Splices & Taps	xxx.140 Splices & Taps
	xxx.22 # of Conductors	xxx.42 # of Conductors	xxx.60 Grounding	xxx.150 Grounding
	xxx.24 Scope			